T0230611

ESSENTIAL DC/DC CONVERTERS

ESSENTIAL DC/DC CONVERTERS

Fang Lin Luo

Nanyang Technological University
Singapore

Hong Ye

Nanyang Technological University
Singapore

Taylor & Francis
Taylor & Francis Group
Boca Raton   London   New York

A CRC title, part of the Taylor & Francis imprint, a member of the
Taylor & Francis Group, the academic division of T&F Informa plc.

The material was previously published in *Advanced DC/DC Converters* © CRC Press LLC 2003

Published in 2006 by
CRC Press
Taylor & Francis Group
6000 Broken Sound Parkway NW, Suite 300
Boca Raton, FL 33487-2742

© 2006 by Taylor & Francis Group, LLC
CRC Press is an imprint of Taylor & Francis Group

International Standard Book Number-10: 0-8493-7238-0 (Hardcover)
International Standard Book Number-13: 978-0-8493-7238-4 (Hardcover)
Library of Congress Card Number 2005050494

Library of Congress Cataloging-in-Publication Data

Luo, Fang Lin.
    Essential DC/DC converters / Fang Lin Luo, Hong Ye.
        p. cm.
    Includes bibliographical references and index.
    ISBN 0-8493-7238-0 (alk. paper)
    1. DC-to-DC converters. I. Ye, Hong, 1973- II. Title.

TK7872.C8L855 2005
621.31'3--dc22   2005050494

Preface

DC/DC conversion technology is the main branch of Power Electronics and is progressing rapidly. Recent reports indicate that the production of DC/DC converters occupies the largest percentage of the total turnover of all conversion equipment productions. Based on incomplete statistics, there are more than 500 topologies of DC/DC converters existing. Many new topologies are still created every year. It is a lofty undertaking to write about the large number of DC/DC converters. The authors have sorted these converters into six generations since 2000. This systematical work is very helpful for understanding DC/DC converter evolution and development. The categories are listed below:

1. The first generation (classical/traditional) converters
2. The second generation (multiple quadrant) converters
3. The third generation (switched component) converters
4. The fourth generation (soft switching) converters
5. The fifth generation (synchronous rectifier) converters
6. The sixth generation (multiple element resonant power) converters

More than 300 prototypes of the first generation converters have been developed in the past 80 years.

The purpose of this book is to provide information about **essential DC/DC converters** that is both concise and useful for engineering students and practicing professionals. It is well organized in 410 pages and 200 diagrams to introduce more than 80 topologies of the essential DC/DC converters originally developed by the authors. All topologies are novel approaches and great contributions to modern power electronics. They are sorted in three groups:

- The voltage-lift converters
- The super-lift converters
- The ultra-lift converter

The voltage lift technique is a popular method that is widely applied in electronic circuit design. Applying this technique effectively overcomes the effects of parasitic elements and greatly increases the output voltage. Therefore, these DC/DC converters can convert the source voltage into a higher output voltage with high power efficiency, high power density and a simple structure. It is applied in the periodical switching circuit. Usually, a capacitor is charged during switching-on by certain voltage. This charged capacitor voltage can be arranged on top-up to output voltage during switching-off. Therefore, the output voltage can be increased. A typical example is the saw-tooth-wave generator with voltage lift circuit.

Voltage lift technique has its output voltage increasing in arithmetic progression, stage by stage. Super lift technique is more powerful than voltage-lift technique. The output voltage transfer gain of super-lift converters can be very high, and increases in geometric progression, stage by stage. It effectively enhances the voltage transfer gain in power series. Four series of super-lift converters created by the authors are introduced in this book. Some industrial applications verified their versatile and powerful characteristics. Super-lift technique is an outstanding achievement in DC/DC conversion technology.

Ultra-lift technique is another outstanding achievement in DC/DC conversion technology. It combines the characteristics of the voltage-lift and super-lift techniques to create the very high voltage transfer gain converter ultra-lift Luo converter. It effectively enhances the voltage transfer gain.

This book is organized in seven chapters. The general knowledge of DC/DC conversion technology is introduced in Chapter 1; and voltage lift converters in Chapter 2. Chapters 3–6 introduce the four series super-lift converters. Chapter 7 introduces the ultra-lift Luo converter.

The authors have worked in this research area for long periods and created a large number of outstanding converters: namely Luo converters, which cover all six generations of converters. Super-lift converters are our favorite achievements in our 20-years' research.

Our acknowledgment goes to the executive editor for this book.

Fang Lin Luo and Hong Ye
Nanyang Technological University
Singapore
May 2005

Contents

Figure Credits

The following figures were reprinted from *Power Electronics Handbook* (M.H. Rashid, Ed.), Chapter 17, Copyright 2001, with kind permission from Elsevier.

Authors

Dr. Fang Lin Luo received his bachelor of science (First Class with Honours) in radio-electronic physics at the Sichuan University, Chengdu, China and his Ph.D. in electrical engineering and computer science from Cambridge University, England, in 1986.

Dr. Luo was with the Chinese Automation Research Institute of Metallurgy (CARIM), Beijing, as a senior engineer after his graduation from Sichuan University. He was with the Entreprises Saunier Duval, Paris, France as a project engineer in 1981 and 1982. He was with Hocking NDT Ltd, Allen-Bradley IAP Ltd. and Simplatroll Ltd. in England as a senior engineer after he earned his Ph.D. from Cambridge University.

He is with the School of Electrical and Electronic Engineering, Nanyang Technological University in Singapore. He is a senior member of IEEE. He has published seven teaching textbooks and 218 technical papers in *IEEE-Transactions, IEEE-Proceedings* and other international journals, and various international conferences. His present research interest is in the digital power electronics and DC and AC motor drives with computerized artificial intelligent control (AIC) and digital signal processing (DSP), and DC/AC inverters, AC/DC rectifiers, AC/AC and DC/DC converters.

Dr. Luo was the chief editor of the international *Power Supply Technologies and Applications Journal* in 1998–2003. He is the international editor of *Advanced Technology of Electrical Engineering and Energy*. He is currently the associate editor of the *IEEE Transactions* on both power electronics and industrial electronics.

Dr. Hong Ye earned a bachelor degree (First Class with Honors) in 1995 and the master of engineering degree from Xi'an Jiaotong University, China in 1999. She completed her Ph.D. at Nanyang Technological University, Singapore.

Dr. Ye was with the R&D Institute, XIYI Company, Ltd., China as a research engineer from 1995 to 1997. She joined Nanyang Technical University in 2003.

Dr. Ye is an IEEE Member and has co-authored seven books. She has published more than 48 technical papers in *IEEE-Transactions, IEEE-Proceedings* and other international journals and various international conferences. Her research interests are in the areas of DC/DC converters, signal processing, operations research and structural biology.

1

Introduction

Conversion technique is a major research area in the field of power electronics. The equipment for conversion techniques have applications in industry,
research and development, government organizations, and daily life. The
equipment can be divided in four technologies:

- AC/AC transformers
- AC/DC rectifiers
- DC/DC converters
- DC/AC inverters

According to incomplete statistics, there have been more than 500 prototypes of DC/DC converters developed in the past six decades. All existing
DC/DC converters were designed to meet the requirements of certain applications. They are usually called by their function, for example, Buck converter, Boost converter and Buck-Boost converter, and zero current switching
(ZCS) and zero voltage switching (ZVS) converters. The large number of
DC/DC converters had not been evolutionarily classified until 2001. The
authors have systematically classified the types of converters into six generations according to their characteristics and development sequence. This
classification grades all DC/DC converters and categorizes new prototypes.
Since 2001, the DC/DC converter family tree has been built and this classification has been recognized worldwide. Following this principle, it is now easy
to sort and allocate DC/DC converters and assess their technical features.

1.1 Historical Review

DC/DC conversion technology is a major subject area in the field of power
engineering and drives, and has been under development for six decades.
DC/DC converters are widely used in industrial applications and computer
hardware circuits. DC/DC conversion techniques have developed very

quickly. Statistics show that the DC/DC converter worldwide market will grow from U.S. $3336 million in 1995 to U.S. $5128 million in the year 2004 with a compound annual growth rate (CAGR) of 9%.* This compares to the AC/DC power supply market, which will have a CAGR of only about 7.5% during the same period. In addition to its higher growth rate, the DC/DC converter market is undergoing dramatic changes as a result of two major trends in the electronics industry: low voltage and high power density. From this investigation it can be seen that the production of DC/DC converters in the world market is much higher than that of AC/DC converters.

The DC/DC conversion technique was established in the 1920s. A simple voltage conversion, the simplest DC/DC converter is a voltage divider (such as rheostat, potential–meter, and so on), but it only transfers output voltage lower than input voltage with poor efficiency. The multiple-quadrant chopper is the second step in DC/DC conversion. Much time has been spent trying to find equipment to convert the DC energy source of one voltage to another DC actuator with another voltage, as does a transformer employed in AC/AC conversion.

Some preliminary types of DC/DC converters were used in industrial applications before the Second World War. Research was blocked during the war, but applications of DC/DC converters were recognized. After the war, communication technology developed very rapidly and required low voltage DC power supplies. This resulted in the rapid development of DC/DC conversion techniques. Preliminary prototypes can be derived from choppers.

1.2   Multiple-Quadrant Choppers

Choppers are the circuits that convert fixed DC voltage to variable DC voltage or pulse-width–modulated (PWM) AC voltage. In this book, we concentrate on its first function.

1.2.1   Multiple-Quadrant Operation

A DC motor can run in forward running or reverse running. During the forward starting process its armature voltage and armature current are both positive. We usually call this forward motoring operation or *quadrant I* operation. During the forward braking process its armature voltage is still positive and its armature current is negative. This state is called the forward regenerating operation or *quadrant II* operation. Analogously, during the reverse starting process the DC motor armature voltage and current are both negative. This reverse motoring operation is called the *quadrant III* operation.

---

* Figures are taken from the Darnell Group News report on *DC-DC Converters: Global Market Forecasts, Demand Characteristics and Competitive Environment* published in February 2000.

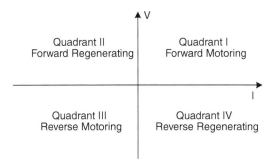

FIGURE 1.1
Four-quadrant operation.

During reverse braking process its armature voltage is still negative and its armature current is positive. This state is called the reverse regenerating operation *quadrant IV* operation.

Referring to the DC motor operation states; we can define the multiple-quadrant operation as below:

Quadrant I operation: forward motoring, voltage is positive, current is positive

Quadrant II operation: forward regenerating, voltage is positive, current is negative

Quadrant III operation: reverse motoring, voltage is negative, current is negative

Quadrant IV operation: reverse regenerating, voltage is negative, current is positive

The operation status is shown in the Figure 1.1. Choppers can convert a fixed DC voltage into various other voltages. The corresponding chopper is usually named according to its quadrant operation chopper, e.g., the first-quadrant chopper or "A"-type chopper. In the following description we use the symbols V_{in} as the fixed voltage, V_p the chopped voltage, and V_O the output voltage.

1.2.2 The First-Quadrant Chopper

The first-quadrant chopper is also called "A"-type chopper and its circuit diagram is shown in Figure 1.2a and corresponding waveforms are shown in Figure 1.2b. The switch S can be some semiconductor devices such as BJT, IGBT, and MOSFET. Assuming all parts are ideal components, the output voltage is calculated by the formula,

$$V_O = \frac{t_{on}}{T} V_{in} = k V_{in} \tag{1.1}$$

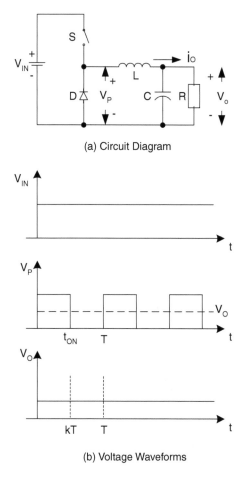

(a) Circuit Diagram

(b) Voltage Waveforms

FIGURE 1.2
The first-quadrant chopper.

where T is the repeating period $T = 1/f$, f is the chopping frequency, t_{on} is the switch-on time, k is the conduction duty cycle $k = t_{on}/T$.

1.2.3   The Second-Quadrant Chopper

The second-quadrant chopper is the called "B"-type chopper and the circuit diagram and corresponding waveforms are shown in Figure 1.3a and b. The The output voltage can be calculated by the formula,

$$V_O = \frac{t_{off}}{T} V_{in} = (1-k)V_{in} \tag{1.2}$$

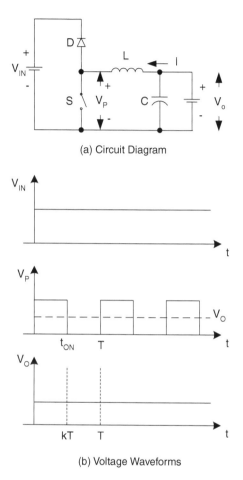

(a) Circuit Diagram

(b) Voltage Waveforms

FIGURE 1.3
The second-quadrant chopper.

where T is the repeating period $T = 1/f$, f is the chopping frequency, t_{off} is the switch-off time $t_{off} = T - t_{on}$, and k is the conduction duty cycle $k = t_{on}/T$.

1.2.4   The Third-Quadrant Chopper

The third-quadrant chopper and corresponding waveforms are shown in Figure 1.4a and b. All voltage polarity is defined in the figure. The output voltage (absolute value) can be calculated by the formula,

$$V_O = \frac{t_{on}}{T} V_{in} = kV_{in} \qquad (1.3)$$

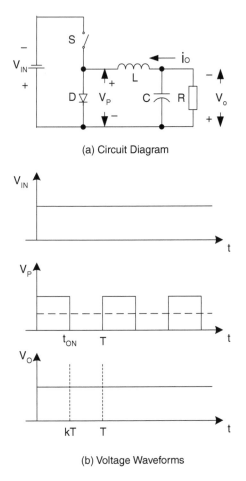

(a) Circuit Diagram

(b) Voltage Waveforms

FIGURE 1.4
The third-quadrant chopper.

where t_{on} is the switch-on time, and k is the conduction duty cycle $k = t_{on}/T$.

1.2.5   The Fourth-Quadrant Chopper

The fourth-quadrant chopper and corresponding waveforms are shown in Figure 1.5a and b. All voltage polarity is defined in the figure. The output voltage (absolute value) can be calculated by the formula,

$$V_O = \frac{t_{off}}{T} V_{in} = (1-k)V_{in} \tag{1.4}$$

where t_{off} is the switch-off time $t_{off} = T - t_{on}$, time, and k is the conduction duty cycle $k = t_{on}/T$.

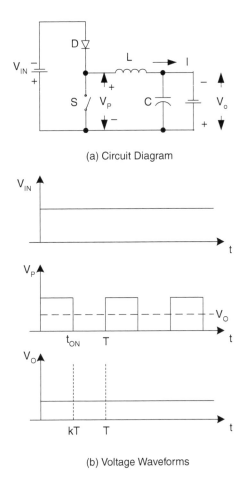

(a) Circuit Diagram

(b) Voltage Waveforms

FIGURE 1.5
The fourth-quadrant chopper.

1.2.6 The First and Second Quadrant Chopper

The first and second quadrant chopper is shown in Figure 1.6. Dual quadrant operation is usually requested in the system with two voltage sources V_1 and V_2. Assume that the condition $V_1 > V_2$, and the inductor L is an ideal component. During quadrant I operation, S_1 and D_2 work, and S_2 and D_1 are idle. Vice versa, during quadrant II operation, S_2 and D_1 work, and S_1 and D_2 are idle. The relation between the two voltage sources can be calculated by the formula,

$$V_2 = \begin{cases} kV_1 & QI\_operation \\ (1-k)V_1 & QII\_operation \end{cases} \tag{1.5}$$

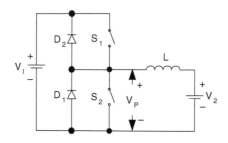

FIGURE 1.6
The first-second quadrant chopper.

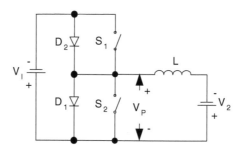

FIGURE 1.7
The third-fourth quadrant chopper.

where k is the conduction duty cycle $k = t_{on}/T$.

1.2.7 The Third and Fourth Quadrant Chopper

The third and fourth quadrant chopper is shown in Figure 1.7. Dual quadrant operation is usually requested in the system with two voltage sources V_1 and V_2. Both voltage polarity is defined in the figure, we just concentrate their absolute values in analysis and calculation. Assume that the condition $V_1 > V_2$, the inductor L is ideal component. During quadrant I operation, S_1 and D_2 work, and S_2 and D_1 are idle. Vice versa, during quadrant II operation, S_2 and D_1 work, and S_1 and D_2 are idle. The relation between the two voltage sources can be calculated by the formula,

$$V_2 = \begin{cases} kV_1 & QIII\_operation \\ (1-k)V_1 & QIV\_operation \end{cases} \qquad (1.6)$$

where k is the conduction duty cycle $k = t_{on}/T$.

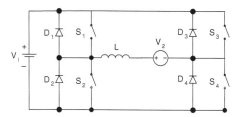

FIGURE 1.8
The four-quadrant chopper.

TABLE 1.1

The Switches and Diode's Status for Four-Quadrant Operation

Switch or Diode	Quadrant I	Quadrant II	Quadrant III	Quadrant IV
S_1	Works	Idle	Idle	Works
D_1	Idle	Works	Works	Idle
S_2	Idle	Works	Works	Idle
D_2	Works	Idle	Idle	Works
S_3	Idle	Idle	On	Idle
D_3	Idle	Idle	Idle	On
S_4	On	Idle	Idle	Idle
D_4	Idle	On	Idle	Idle
Output	$V_2 +, I_2 +$	$V_2 +, I_2 -$	$V_2 -, I_2 -$	$V_2 -, I_2 +$

1.2.8 The Four-Quadrant Chopper

The four-quadrant chopper is shown in Figure 1.8. The input voltage is positive, output voltage can be either positive or negative. The switches and diode status for the operation are shown in Table 1.1. The output voltage can be calculated by the formula,

$$V_2 = \begin{cases} kV_1 & QI\_operation \\ (1-k)V_1 & QII\_operation \\ -kV_1 & QIII\_operation \\ -(1-k)V_1 & QIV\_operation \end{cases} \tag{1.7}$$

1.3 Pump Circuits

The electronic pump is a major component of all DC/DC converters. Historically, they can be sorted into four groups:

- Fundamental pumps
- Developed pumps
- Transformer-type pumps
- Super-lift pumps

1.3.1 Fundamental Pumps

Fundamental pumps are developed from fundamental DC/DC converters just like their name:

- Buck pump
- Boost pump
- Buck-boost pump

All fundamental pumps consist of three components: a switch S, a diode D, and an inductor L.

1.3.1.1 Buck Pump

The circuit diagram of the buck pump, and some current waveforms are shown in Figure 1.9. Switch S and diode D are alternately on and off. Usually, the buck pump works in continuous operation mode, inductor current is continuous in this case.

1.3.1.2 Boost Pump

The circuit diagram of the boost pump, and some current waveforms are shown in Figure 1.10. Switch S and diode D are alternately on and off. The inductor current is usually continuous.

1.3.1.3 Buck-Boost Pump

The circuit diagram of the buck-boost pump and some current waveforms are shown in Figure 1.11. Switch S and diode D are alternately on and off. Usually, the buck-boost pump works in continuous operation mode, inductor current is continuous in this case.

1.3.2 Developed Pumps

Developed pumps are created from the developed DC/DC converters just like their name:

- Positive Luo-pump
- Negative Luo-pump
- Cúk-pump

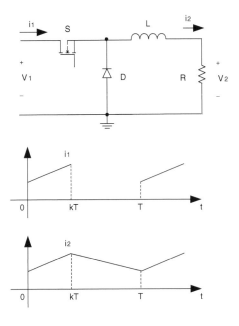

FIGURE 1.9
Buck pump.

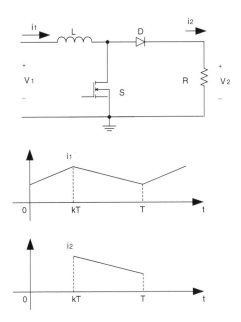

FIGURE 1.10
Boost pump.

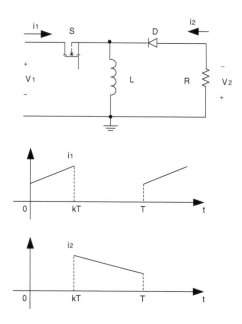

FIGURE 1.11
Buck-boost pump.

All developed pumps consist of four components: a switch S, a diode D, a capacitor C, and an inductor L.

1.3.2.1 *Positive Luo-Pump*

The circuit diagram of the positive Luo-pump and some current and voltage waveforms are shown in Figure 1.12. Switch S and diode D are alternately on and off. Usually, this pump works in continuous operation mode, inductor current is continuous in this case. The output terminal voltage and current is usually positive.

1.3.2.2 *Negative Luo-Pump*

The circuit diagram of the negative Luo-pump and some current and voltage waveforms are shown in Figure 1.13. Switch S and diode D are alternately on and off. Usually, this pump works in continuous operation mode, inductor current is continuous in this case. The output terminal voltage and current is usually negative.

1.3.2.3 *Cúk-Pump*

The circuit diagram of the Cúk pump and some current and voltage waveforms are shown in Figure 1.14. Switch S and diode D are alternately on and off. Usually, the Cúk pump works in continuous operation mode, inductor current is continuous in this case. The output terminal voltage and current is usually negative.

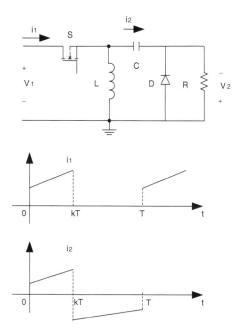

FIGURE 1.12
Positive Luo-pump.

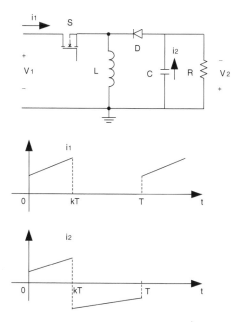

FIGURE 1.13
Negative Luo-pump.

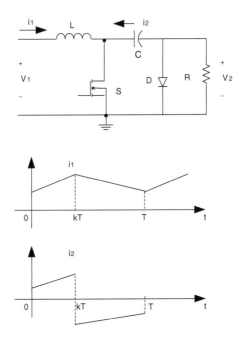

FIGURE 1.14
Cúk-pump.

1.3.3   Transformer-Type Pumps

Transformer-type pumps are developed from transformer-type DC/DC converters just like their name:

- Forward pump
- Fly-Back pump
- ZETA pump

All transformer-type pumps consist of a switch S, a transformer with the turn ratio N and other components such as diode D (one or more) and capacitor C.

1.3.3.1   Forward Pump

The circuit diagram of the forward pump and some current waveforms are shown in Figure 1.15. Switch S and diode D_1 are synchronously on and off, and diode D_2 is alternately off and on. Usually, the forward pump works in discontinuous operation mode, input current is discontinuous in this case.

1.3.3.2   Fly-Back Pump

The circuit diagram of the fly-back pump and some current waveforms are in Figure 1.16. Since the primary and secondary windings of the transformer

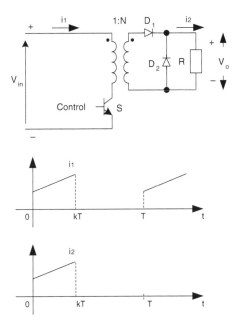

FIGURE 1.15
Forward pump.

are purposely arranged in inverse polarities, switch S and diode D are alternately on and off. Usually, the fly-back pump works in discontinuous operation mode, input current is discontinuous in this case.

1.3.3.3 ZETA Pump

The circuit diagram of the ZETA pump and some current waveforms are shown in Figure 1.17. Switch S and diode D are alternately on and off. Usually, the ZETA pump works in discontinuous operation mode, input current is discontinuous in this case.

1.3.4 Super-Lift Pumps

Super-lift pumps are developed from super-lift DC/DC converters:

- Positive super Luo-pump
- Negative super Luo-pump
- Positive push-pull pump
- Negative push-pull pump
- Double/Enhanced circuit (DEC)

All super-lift pumps consist of switches, diodes, capacitors, and sometimes an inductor.

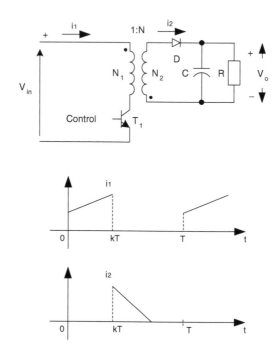

FIGURE 1.16
Fly-back pump.

1.3.4.1   Positive Super Luo-Pump

The circuit diagram of the positive super-lift pump and some current waveforms are shown in Figure 1.18. Switch S and diode D_1 are synchronously on and off, but diode D_2 is alternately off and on. Usually, the positive super-lift pump works in continuous conduction mode (CCM), inductor current is continuous in this case.

1.3.4.2   Negative Super Luo-Pump

The circuit diagram of the negative super-lift pump and some current waveforms are shown in Figure 1.19. Switch S and diode D_1 are synchronously on and off, but diode D_2 is alternately off and on. Usually, the negative super-lift pump works in CCM, but input current is discontinuous in this case.

1.3.4.3   Positive Push-Pull Pump

All push-pull pumps consist of two switches without any inductor. They can be employed in multiple-lift switched capacitor converters. The circuit diagram of positive push-pull pump and some current waveforms are shown in Figure 1.20. Since there is no inductor in the pump, it is applied in

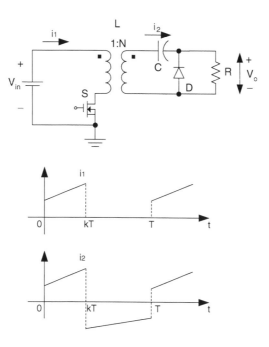

FIGURE 1.17
ZETA pump.

switched-capacitor converters. The main switch S and diode D_1 are synchronously on and off, but the slave switch S_1 and diode D_2 are alternately off and on. Usually, the positive push-pull pump works in push-pull state continuous operation mode.

1.3.4.4 Negative Push-Pull Pump

The circuit diagram of this push-pull pump and some current waveforms are shown in Figure 1.21. Since there is no inductor in the pump, it is often used in switched-capacitor converters. The main switch S and diode D are synchronously on and off, but the slave switch S_1 is alternately off and on. Usually, the super-lift pump works in push-pull state continuous operation mode, inductor current is continuous in this case.

1.3.4.5 Double/Enhanced Circuit (DEC)

The circuit diagram of the double/enhanced circuit and some current waveforms are in Figure 1.22. The switch is the only other existing circuit part. This circuit is usually applied in lift, super-lift, and push-pull converters. These two diodes are alternately on and off, so that two capacitors are alternately charging and discharging. Usually, this circuit can enhance the voltage doubly or at certain times.

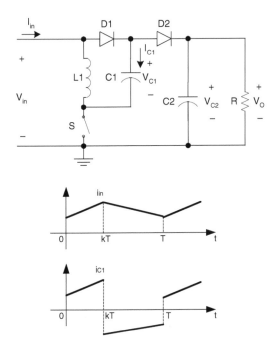

FIGURE 1.18
Positive super Luo-pump.

1.4   Development of DC/DC Conversion Technique

According to incomplete statistics, there are more than 500 existing proto-types of DC/DC converters. The main purpose of this book is to catorgorize all existing prototypes of DC/DC converters. This job is of vital importance for future development of DC/DC conversion techniques. The authors have devoted 20 years to this subject area, their work has been recognized and assessed by experts worldwide. The authors classify all existing prototypes of DC/DC converters into six generations. They are

- First generation (classical/traditional) converters
- Second generation (multi-quadrant) converters
- Third generation (switched-component **SI/SC**) converters
- Fourth generation (soft-switching: **ZCS/ZVS/ZT**) converters
- Fifth generation (synchronous rectifier **SR**) converters
- Sixth generation (multiple energy-storage elements resonant **MER**) converters

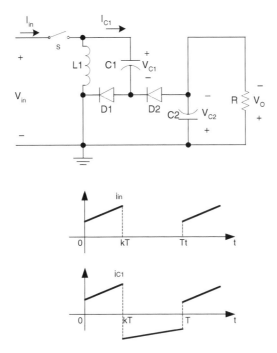

FIGURE 1.19
Negative super Luo-pump.

1.4.1 The First Generation Converters

The first-generation converters perform in a single quadrant mode and in low power range (up to around 100 W). Since its development lasts a long time, it has, briefly, five categories:

- Fundamental converters
- Transformer-type converters
- Developed converters
- Voltage-lift converters
- Super-lift converters

1.4.1.1 Fundamental Converters

Three types of fundamental DC/DC classifications were constructed, these are **buck** converter, **boost** converter, and **buck-boost** converter. They can be derived from single quadrant operation choppers. For example, the buck converter was derived from an A-type chopper. These converters have two main problems: linkage between input and output and very large output voltage ripple.

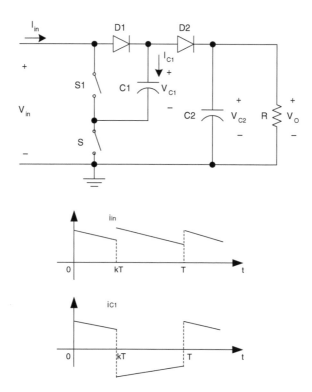

FIGURE 1.20
Positive push-pull pump.

1.4.1.1.1   Buck Converter

The buck converter is a step-down DC/DC converter. It works in first-quadrant operation. It can be derived from a quadrant I chopper. Its circuit diagram, and switch-on and -off equivalent circuit are shown in Figure 1.23. The output voltage is calculated by the formula,

$$V_O = \frac{t_{on}}{T} V_{in} = kV_{in} \tag{1.8}$$

where T is the repeating period $T = 1/f$, f is the chopping frequency, t_{on} is the switch-on time, and k is the conduction duty cycle $k = t_{on}/T$.

1.4.1.1.2   Boost Converter

The boost converter is a step-up DC/DC converter. It works in second-quadrant operation. It can be derived from quadrant II chopper. Its circuit diagram, and switch-on and -off equivalent circuit are shown in Figure 1.24. The output voltage is calculated by the formula,

$$V_O = \frac{T}{T - t_{on}} V_{in} = \frac{1}{1 - k} V_{in} \tag{1.9}$$

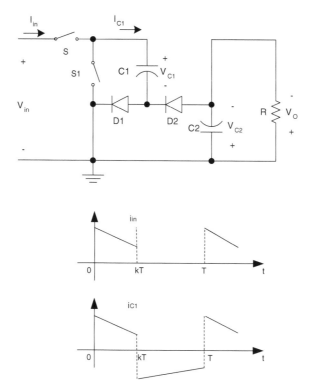

FIGURE 1.21
Negative push-pull pump.

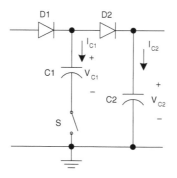

FIGURE 1.22
Double/enhanced circuit (DEC).

where T is the repeating period $T = 1/f$, f is the chopping frequency, t_{on} is the switch-on time, k is the conduction duty cycle $k = t_{on}/T$.

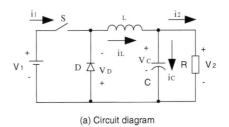

(a) Circuit diagram

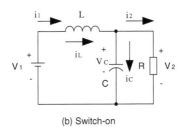

(b) Switch-on

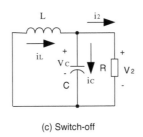

(c) Switch-off

FIGURE 1.23
Buck converter.

1.4.1.1.3   *Buck-Boost Converter*

The buck-boost converter is a step down/up DC/DC converter. It works in third-quadrant operation. Its circuit diagram, switch-on and -off equivalent circuit, and waveforms are shown in Figure 1.25. The output voltage is calculated by the formula,

$$V_O = \frac{t_{on}}{T - t_{on}} V_{in} = \frac{k}{1 - k} V_{in} \tag{1.10}$$

where T is the repeating period $T = 1/f$, f is the chopping frequency, t_{on} is the switch-on time, and k is the conduction duty cycle $k = t_{on}/T$. By using this converter it is easy to obtain the random output voltage, which can be higher or lower than the input voltage. It provides great convenience for industrial applications.

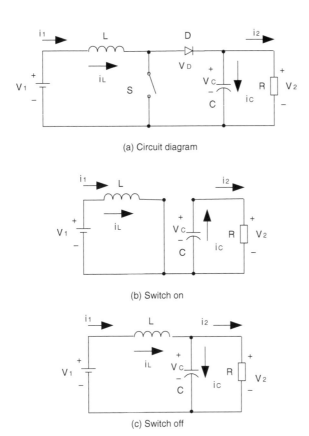

(a) Circuit diagram

(b) Switch on

(c) Switch off

FIGURE 1.24
Boost converter.

1.4.1.2 Transformer-Type Converters

Since all fundamental DC/DC converters keep the linkage from input side to output side and the voltage transfer gain is comparably low, transformer-type converters were developed in the 1960s to 1980s. There are a large group of converters such as the **forward** converter, **push-pull** converter, **fly-back** converter, **half-bridge** converter, **bridge** converter, and **Zeta** (or **ZETA**) converter. Usually, these converters have high transfer voltage gain and high insulation between both sides. Their gain usually depends on the transformer's turn ratio N, which can be thousands times.

1.4.1.2.1 Forward Converter

A forward converter is a transformer-type buck converter with a turn ratio N. It works in first quadrant operation. Its circuit diagram is shown in Figure 1.26. The output voltage is calculated by the formula,

$$V_O = kNV_{in} \tag{1.12}$$

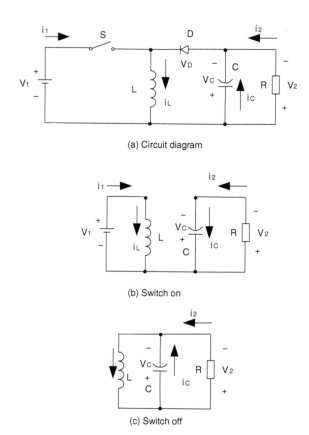

FIGURE 1.25
Buck-boost converter.

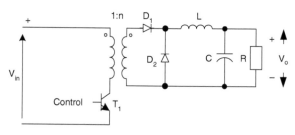

FIGURE 1.26
Forward converter.

where N is the transformer turn ratio, and k is the conduction duty cycle $k = t_{on}/T$.

In order to exploit the magnetic ability of the transformer iron core, a tertiary winding can be employed in the transformer. Its corresponding circuit diagram is shown in Figure 1.27.

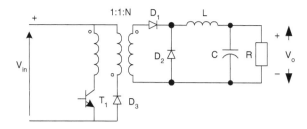

FIGURE 1.27
Forward converter with tertiary winding.

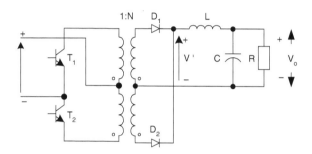

FIGURE 1.28
Push-pull converter.

1.4.1.2.2 Push-Pull Converter

The boost converter works in push-pull state, which effectively avoids the iron core saturation. Its circuit diagram is shown in Figure 1.28. Since there are two switches, which work alternately, the output voltage is doubled. The output voltage is calculated by the formula,

$$V_O = 2kNV_{in} \tag{1.13}$$

where N is the transformer turn ratio, and k is the conduction duty cycle $k = t_{on}/T$.

1.4.1.2.3 Fly-Back Converter

The fly-back converter is a transformer type converter using the demagnetizing effect. Its circuit diagram is shown in Figure 1.29. The output voltage is calculated by the formula,

$$V_O = \frac{k}{1-k} NV_{in} \tag{1.14}$$

where N is the transformer turn ratio, and k is the conduction duty cycle $k = t_{on}/T$.

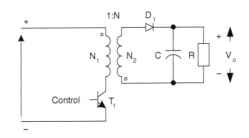

FIGURE 1.29
Fly-back converter.

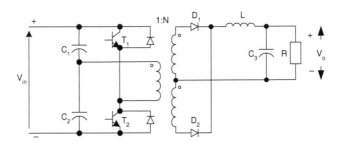

FIGURE 1.30
Half-bridge converter.

1.4.1.2.4 Half-Bridge Converter

In order to reduce the primary side in one winding, the half-bridge converter was constructed. Its circuit diagram is shown in Figure 1.30. The output voltage is calculated by the formula,

$$V_O = kNV_{in} \tag{1.15}$$

where N is the transformer turn ratio, and k is the conduction duty cycle $k = t_{on}/T$.

1.4.1.2.5 Bridge Converter

The bridge converter employs more switches and therefore gains double output voltage. Its circuit diagram is shown in Figure 1.31. The output voltage is calculated by the formula,

$$V_O = 2kNV_{in} \tag{1.16}$$

where N is the transformer turn ratio, and k is the conduction duty cycle $k = t_{on}/T$.

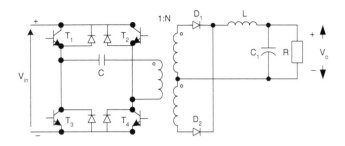

FIGURE 1.31
Bridge converter.

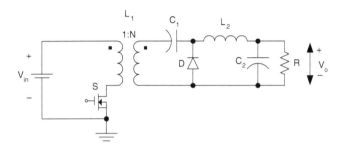

FIGURE 1.32
Zeta converter.

1.4.1.2.6 *ZETA Converter*

The ZETA converter is a transformer type converter with a low-pass filter. Its output voltage ripple is small. Its circuit diagram is shown in Figure 1.32. The output voltage is calculated by the formula,

$$V_O = \frac{k}{1-k} N V_{in} \tag{1.17}$$

where N is the transformer turn ratio, and k is the conduction duty cycle $k = t_{on}/T$.

1.4.1.2.7 *Forward Converter with Tertiary Winding and Multiple Outputs*

Some industrial applications require multiple outputs. This requirement is easily realized by constructing multiple secondary windings and the corresponding conversion circuit. For example, a forward converter with tertiary winding and three outputs is shown in Figure 1.33. The output voltage is calculated by the formula,

$$V_O = k N_i V_{in} \tag{1.18}$$

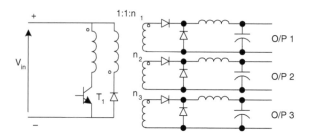

FIGURE 1.33
Forward converter with tertiary winding and three outputs.

where N_i is the transformer turn ratio to the secondary winding, $i = 1, 2$, and 3 respectively, and k is the conduction duty cycle $k = t_{on}/T$. In principle, this structure is available for all transformer-type DC/DC converters for multiple outputs applications.

1.4.1.3   Developed Converters

Developed-type converters overcome the second fault of the fundamental DC/DC converters. They are derived from fundamental converters by the addition of a low-pass filter. The preliminary design was published in a conference in 1977 (Massey and Snyder, 1977). The author designed three types of converters that derived from fundamental DC/DC converters plus a low-pass filter. This conversion technique was very popular between 1970 and 1990. Typical prototype converters are positive output (P/O) **Luo**-converter, negative output (N/O) **Luo**-converter, double output (D/O) **Luo**-converter, **Cúk**-converter, **SEPIC** (single-ended primary inductance converter) and Watkins–Johnson converters. The output voltage ripple of all developed-type converters is usually small and can be lower than 2%.

In order to obtain the random output voltage, which can be higher or lower input voltage. All developed converters provide ease of application for industry. Therefore, the output voltage gain of all developed converters is

$$V_O = \frac{k}{1-k} V_{in} \qquad (1.19)$$

1.4.1.3.1   Positive Output (P/O) Luo-Converter

The positive output **Luo**-converter is the elementary circuit of the series *positive output Luo-converters*. It can be derived from the buck-boost converter. Its circuit diagram is shown in Figure 1.34. The output voltage is calculated by the Equation (1.19).

1.4.1.3.2   Negative Output (N/O) Luo-Converter

The negative output Luo-converter is the elementary circuit of the series *negative output Luo-converters*. It can also be derived from buck-boost converters. Its

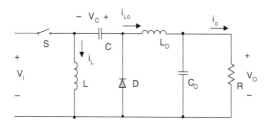

FIGURE 1.34
Positive output Luo-converter.

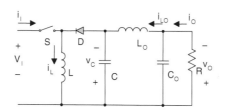

FIGURE 1.35
Negative output Luo-converter.

circuit diagram is shown in Figure 1.35. The output voltage is calculated by the Equation (1.19).

1.4.1.3.3 Double Output (D/O) Luo-Converter

In order to obtain mirror symmetrical positive plus negative output voltage double output (D/O) Luo-converters were constructed. The double output Luo-converter is the elementary circuit of the series *double output Luo-converters*. It can also be derived from the buck-boost converter. Its circuit diagram is shown in Figure 1.36. The output voltage is calculated by Equation (1.19).

1.4.1.3.4 Cúk-Converter

The Cúk-converter is derived from boost converter. Its circuit diagram is shown in Figure 1.37. The output voltage is calculated by Equation (1.19).

1.4.1.3.5 Single-Ended Primary Inductance Converter

The single-ended primary inductance converter (**SEPIC**) is derived from the boost converter. Its circuit diagram is shown in Figure 1.38. The output voltage is calculated by Equation (1.19).

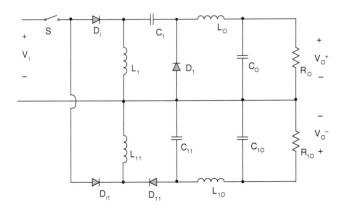

FIGURE 1.36
Double output Luo-converter.

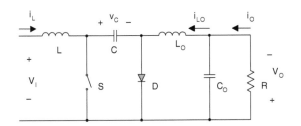

FIGURE 1.37
Cúk-converter.

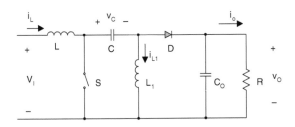

FIGURE 1.38
SEPIC.

1.4.1.3.6 *Tapped Inductor Converter*

These converters are derived from fundamental converters. The circuit diagrams are shown in Table 1.2. The voltage transfer gains are shown in Table 1.3. Here the tapped inductor ratio is $n = n1/(n1 + n2)$.

TABLE 1.2

The Circuit Diagrams of the Tapped Inductor Fundamental Converters

TABLE 1.3

The Voltage Transfer Gains of the Tapped Inductor Fundamental Converters

Converter	No tap	Switched to tap	Diode to tap	Rail to tap
Buck	k	$\dfrac{k}{n+k(1-n)}$	$\dfrac{nk}{1+k(n-1)}$	$\dfrac{k-n}{k(1-n)}$
Boost	$\dfrac{1}{1-k}$	$\dfrac{n+k(1-n)}{n(1-k)}$	$\dfrac{1+k(n-1)}{1-k}$	$\dfrac{n-k}{n(1-k)}$
Buck-Boost	$\dfrac{k}{1-k}$	$\dfrac{k}{n(1-k)}$	$\dfrac{nk}{1-k}$	$\dfrac{k}{1-k}$

1.4.1.4   Voltage Lift Converters

Voltage lift technique is a good method to lift the output voltage, and is widely applied in electronic circuit design. After long-term industrial application and research this method has been successfully used in DC/DC conversion technique. Using this method the output voltage can be easily lifted by tens to hundreds of times. Voltage lift converters can be classed into **self**-lift, **re**-lift, **triple**-lift, **quadruple**-lift, and **high-stage** lift converters. The main contributors in this area are Dr. Fang Lin Luo and Dr. Hong Ye. These circuits will be introduced in Chapter 2 in detail.

1.4.1.5   Super Lift Converters

Voltage lift (VL) technique is a popular method that is widely used in electronic circuit design. It has been successfully employed in DC/DC converter applications in recent years, and has opened a way to design high voltage gain converters. Three series Luo-converters are examples of voltage lift technique implementations. However, the output voltage increases in stage by stage just along the arithmetic progression. A novel approach — super lift (SL) technique — has been developed, which implements the output voltage increasing stage by stage along in geometric progression. It effectively enhances the voltage transfer gain in power-law. The typical circuits are sorted into four series: positive output super-lift Luo-converters, negative output super-lift Luo-converters, positive output cascade boost converters, and negative output cascade boost converters. These circuits will be introduced in Chapters 3 to 6 in detail.

1.4.2   The Second Generation Converters

The second generation converters are called multiple quadrant operation converters. These converters perform in two-quadrant operation and four-quadrant operation with medium output power range (hundreds of Watts or higher). The topologies can be sorted into two main categories: first are the converters derived from the multiple-quadrant choppers and/or from the first generation converters. Second are constructed with transformers. Usually, one quadrant operation requires at least one switch. Therefore, a two-quadrant operation converter has at least two switches, and a four-quadrant operation converter has at least four switches. Multiple-quadrant choppers were employed in industrial applications for a long time. They can be used to implement the DC motor multiple-quadrant operation. As the chopper titles indicate, there are class-A converters (one-quadrant operation), class-B converters (two-quadrant operation), class-C converters, class-D converters, and class-E (four-quadrant operation) converters. These converters are derived from multi-quadrant choppers, for example, class B converters are derived from B-type choppers and class E converters are derived from E-type choppers. The class-A converter works in quadrant I, which corresponds to the forward-motoring operation of a DC motor drive. The class-B converter works in quadrant I and II operation, which corresponds to the forward-running motoring and regenerative braking operation of a DC motor drive. The class-C converter works in quadrant I and VI operation. The class-D converter works in quadrant III and VI operation, which corresponds to the reverse-running motoring and regenerative braking operation of a DC motor drive. The class-E converter works in four-quadrant operation, which corresponds to the four-quadrant operation of a DC motor drive. In recent years many papers have investigated the class-E converters for industrial applications. Multi-quadrant operation converters can be derived from the first generation converters. For example, multi-quadrant Luo-converters are derived from positive-output Luo-converters and negative-output Luo-converters. The transformer-type multi-quadrant converters easily change the current direction by transformer polarity and diode rectifier. The main types of such converters can be derived from the **forward** converter, **half-bridge** converter, and **bridge** converter.

1.4.3   The Third Generation Converters

The third generation converters are called switched component converters, and are made of either inductors or capacitors, so-called switched-inductor and switched-capacitors. They can perform in two- or four-quadrant operation with high output power range (thousands of Watts). Since they are made of only inductor or capacitors, they are small. Consequently, the power density and efficiency are high.

1.4.3.1   Switched Capacitor Converters

Switched-capacitor DC/DC converters consist of only capacitors. Because there is no inductor in the circuit, their size is small. They have outstanding advantages such as low power losses and low electromagnetic interference

(EMI). Since its electromagnetic radiation is low, switched-capacitor DC/DC converters are required in certain equipment. The switched-capacitor can be integrated into an integrated-chip (IC). Hence, its size is largely reduced. Much attention has been drawn to the switched-capacitor converter since its development. Many papers have been published discussing its characteristics and advantages. However, most of the converters in the literature perform a single-quadrant operation. Some of them work in the push-pull status. In addition, their control circuit and topologies are very complex, especially, for the large difference between input and output voltages.

1.4.3.2   Multiple-Quadrant Switched Capacitor Luo-Converters

Switched-capacitor DC/DC converters consist of only capacitors. Since its power density is very high it is widely applied in industrial applications. Some industrial applications require multiple quadrant operation, so that, multiple-quadrant switched-capacitor Luo-converters have been developed. There are two-quadrant operation type and four-quadrant operation type, which will be discussed in detail.

1.4.3.3   Multiple-Lift Push-Pull Switched Capacitor Converters

Voltage lift (VL) technique is a popular method widely used in electronic circuit design. It has been successfully employed in DC/DC converter applications in recent years, and has opened a way to design high voltage gain converters. Three series Luo-converters are examples of voltage lift technique implementation. However, the output voltage increases stage-by-stage just along the arithmetic progression. A novel approach — multiple-lift push-pull (ml-pp) technique — has been developed that implements the output voltage, which increases stage by stage along the arithmetic progression. It effectively enhances the voltage transfer gain. The typical circuits are sorted into two series: *positive output multiple-lift push-pull switched capacitor Luo-converters* and *negative output multiple-lift push-pull switched capacitor Luo-converters*.

1.4.3.4   Multiple-Quadrant Switched Inductor Converters

The switched-capacitors have many advantages, but their circuits are not simple. If the difference of input and output voltages is large, many capacitors must be required. The switched-inductor has the outstanding advantage that only one inductor is required for one switched inductor converter no matter how large the difference between input and output voltages is. This characteristic is very important for large power conversion. At the present time, large power conversion equipment is close to using switched-inductor converters. For example, the MIT DC/DC converter designed by Prof. John G. Kassakian for his new system in the 2005 automobiles is a two-quadrant switched-inductor DC/DC converter.

1.4.4 The Fourth Generation Converters

The fourth generation DC/DC converters are called soft-switching converters. There are four types of soft-switching methods:

1. Resonant-switch converters
2. Load-resonant converters
3. Resonant-dc-link converters
4. High-frequency-link integral-half-cycle converters

Until now attention has been paid only to the resonant-switch conversion method. This resonance method is available for working independently to load. There are three main categories: zero-current-switching (ZCS), zero-voltage-switching (ZVS), and zero-transition (ZT) converters. Most topologies usually perform in single quadrant operation in the literature. Actually, these converters can perform in two- and four-quadrant operation with high output power range (thousands of Watts). According to the transferred power becomes large, the power losses increase largely. Main power losses are produced during the switch-on and switch-off period. How to reduce the power losses across the switch is the clue to increasing the power transfer efficiency. Soft-switching technique successfully solved this problem. Professor Fred Lee is the pioneer of the soft-switching technique. He established a research center and manufacturing base to realize the zero-current-switching (ZCS) and zero-voltage-switching (ZVS) DC/DC converters. His first paper introduced his research in 1984. ZCS and ZVS converters have three resonant states: over resonance (completed resonance), optimum resonance (critical resonance) and quasi resonance (subresonance). Only the quasi-resonance state has two clear cross-zero points in a repeating period. Many papers after 1984 have been published that develop the ZCS quasi-resonant-converters (QRCs) and ZVS-QRCs.

1.4.4.1 Zero-Current-Switching Quasi-Resonant Converters

ZCS-QRC equips the resonant circuit on the switch side to keep the switch-on and switch-off at zero-current condition. There are two states: full-wave state and half wave state. Most engineers use the half-wave state. This technique has half-wave current resonance waveform with two zero-cross points.

1.4.4.2 Zero-Voltage-Switching Quasi-Resonant Converters

ZVS-QRC equips the resonant circuit on the switch side to keep the switch-on and switch-off at zero-current condition. There are two states: full-wave state and half wave state. Most engineers use the half-wave state. This technique has half-wave current resonance waveform with two zero-cross points.

1.4.4.3   Zero-Transition Converters

Using ZCS-QRC and ZVS-QRC largely reduces the power losses across the switches. Consequently, the switch device power rates become lower and converter power efficiency is increased. However, ZCS-QRC and ZVS-QRC have large current and voltage stresses. Therefore the device's current and voltage peak rates usually are 3 to 5 times higher than the working current and voltage. It is not only costly, but also ineffective. Zero-Transition (ZT) technique overcomes this fault. It implements zero-voltage plus zero-current-switching (ZV-ZCS) technique without significant current and voltage stresses.

1.4.5   The Fifth Generation Converters

The fifth generation converters are called synchronous rectifier (SR) DC/DC converters. This type of converter was required by development of computing technology. Corresponding to the development of the micro-power consumption technique and high-density IC manufacture, the power supplies with low output voltage and strong current are widely used in communications, computer equipment, and other industrial applications. Intel , which developed the Zelog-type computers, governed the world market for a long time. Inter-80 computers used the 5 V power supply. In order to increase the memory size and operation speed, large-scale integrated chip (LSIC) technique has been quickly developed. As the amount of IC manufacturing increased, the gaps between the layers became narrower. At the same time, the micro-power-consumption technique was completed. Therefore new computers such as those using Pentium I, II, III, and IV, use a 3.3 V power supply. Future computers will have larger memory and will require lower power supply voltages, e.g. 2.5, 1.8, 1.5, even if 1.1 V. Such low power supply voltage cannot be obtained by the traditional diode rectifier bridge because the diode voltage drop is too large. Because of this requirement, new types of MOSFET were developed. They have very low conduction resistance (6 to 8 m Ω,) and forward voltage drop (0.05 to 0.2 V). Many papers have been published since 1990 and many prototypes have been developed. The fundamental topology is derived from the forward converter. Active-clamped circuit, flat-transformers, double current circuit, soft-switching methods, and multiple current methods can be used in SR DC/DC converters.

1.4.6   The Sixth Generation Converters

The sixth generation converters are called multiple energy-storage elements resonant (MER) converters. Current source resonant inverters are the heart of many systems and equipment, e.g., uninterruptible power supply (UPS) and high-frequency annealing (HFA) apparatus. Many topologies shown in the literature are the series resonant converters (SRC) and parallel resonant converters (PRC) that consist of two or three or four energy storage elements. However, they have limitations. These limitations of two-, three-, and/or four-element resonant topologies can be overcome by special design. These converters have been catagorized into three main types:

- Two energy-storage elements resonant DC/AC and DC/AC/DC converters
- Three energy-storage elements resonant DC/AC and DC/AC/DC converters
- Four energy-storage elements (2L-2C) resonant DC/AC and DC/AC/DC converters

By mathematical calculation there are eight prototypes of two-element converters, 38 prototypes of three-element converters, and 98 prototypes of four-element (2L-2C) converters. By careful analysis of these prototypes we find that few circuits can be realized. If we keep the output in low-pass bandwidth, the series components must be inductors and shunt components must be capacitors. Through further analysis, the first component of the resonant-filter network can be an inductor in series, or a capacitor in shunt. In the first case, only alternate (square wave) voltage sources can be applied to the network. In the second case, only alternate (square wave) current sources can be applied to the network.

1.5   Categorize Prototypes and DC/DC Converter Family Tree

There are more than 500 topologies of DC/DC converters. It is urgently necessary to categorize all prototypes. From all accumulated knowledge we can build a DC/DC converter family tree, which is shown in Figure 1.39. In each generation we introduce some circuits to readers to promote understanding of the characteristics.

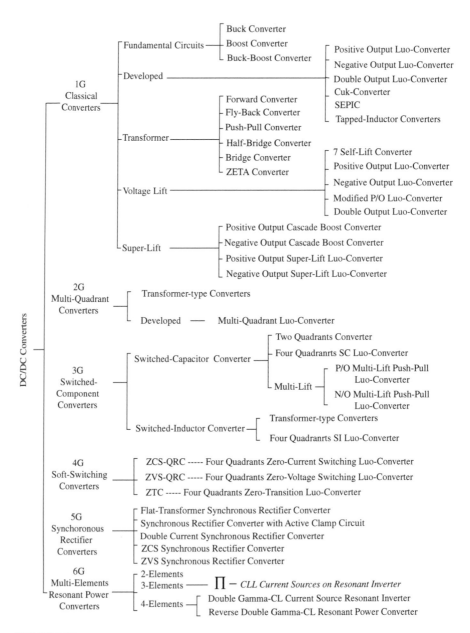

FIGURE 1.39
DC/DC converter family.

Bibliography

Cúk, S., Basics of switched-mode power conversion: topologies, magnetics, and control, in Cúk, S., Ed., *Advances in Switched-Mode Power Conversion*, Irvine, CA, Teslaco, 1995, vol. 2.

Erickson, R.W. and Maksimovic, D., *Fundamentals of Power Electronics*, 2nd ed., Kluwer Academic Publishers, Norwell, MA, 1999.

Kassakian, J., Schlecht, M., and Vergese, G., *Principles of Power Electronics*, Addison-Wesley, Reading, MA, 1991.

Kazimierczuk, M.K. and Bui, X.T., Class-E DC-DC converters with an inductive impedance inverter, *IEEE-Transactions on Power Electronics*, 4, 124, 1989.

Liu, Y. and Sen, P.C., New class-E DC-DC converter topologies with constant switching frequency, *IEEE-Transactions on Industry Applications*, 32, 961, 1996.

Luo, F.L., Re-lift circuit, a new DC-DC step-up (boost) converter, *IEE-Electronics Letters*, 33, 5, 1997.

Luo, F.L., Positive output Luo-converters, a series of new DC-DC step-up (boost) conversion circuits, in *Proceedings of the IEEE International Conference PEDS 1997*, Singapore, p. 882.

Luo, F.L., Re-lift converter: design, test, simulation and stability analysis, *IEE Proceedings on Electric Power Applications*, 145, 315, 1998.

Luo, F.L., Negative output Luo-converters, implementing the voltage lift technique, in *Proceedings of the Second World Energy System International Conference WES 1998*, Toronto, Canada, 1998, 253.

Luo, F.L. and Ye, H., DC/DC conversion techniques and nine series Luo-converters, in Rashid M.H., Luo F.L., et al., Eds., *Power Electronics Handbook*, Academic Press, San Diego, 2001, chap. 17.

Maksimovic, D. and Cúk, S., Switching converters with wide DC conversion range, *IEEE Transactions on Power Electronics*, 6, 151, 1991.

Massey, R.P. and Snyder, E.C., High voltage single-ended DC/DC converter, in *Proceedings of IEEE PESC '77*, Record. 1977. p. 156.

Middlebrook, R.D. and Cúk, S., *Advances in Switched-Mode Power Conversion*, TESLAco, Pasadena, CA, 1981, vols. I and II.

Mohan, N., Undeland, T.M., and Robbins, W.P., *Power Electronics: Converters, Applications and Design*, 3rd ed., John Wiley & Sons, New York, 2003.

Oxner, E., *Power FETs and Their Applications*, Prentice Hall, NJ, 1982.

Rashid, M.H., *Power Electronics: Circuits, Devices and Applications*, 2nd ed., Prentice-Hall, New York, 1993.

Redl, R., Molnar, B., and Sokal, N.O., Class-E resonant DC-DC power converters: analysis of operations, and experimental results at 1.5 Mhz, *IEEE-Transactions on Power Electronics*, 1, 111, 1986.

Severns, R.P. and Bloom, E., *Modern DC-to-DC Switch Mode Power Converter Circuits*, Van Nostrand Reinhold Company, New York, 1985.

Smedley, K.M. and Cúk, S., One-cycle control of switching converters, *IEEE Transactions on Power Electronics*, 10, 625, 1995.

2

Voltage-Lift Converters

The voltage-lift (VL) technique is a popular method that is widely applied in electronic circuit design. Applying this technique effectively overcomes the effects of parasitic elements and greatly increases the output voltage. Therefore, these DC/DC converters can convert the source voltage into a higher output voltage with high power efficiency, high power density, and a simple structure.

2.1 Introduction

VL technique is applied in the periodical switching circuit. Usually, a capacitor is charged during switch-on by certain voltages, e.g., source voltage. This charged capacitor voltage can be arranged on top-up to some parameter, e.g., output voltage during switch-off. Therefore, the output voltage can be lifted higher. Consequently, this circuit is called a self-lift circuit. A typical example is the saw-tooth-wave generator with a self-lift circuit.

Repeating this operation, another capacitor can be charged by a certain voltage, which may possibly be the input voltage or other equivalent voltage. The second capacitor charged voltage is also possibly arranged on top-up to some parameter, especially output voltage. Therefore, the output voltage can be higher than that of the self-lift circuit. As usual, this circuit is called re-lift circuit.

Analogously, this operation can be repeated many times. Consequently, the series circuits are called triple-lift circuits, quadruple-lift circuits, and so on.

Because of the effect of parasitic elements the output voltage and power transfer efficiency of DC-DC converters are limited. Voltage lift technique opens a way to improve circuit characteristics. After long-term research, this technique has been successfully applied to DC-DC converters. Three series Luo-converters are the DC-DC converters, which were developed from prototypes using VL technique. These converters perform DC-DC voltage increasing conversion with high power density, high efficiency, and cheap topology in simple structure. They are different from any other DC-DC step-up converters and possess many advantages including a high output voltage with small ripples. Therefore, these converters are widely used in computer peripheral equipment and industrial applications, especially for high output voltage projects. This chapter's contents are arranged thusly:

1. Seven types of self-lift converters
2. Positive output Luo-converters
3. Negative output Luo-converters
4. Modified positive output Luo-converters
5. Double output Luo-converters

2.2 Seven Self-Lift Converters

All self-lift converters introduced here are derived from developed converters such as Luo-converters, Cúk-converters, and single-ended primary inductance converters (SEPICs) discussed in Section 1.3. Since all circuits are simple, usually only one more capacitor and diode required that the output voltage be higher by an input voltage. The output voltage is calculated by the formula

$$V_O = (\frac{k}{1-k} + 1)V_{in} = \frac{1}{1-k} V_{in} \qquad (2.1)$$

There are seven circuits:

- Self-lift Cúk converter
- Self-lift P/O Luo-converter
- Reverse self-lift P/O Luo-converter
- Self-lift N/O Luo-converter
- Reverse self-lift Luo-converter
- Self-lift SEPIC
- Enhanced self-lift P/O Luo-converter

These converters perform DC-DC voltage increasing conversion in simple structures. In these circuits the switch S is a semiconductor device (MOSFET, BJT, IGBT and so on). It is driven by a pulse-width-modulated (PWM) switching signal with variable frequency f and conduction duty k. For all circuits, the load is usually resistive, i.e.,

$$R = V_O / I_O$$

The normalized load is

$$z_N = \frac{R}{fL_{eq}} \tag{2.2}$$

where L_{eq} is the equivalent inductance.

We concentrate on the absolute values rather than polarity in the following description and calculations. The directions of all voltages and currents are defined and shown in the corresponding figures. We also assume that the semiconductor switch and the passive components are all ideal. All capacitors are assumed to be large enough that the ripple voltage across the capacitors can be negligible in one switching cycle for the average value discussions.

For any component X (e.g., C, L and so on): its instantaneous current and voltage are expressed as i_X and v_X. Its average current and voltage values are expressed as I_x and V_x. The input voltage and current are V_O and I_O; the output voltage and current are V_I and I_I. T and f are the switching period and frequency.

The voltage transfer gain for the continuous conduction mode (CCM) is

$$M = \frac{V_o}{V_I} = \frac{I_I}{I_o} \tag{2.3}$$

Variation of current i_L:
$$\zeta_1 = \frac{\Delta i_L / 2}{I_L} \tag{2.4}$$

Variation of current i_{LO}:
$$\zeta_2 = \frac{\Delta i_{Lo} / 2}{I_{Lo}} \tag{2.5}$$

Variation of current i_D:
$$\xi = \frac{\Delta i_D / 2}{I_D} \tag{2.6}$$

Variation of voltage v_C:
$$\rho = \frac{\Delta v_C / 2}{V_C} \tag{2.7}$$

Variation of voltage v_{C1}:
$$\sigma_1 = \frac{\Delta v_{C1} / 2}{v_{C1}} \tag{2.8}$$

Variation of voltage v_{C2}:
$$\sigma_2 = \frac{\Delta v_{C2} / 2}{v_{C2}} \tag{2.9}$$

Variation of output voltage v_O: $\varepsilon = \dfrac{\Delta V_o / 2}{V_o}$ (2.10)

Here I_D refers to the average current i_D which flows through the diode D during the switch-off period, not its average current over the whole period.

Detailed analysis of the seven self-lift DC-DC converters will be given in the following sections. Due to the length limit of this book, only the simulation and experimental results of the self-lift Cúk converter are given. However, the results and conclusions of other self-lift converters should be quite similar to those of the self-lift Cúk-converter.

2.2.1   Self-Lift Cúk Converter

Self-lift Cúk converters and their equivalent circuits during switch-on and switch-off period are shown in Figure 2.1. It is derived from the Cúk converter. During switch-on period, S and D_1 are on, D is off. During switch-off period, D is on, S and D_1 are off.

2.2.1.1   *Continuous Conduction Mode*

In steady state, the average inductor voltages over a period are zero. Thus

$$V_{C1} = V_{CO} = V_O \qquad (2.11)$$

During switch-on period, the voltage across capacitors C and C_1 are equal. Since we assume that C and C_1 are sufficiently large, so

$$V_C = V_{C1} = V_O \qquad (2.12)$$

The inductor current i_L increases during switch-on and decreases during switch-off. The corresponding voltages across L are V_I and $-(V_C - V_I)$.

Therefore,

$$kTV_I = (1-k)T(V_C - V_I)$$

Hence,

$$V_O = V_C = V_{C1} = V_{CO} = \frac{1}{1-k}V \qquad (2.13)$$

The voltage transfer gain in the CCM is

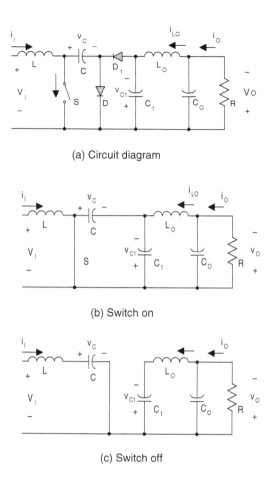

(a) Circuit diagram

(b) Switch on

(c) Switch off

FIGURE 2.1
Self-lift Cúk converter and equivalent circuits. (a) The self-lift Cúk converter. (b) The equivalent circuit during switch-on. (c) The equivalent circuit during switch-off.

$$M = \frac{V_O}{V_I} = \frac{I_I}{I_O} = \frac{1}{1-k} \tag{2.14}$$

The characteristics of M vs. conduction duty cycle k are shown in Figure 2.2.

Since all the components are considered ideal, the power loss associated with all the circuit elements are neglected. Therefore the output power P_O is considered to be equal to the input power P_{IN}:

$$V_O I_O = V_I I_I$$

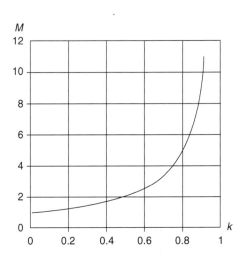

FIGURE 2.2
Voltage transfer gain M vs. k.

Thus,

$$I_L = I_I = \frac{1}{1-k} I_O$$

During switch-off,

$$i_D = i_L \qquad I_D = \frac{1}{1-k} I_O \qquad (2.15)$$

The capacitor C_O acts as a low pass filter so that

$$I_{LO} = I_O$$

The current i_L increases during switch-on. The voltage across it during switch-on is V_I, therefore its peak-to-peak current variation is

$$\Delta i_L = \frac{kTV_I}{L}$$

The variation ratio of the current i_L is

$$\zeta_1 = \frac{\Delta i_L / 2}{I_L} = \frac{kTV_I}{2I_L} = \frac{k(1-k)^2 R}{2fL} = \frac{kR}{2M^2 fL} \qquad (2.16)$$

The variation of current i_D is

$$\xi = \zeta_1 = \frac{kR}{2M^2 fL} \tag{2.17}$$

The peak-to-peak variation of voltage v_C is

$$\Delta v_C = \frac{I_L(1-k)T}{C} = \frac{I_O}{fC} \tag{2.18}$$

The variation ratio of the voltage v_C is

$$\rho = \frac{\Delta v_C / 2}{V_C} = \frac{I_O}{2fCV_O} = \frac{1}{2fRC} \tag{2.19}$$

The peak-to-peak variation of the voltage v_{C1} is

$$\Delta v_{C1} = \frac{I_{LO}(1-k)T}{C_1} = \frac{I_O(1-k)}{fC_1} \tag{2.20}$$

The variation ratio of the voltage v_{C1} is

$$\sigma_1 = \frac{\Delta v_{C1} / 2}{V_{C1}} = \frac{I_O(1-k)}{2fC_1V_O} = \frac{1}{2MfRC_1} \tag{2.21}$$

The peak-to-peak variation of the current i_{LO} is approximately:

$$\Delta i_{LO} = \frac{\frac{1}{2}\frac{\Delta v_{C1}}{2}\frac{T}{2}}{L_O} = \frac{I_O(1-k)}{8f^2 L_O C_1} \tag{2.22}$$

The variation ratio of the current i_{LO} is approximately:

$$\zeta_2 = \frac{\Delta i_{LO} / 2}{I_{LO}} = \frac{I_O(1-k)}{16f^2 L_O C_1 I_O} = \frac{1}{16Mf^2 L_O C_1} \tag{2.24}$$

The peak-to-peak variation of voltage v_O and v_{CO} is

$$\Delta v_O = \Delta v_{CO} = \frac{\frac{1}{2}\frac{\Delta i_{LO}}{2}\frac{T}{2}}{C_O} = \frac{I_O(1-k)}{64f^3 L_O C_1 C_O} \tag{2.25}$$

The variation ratio of the output voltage is

$$\varepsilon = \frac{\Delta v_O / 2}{V_O} = \frac{I_O(1-k)}{128 f^3 L_O C_1 C_O V_O} = \frac{1}{128 M f^3 L_O C_1 C_O R} \qquad (2.26)$$

The voltage transfer gain of the self-lift Cúk converter is the same as the original boost converter. However, the output current of the self-lift Cúk converter is continuous with small ripple.

The output voltage of the self-lift Cúk converter is higher than the corresponding Cúk converter by an input voltage. It retains one of the merits of the Cúk converter. They both have continuous input and output current in CCM. As for component stress, it can be seen that the self-lift converter has a smaller voltage and current stresses than the original Cúk converter.

2.2.1.2 Discontinuous Conduction Mode

Self-lift Cúk converters operate in the discontinuous conduction mode (DCM) if the current i_D reduces to zero during switch-off. As a special case, when i_D decreases to zero at $t = T$, then the circuit operates at the boundary of CCM and DCM. The variation ratio of the current i_D is 1 when the circuit works in the boundary state.

$$\xi = \frac{k}{2} \frac{R}{M^2 fL} = 1 \qquad (2.27)$$

Therefore the boundary between CCM and DCM is

$$M_B = \sqrt{k} \sqrt{\frac{R}{2fL}} = \sqrt{\frac{k z_N}{2}} \qquad (2.28)$$

where z_N is the normalized load $R/(fL)$.

The boundary between CCM and DCM is shown in Figure 2.3a. The curve that describes the relationship between M_B and z_N has the minimum value $M_B = 1.5$ and $k = 1/3$ when the normalized load z_N is 13.5.

When $M > M_B$, the circuit operates in the DCM. In this case the diode current i_D decreases to zero at $t = t_1 = [k + (1-k)m]T$ where $kT < t1 < T$ and $0 < m < 1$.

Define m as the current filling factor (FF). After mathematical manipulation:

$$m = \frac{1}{\xi} = \frac{M^2}{k \dfrac{R}{2fL}} \qquad (2.29)$$

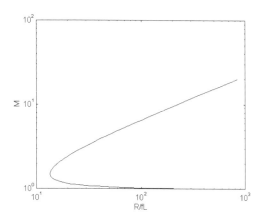

a) Boundary between CCM and DCM

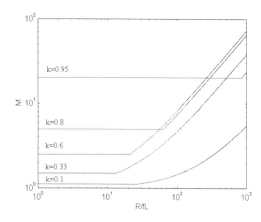

b) The voltage transfer gain M vs. the
normalized load at various k

FIGURE 2.3
Boundary between CCM and DCM and DC voltage transfer gain M vs. the normalized load at various k. (a) Boundary between CCM and DCM. (b) The voltage transfer gain M vs. the normalized load at various k.

From the above equation we can see that the discontinuous conduction mode is caused by the following factors:

- Switch frequency f is too low
- Duty cycle k is too small
- Inductance L is too small
- Load resistor R is too big

In the discontinuous conduction mode, current i_L increases during switch-on and decreases in the period from kT to $(1-k)mT$. The corresponding voltages across L are V_I and $-(V_C - V_I)$. Therefore, $kTV_I = (1-k)mT(V_C - V_I)$ Hence,

$$V_C = [1 + \frac{k}{(1-k)m}]V_I \tag{2.30}$$

Since we assume that C, C_1, and C_O are large enough,

$$V_O = V_C = V_{CO} = [1 + \frac{k}{(1-k)m}]V_I \tag{2.31}$$

or

$$V_O = [1 + k^2(1-k)\frac{R}{2fL}]V_I \tag{2.32}$$

The voltage transfer gain in the DCM is

$$M_{DCM} = 1 + k^2(1-k)\frac{R}{2fL} \tag{2.33}$$

The relation between DC voltage transfer gain M and the normalized load at various k in the DCM is also shown Figure 2.3b. It can be seen that in DCM, the output voltage increases as the load resistance R is increasing.

2.2.2   Self-Lift P/O Luo-Converter

Self-lift positive output Luo-converters and the equivalent circuits during switch-on and switch-off period are shown in Figure 2.4. It is the self-lift circuit of the positive output Luo-converter. It is derived from the elementary circuit of positive output Luo-converter. During switch-on period, S and D_1 are switch-on, D is switch-off. During switch-off period, D is on, S and D_1 are off.

2.2.2.1   Continuous Conduction Mode

In steady state, the average inductor voltages over a period are zero. Thus

$$V_C = V_{CO} = V_O$$

During switch-on period, the voltage across capacitor C_1 is equal to the source voltage. Since we assume that C and C_1 are sufficiently large,

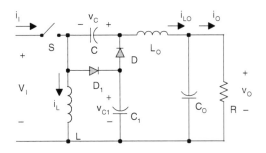

a) Self–Lift Positive Output Luo–Converter

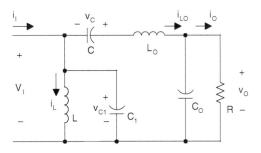

b) The equivalent circuit during switch–on

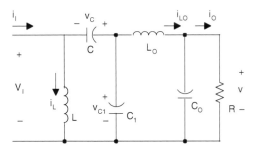

c) The equivalent circuit during switch–off

FIGURE 2.4
Self-lift positive output Luo-converter and its equivalent circuits. (a) Self-lift positive output Luo-converter. (b) The equivalent circuit during switch-on. (c) The equivalent circuit during switch-off.

$$V_{C1} = V_I$$

The inductor current i_L increases in the switch-on period and decreases in the switch-off period. The corresponding voltages across L are V_I and $-(V_C - V_{C1})$.

Therefore

$$kTV_I = (1-k)T(V_C - V_{C1})$$

Hence,

$$V_O = \frac{1}{1-k} V_I$$

The voltage transfer gain in the CCM is

$$M = \frac{V_O}{V_I} = \frac{1}{1-k} \tag{2.34}$$

Since all the components are considered ideal, the power loss associated with all the circuit elements are neglected. Therefore the output power P_O is considered to be equal to the input power P_{IN}, $V_O I_O = V_I I_I$. Thus,

$$I_I = \frac{1}{1-k} I_O$$

The capacitor C_O acts as a low pass filter so that

$$I_{LO} = I_O$$

The charge of capacitor C increases during switch-on and decreases during switch-off.

$$Q_+ = I_{C-ON} kT = I_O kT \qquad Q_- = I_{C-OFF}(1-k)T = I_L(1-k)T$$

In a switch period,

$$Q_+ = Q_- \qquad I_L = \frac{k}{1-k} I_O$$

during switch-off period,

$$i_D = i_L + i_{LO}$$

Therefore,

$$I_D = I_L + I_{LO} = \frac{1}{1-k} I_O$$

For the current and voltage variations and boundary condition, we can get the following equations using a similar method that was used in the analysis of self-lift Cúk converter.
Current variations:

$$\zeta_1 = \frac{1}{2M^2}\frac{R}{fL} \quad \zeta_2 = \frac{k}{2M}\frac{R}{fL_O} \quad \xi = \frac{k}{2M^2}\frac{R}{fL_{eq}}$$

where L_{eq} refers to

$$L_{eq} = \frac{LL_O}{L+L_O}$$

Voltage variations:

$$\rho = \frac{k}{2}\frac{1}{fCR} \quad \sigma_1 = \frac{M}{2}\frac{1}{fC_1R} \quad \varepsilon = \frac{k}{8M}\frac{1}{f^2L_OC_O}$$

2.2.2.2 Discontinuous Conduction Mode

Self-lift positive output Luo-converters operate in the DCM if the current i_D reduces to zero during switch-off. As the critical case, when i_D decreases to zero at $t = T$, then the circuit operates at the boundary of CCM and DCM.

The variation ratio of the current i_D is 1 when the circuit works in the boundary state.

$$\xi = \frac{k}{2M^2}\frac{R}{fL_{eq}} = 1$$

Therefore the boundary between CCM and DCM is

$$M_B = \sqrt{k}\sqrt{\frac{R}{2fL_{eq}}} = \sqrt{\frac{kz_N}{2}} \tag{2.35}$$

where z_N is the normalized load $R/(fL_{eq})$ and L_{eq} refers to $L_{eq} = LL_O/L+L_O$.

When $M > M_B$, the circuit operates at the DCM. In this case the circuit operates in the diode current i_D decreases to zero at $t = t_1 = [k + (1 - k) m]$ T, where $KT < t_1 < T$ and $0 < m < 1$, with m as the current filling factor. We define m as:

$$m = \frac{1}{\xi} = \frac{M^2}{k\dfrac{R}{2fL_{eq}}} \tag{2.36}$$

In the discontinuous conduction mode, current i_L increases in the switch-on period kT and decreases in the period from kT to $(1 - k)mT$. The corresponding voltages across L are V_I and $-(V_C - V_{C1})$. Therefore,

$$kTV_I = (1 - k)mT(V_C - V_{C1})$$

and

$$V_C = V_{CO} = V_O \qquad V_{C1} = V_I$$

Hence,

$$V_O = [1 + \frac{k}{(1-k)m}]V_I$$

or

$$V_O = [1 + k^2(1-k)\frac{R}{2fL_{eq}}]V_I \qquad (2.37)$$

So the real DC voltage transfer gain in the DCM is

$$M_{DCM} = 1 + k^2(1-k)\frac{R}{2fL_{eq}} \qquad (2.38)$$

In DCM, the output voltage increases as the load resistance R is increasing.

2.2.3   Reverse Self-Lift P/O Luo-Converter

Reverse self-lift positive output Luo-converters and their equivalent circuits during switch-on and switch-off period are shown in Figure 2.5. It is derived from the elementary circuit of positive output Luo-converters. During switch-on period, S and D_1 are on, D is off. During switch-off period, D is on, S and D_1 are off.

2.2.3.1   *Continuous Conduction Mode*

In steady state, the average inductor voltages over a period are zero. Thus

$$V_{C1} = V_{CO} = V_O$$

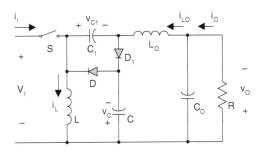

a) Reverse Self–Lift Positive Output Luo–Converter

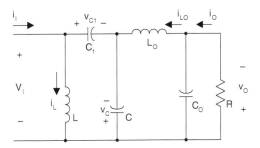

b) The equivalent circuit during switch–on

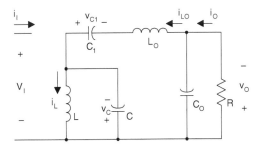

c) The equivalent circuit during switch–off

FIGURE 2.5
Reverse self-lift positive output Luo-converter and its equivalent circuits. (a) Reverse self-lift positive output Luo-converter. (b) The equivalent circuit during switch-on. (c) The equivalent circuit during switch-off.

During switch-on period, the voltage across capacitor C is equal to the source voltage plus the voltage across C_1. Since we assume that C and C_1 are sufficiently large,

$$V_{C1} = V_I + V_C$$

Therefore,

$$V_{C1} = V_I + \frac{k}{1-k}V_I = \frac{1}{1-k}V_I \qquad V_O = V_{CO} = V_{C1} = \frac{1}{1-k}V_I \qquad (2.39)$$

The voltage transfer gain in the CCM is

$$M = \frac{V_O}{V_I} = \frac{1}{1-k} \qquad (2.40)$$

Since all the components are considered ideal, the power losses on all the circuit elements are neglected. Therefore the output power P_O is considered to be equal to the input power P_{IN},

$$V_O I_O = V_I I_I$$

Thus,

$$I_I = \frac{1}{1-k}I_O$$

The capacitor C_O acts as a low pass filter so that

$$I_{LO} = I_O$$

The charge of capacitor C_1 increases during switch-on and decreases during switch-off

$$Q_+ = I_{C1-ON}kT$$

$$Q_- = I_{LO}(1-k)T = I_O(1-k)T$$

In a switch period,

$$Q_+ = Q_- \qquad I_{C1-ON} = \frac{1-k}{k}I_O$$

$$I_{C-ON} = I_{LO} + I_{C1-ON} = I_O + \frac{1-k}{k}I_O = \frac{1}{k}I_O \qquad (2.41)$$

The charge of capacitor C increases during switch-off and decreases during switch-on.

$$Q_+ = I_{C-OFF}(1-k)T \quad Q_- = I_{C-ON}kT = \frac{1}{k}I_O kT$$

In a switch period,

$$Q_+ = Q_- \quad I_{C-OFF} = \frac{1-k}{k}I_{C-ON} = \frac{1}{1-k}I_O \tag{2.42}$$

Therefore,

$$I_L = I_{LO} + I_{C-OFF} = I_O + \frac{1}{1-k}I_O = \frac{2-k}{1-k}I_O = I_O + I_I$$

During switch-off,

$$i_D = i_L - i_{LO}$$

Therefore,

$$I_D = I_L - I_{LO} = I_O$$

The following equations are used for current and voltage variations and boundary conditions.

Current variations:

$$\zeta_1 = \frac{k}{(2-k)M^2}\frac{R}{fL}, \quad \zeta_2 = \frac{k}{2M}\frac{R}{fL_O} \quad \xi = \frac{1}{2M^2}\frac{R}{fL_{eq}}$$

where L_{eq} refers to

$$L_{eq} = \frac{LL_O}{L+L_O}$$

Voltage variations:

$$\rho = \frac{1}{2k}\frac{1}{fCR} \quad \sigma_1 = \frac{1}{2M}\frac{1}{fC_1R} \quad \varepsilon = \frac{k}{16M}\frac{1}{f^2C_OL_O}$$

2.2.3.2   Discontinuous Conduction Mode

Reverse self-lift positive output Luo-converter operates in the DCM if the current i_D reduces to zero during switch-off at $t = T$, then the circuit operates at the boundary of CCM and DCM. The variation ratio of the current i_D is 1 when the circuit works in the boundary state.

$$\xi = \frac{k}{2M^2} \frac{R}{fL_{eq}} = 1$$

Therefore the boundary between CCM and DCM is

$$M_B = \sqrt{k} \sqrt{\frac{R}{2fL_{eq}}} = \sqrt{\frac{kz_N}{2}} \tag{2.43}$$

where z_N is the normalized load $R/(fL_{eq})$ and L_{eq} refers to $L_{eq} = LL_O/L + L_O$.

When $M > M_B$, the circuit operates in the DCM. In this case the diode current i_D decreases to zero at $t = t_1 = [k + (1 - k)\,m]\,T$, where $KT < t1 < T$ and $0 < m < 1$. m is the current filling factor.

$$m = \frac{1}{\xi} = \frac{M^2}{k\dfrac{R}{2fL_{eq}}} \tag{2.44}$$

In the discontinuous conduction mode, current i_L increases during switch-on and decreases in the period from kT to $(1 - k)mT$. The corresponding voltages across L are V_I and $-V_C$.

Therefore,

$$kTV_I = (1 - k)mTV_C$$

and

$$V_{C1} = V_{CO} = V_O \qquad V_{C1} = V_I + V_C$$

Hence,

$$V_O = [1 + \frac{k}{(1 - k)m}]V_I$$

or

$$V_O = \left[1 + k^2(1 - k)\frac{R}{2fL_{eq}}\right]V_I \tag{2.45}$$

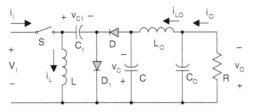

a) Self-Lift Negative Output Luo-Converter

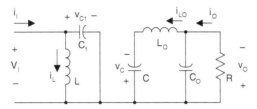

b) The equivalent circuit during switch-on

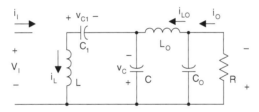

c) The equivalent circuit during switch-off

FIGURE 2.6
Self-lift negative output Luo-converter and its equivalent circuits. (a) Self-lift negative output Luo-converter. (b) The equivalent circuit during switch-on. (c) The equivalent circuit during switch-off.

So the real DC voltage transfer gain in the DCM is

$$M_{DCM} = 1 + k^2(1-k)\frac{R}{2fL} \tag{2.46}$$

In DCM the output voltage increases as the load resistance R increases.

2.2.4   Self-Lift N/O Luo-Converter

Self-lift negative output Luo-converters and their equivalent circuits during switch-on and switch-off period are shown in Figure 2.6. It is the self-lift circuit of the negative output Luo-converter. The function of capacitor C_1 is to lift the voltage V_C by a source voltage V_I. S and D_1 are on, and D is off during switch-on period. D is on, and S and D_1 are off during switch-off period.

2.2.4.1   Continuous Conduction Mode

In the steady state, the average inductor voltages over a period are zero. Thus

$$V_C = V_{CO} = V_O$$

During switch-on period, the voltage across capacitor C_1 is equal to the source voltage. Since we assume that C and C_1 are sufficiently large, $V_{C1} = V_I$.

Inductor current i_L increases in the switch-on period and decreases in the switch-off period. The corresponding voltages across L are V_I and $-(V_C - V_{C1})$.

Therefore,

$$kTV_I = (1-k)T(V_C - V_{C1})$$

Hence,

$$V_O = V_C = V_{CO} = \frac{1}{1-k} V_I \tag{2.47}$$

The voltage transfer gain in the CCM is

$$M = \frac{V_O}{V_I} = \frac{1}{1-k} \tag{2.48}$$

Since all the components are considered ideal, the power loss associated with all the circuit elements are neglected. Therefore the output power P_O is considered to be equal to the input power P_{IN}, $V_O I_O = V_I I_I$.

Thus,

$$I_I = \frac{1}{1-k} I_O$$

The capacitor C_O acts as a low pass filter so that $I_{LO} = I_O$.

For the current and voltage variations and boundary condition, the following equations can be obtained by using a similar method that was used in the analysis of self-lift Cúk converter.

Current variations:

$$\zeta_1 = \frac{k}{2M^2} \frac{R}{fL}, \quad \zeta_2 = \frac{k}{16} \frac{1}{f^2 L_O C} \quad \xi = \frac{k}{2M^2} \frac{R}{fL}$$

Voltage variations:

$$\rho = \frac{k}{2} \frac{1}{fCR} \qquad \sigma_1 = \frac{M}{2} \frac{1}{fC_1R} \qquad \varepsilon = \frac{k}{128} \frac{1}{f^3 L_O CC_O R}$$

2.2.4.2 Discontinuous Conduction Mode

Self-lift negative output Luo-converters operate in the DCM if the current i_D reduces to zero at $t = T$, then the circuit operates at the boundary of CCM and DCM. The variation ratio of the current i_D is 1 when the circuit works at the boundary state.

$$\xi = \frac{k}{2M^2} \frac{R}{fL} = 1$$

Therefore the boundary between CCM and DCM is

$$M_B = \sqrt{k} \sqrt{\frac{R}{2fL_{eq}}} = \sqrt{\frac{kz_N}{2}} \qquad (2.49)$$

where L_{eq} refers to $L_{eq} = L$ and z_N is the normalized load $R/(fL_{eq})$.

When $M > M_B$, the circuit operates in the DCM. In this case the diode current i_D decreases to zero at $t = t_1 = [k + (1 - k)m]T$, where $KT < t1 < T$ and $0 < m < 1$. m is the current filling factor and is defined as:

$$m = \frac{1}{\xi} = \frac{M^2}{k \dfrac{R}{2fL}} \qquad (2.50)$$

In the discontinuous conduction mode, current i_L increases during switch-on and decreases during period from kT to $(1 - k)mT$. The voltages across L are V_I and $-(V_C - V_{C1})$.

$$kTV_I = (1 - k)mT(V_C - V_{C1})$$

and

$$V_{C1} = V_I \qquad V_C = V_{CO} = V_O$$

Hence,

$$V_O = [1 + \frac{k}{(1-k)m}]V_I \quad \text{or} \quad V_O = [1 + k^2(1-k)\frac{R}{2fL}]V_I$$

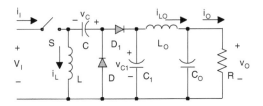

a) Reverse Self-Lift Negative Output Luo-Converter

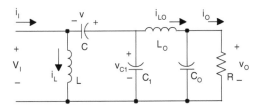

b) The equivalent circuit during switch-on

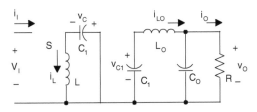

c) The equivalent circuit during switch-off

FIGURE 2.7
Reverse self-lift negative output Luo-converter and its equivalent circuits. (a) Reverse self-lift negative output Luo-converter. (b) The equivalent circuit during switch-on. (c) The equivalent circuit during switch-off.

So the real DC voltage transfer gain in the DCM is

$$M_{DCM} = 1 + k^2(1-k)\frac{R}{2fL} \qquad (2.51)$$

We can see that in DCM, the output voltage increases as the load resistance R is increasing.

2.2.5   Reverse Self-Lift N/O Luo-Converter

Reverse self-lift negative output Luo-converters and their equivalent circuits during switch-on and switch-off period are shown in Figure 2.7. It is derived from the Zeta converter. During switch-on period, S and D_1 are on, D is off. During switch-off period, D is on, S and D_1 are off.

2.2.5.1 Continuous Conduction Mode

In steady state, the average inductor voltages over a period are zero. Thus

$$V_{C1} = V_{CO} = V_O$$

The inductor current i_L increases in the switch-on period and decreases in the switch-off period. The corresponding voltages across L are V_I and $-V_C$. Therefore

$$kTV_I = (1-k)TV_C$$

Hence,

$$V_C = \frac{k}{1-k} V_I \tag{2.52}$$

voltage across C. Since we assume that C and C_1 are sufficiently large,

$$V_{C1} = V_I + V_C$$

Therefore,

$$V_{C1} = V_I + \frac{k}{1-k} V_I = \frac{1}{1-k} V_I \qquad V_O = V_{CO} = V_{C1} = \frac{1}{1-k} V_I$$

The voltage transfer gain in the CCM is

$$M = \frac{V_O}{V_I} = \frac{1}{1-k} \tag{2.53}$$

Since all the components are considered ideal, the power loss associated with all the circuit elements is neglected. Therefore the output power P_O is considered to be equal to the input power P_{IN}, $V_O I_O = V_I I_I$. Thus,

$$I_I = \frac{1}{1-k} I_O$$

The capacitor C_O acts as a low pass filter so that $I_{LO} = I_O$.
The charge of capacitor C_1 increases during switch-on and decreases during switch-off.

$$Q_+ = I_{C1-ON} kT \qquad Q_- = I_{C1-OFF}(1-k)T = I_O(1-k)T$$

In a switch period,

$$Q_+ = Q_- \quad I_{C1-ON} = \frac{1-k}{k} I_{C-OFF} = \frac{1-k}{k} I_o$$

The charge of capacitor C increases during switch-on and decreases during switch-off.

$$Q_+ = I_{C-ON} kT \quad Q_- = I_{C-OFF}(1-k)T$$

In a switch period,

$$Q_+ = Q_-$$

$$I_{C-ON} = I_{C1-ON} + I_{LO} = \frac{1-k}{k} I_o + I_o = \frac{1}{k} I_o$$

$$I_{C-OFF} = \frac{k}{1-k} I_{C-ON} = \frac{k}{1-k} \frac{1}{k} I_o = \frac{1}{1-k} I_o$$

Therefore,

$$I_L = I_{C-OFF} = \frac{1}{1-k} I_o$$

During switch-off period,

$$i_D = i_L \quad I_D = I_L = \frac{1}{1-k} I_o$$

For the current and voltage variations and the boundary condition, we can get the following equations using a similar method that was used in the analysis of self-lift Cúk converter.
Current variations:

$$\zeta_1 = \frac{k}{2M^2} \frac{R}{fL} \quad \zeta_2 = \frac{1}{16M} \frac{R}{f^2 L_0 C_1} \quad \xi = \frac{k}{2M^2} \frac{R}{fL}$$

Voltage variations:

$$\rho = \frac{1}{2k} \frac{1}{fCR} \quad \sigma_1 = \frac{1}{2M} \frac{1}{fC_1 R} \quad \varepsilon = \frac{1}{128M} \frac{1}{f^3 L_0 C_1 C_0 R}$$

2.2.5.2   Discontinuous Conduction Mode

Reverse self-lift negative output Luo-converters operate in the DCM if the current i_D reduces to zero during switch-off. As a special case, when i_D decreases to zero at $t = T$, then the circuit operates at the boundary of CCM and DCM.

The variation ratio of the current i_D is 1 when the circuit works in the boundary state.

$$\xi = \frac{k}{2M^2} \frac{R}{fL_{eq}} = 1$$

The boundary between CCM and DCM is

$$M_B = \sqrt{k} \sqrt{\frac{R}{2fL_{eq}}} = \sqrt{\frac{kz_N}{2}}$$

where z_N is the normalized load $R/(fL_{eq})$ and L_{eq} refers to $L_{eq} = L$.

When $M > M_B$, the circuit operates at the DCM. In this case, diode current i_D decreases to zero at $t = t_1 = [k + (1 - k) m] T$ where $KT < t_1 < T$ and $0 < m < 1$ with m as the current filling factor.

$$m = \frac{1}{\xi} = \frac{M^2}{k \dfrac{R}{2fL_{eq}}} \tag{2.54}$$

In the discontinuous conduction mode, current i_L increases in the switch-on period kT and decreases in the period from kT to $(1 - k)mT$. The corresponding voltages across L are V_I and $-V_C$.

Therefore,

$$kTV_I = (1 - k)mTV_C$$

and

$$V_{C1} = V_{CO} = V_O \qquad V_{C1} = V_I + V_C$$

Hence,

$$V_O = [1 + \frac{k}{(1 - k)m}]V_I$$

or

$$V_O = \left[1 + k^2(1-k)\frac{R}{2fL}\right]V_I \tag{2.55}$$

The voltage transfer gain in the DCM is

$$M_{DCM} = 1 + k^2(1-k)\frac{R}{2fL} \tag{2.56}$$

It can be seen that in DCM, the output voltage increases as the load resistance R is increasing.

2.2.6 Self-Lift SEPIC

Self-lift SEPIC and the equivalent circuits during switch-on and switch-off period are shown in Figure 2.8. It is derived from SEPIC (with output filter). S and D_1 are on, and D is off during switch-on period. D is on, and S and D_1 are off during switch-off period.

2.2.6.1 *Continuous Conduction Mode*

In steady state, the average voltage across inductor L over a period is zero. Thus $V_C = V_I$.

During switch-on period, the voltage across capacitor C_1 is equal to the voltage across C. Since we assume that C and C_1 are sufficiently large,

$$V_{C1} = V_C = V_I$$

In steady state, the average voltage across inductor L_O over a period is also zero.

Thus $$V_{C2} = V_{CO} = V_O$$

The inductor current i_L increases in the switch-on period and decreases in the switch-off period. The corresponding voltages across L are V_I and $-(V_C - V_{C1} + V_{C2} - V_I)$.
Therefore

$$kTV_I = (1-k)T(V_C - V_{C1} + V_{C2} - V_I)$$

or

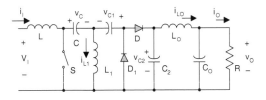

a) Self-Lift Sepic Converter

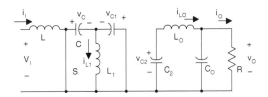

b) The equivalent circuit during switch-on

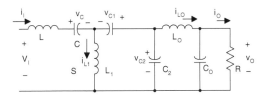

c) The equivalent circuit during switch-off

FIGURE 2.8
Self-lift sepic converter and its equivalent circuits. (a) Self-lift sepic converter. (b) The equivalent circuit during switch-on. (c) The equivalent circuit during switch-off.

$$kTV_I = (1-k)T(V_O - V_I)$$

Hence,

$$V_O = \frac{1}{1-k}V_I = V_{CO} = V_{C2} \tag{2.57}$$

The voltage transfer gain in the CCM is

$$M = \frac{V_O}{V_I} = \frac{1}{1-k} \tag{2.58}$$

Since all the components are considered ideal, the power loss associated with all the circuit elements is neglected. Therefore the output power P_O is considered to be equal to the input power P_{IN}, $V_O I_O = V_I I_I$.
Thus,

$$I_I = \frac{1}{1-k}I_O = I_L$$

The capacitor C_O acts as a low pass filter so that

$$I_{LO} = I_O$$

The charge of capacitor C increases during switch-off and decreases during switch-on.

$$Q_- = I_{C-ON}kT \qquad Q_+ = I_{C-OFF}(1-k)T = I_I(1-k)T$$

In a switch period,

$$Q_+ = Q_- \qquad I_{C-ON} = \frac{1-k}{k}I_{C-OFF} = \frac{1-k}{k}I_I$$

The charge of capacitor C_2 increases during switch-off and decreases during switch-on.

$$Q_- = I_{C2-ON}kT = I_OkT \qquad Q_+ = I_{C2-OFF}(1-k)T$$

In a switch period,

$$Q_+ = Q_- \qquad I_{C2-OFF} = \frac{k}{1-k}I_{C-ON} = \frac{k}{1-k}I_O$$

The charge of capacitor C_1 increases during switch-on and decreases during switch-off.

$$Q_+ = I_{C1-ON}kT \qquad Q_- = I_{C1-OFF}(1-k)T$$

In a switch period,

$$Q_+ = Q_- \qquad I_{C1-OFF} = I_{C2-OFF} + I_{LO} = \frac{k}{1-k}I_O + I_O = \frac{1}{1-k}I_O$$

Therefore

$$I_{C1-ON} = \frac{1-k}{k}I_{C1-OFF} = \frac{1}{k}I_O \qquad I_{L1} = I_{C1-ON} - I_{C-ON} = 0$$

During switch-off,

$$i_D = i_L - i_{L1}$$

Therefore,

$$I_D = I_I = \frac{1}{1-k} I_O$$

For the current and voltage variations and the boundary condition, we can get the following equations using a similar method that is used in the analysis of self-lift Cúk converter.

Current variations:

$$\zeta_1 = \frac{k}{2M^2} \frac{R}{fL} \qquad \zeta_2 = \frac{k}{16} \frac{R}{f^2 L_O C_2} \qquad \xi = \frac{k}{2M^2} \frac{R}{fL_{eq}}$$

where L_{eq} refers to

$$L_{eq} = \frac{LL_O}{L + L_O}$$

Voltage variations:

$$\rho = \frac{M}{2} \frac{1}{fCR} \qquad \sigma_1 = \frac{M}{2} \frac{1}{fC_1 R} \qquad \sigma_2 = \frac{k}{2} \frac{1}{fC_2 R} \qquad \varepsilon = \frac{k}{128} \frac{1}{f^3 L_O C_2 C_O R}$$

2.2.6.2 *Discontinuous Conduction Mode*

Self-lift Sepic converters operate in the DCM if the current i_D reduces to zero during switch-off. As a special case, when i_D decreases to zero at $t = T$, then the circuit operates at the boundary of CCM and DCM.

The variation ratio of the current i_D is 1 when the circuit works in the boundary state.

$$\xi = \frac{k}{2M^2} \frac{R}{fL_{eq}} = 1$$

Therefore the boundary between CCM and DCM is

$$M_B = \sqrt{k} \sqrt{\frac{R}{2fL_{eq}}} = \sqrt{\frac{kz_N}{2}} \qquad (2.59)$$

where z_N is the normalized load $R/(fL_{eq})$ and L_{eq} refers to $L_{eq} = \dfrac{LL_O}{L+L_O}$. ie
When $M > M_B$, the circuit operates in the DCM. In this c.... current i_D decreases to zero at $t = t_1 = [k + (1-k)m]T$ where $KT < t_1 < T$ and $0 < m < 1$. m is defined as:

$$m = \frac{1}{\xi} = \frac{M^2}{k\dfrac{R}{2fL_{eq}}} \tag{2.60}$$

In the discontinuous conduction mode, current i_L increases during switch-on and decreases in the period from kT to $(1-k)mT$. The corresponding voltages across L are V_I and $-(V_C - V_{C1} + V_{C2} - V_I)$.
Thus,

$$kTV_I = (1-k)T(V_C - V_{C1} + V_{C2} - V_I)$$

and

$$V_C = V_I \qquad V_{C1} = V_C = V_I \qquad V_{C2} = V_{CO} = V_O$$

Hence,

$$V_O = [1 + \frac{k}{(1-k)m}]V_I$$

or

$$V_O = \left[1 + k^2(1-k)\frac{R}{2fL_{eq}}\right]V_I$$

So the real DC voltage transfer gain in the DCM is

$$M_{DCM} = 1 + k^2(1-k)\frac{R}{2fL_{eq}} \tag{2.61}$$

In DCM, the output voltage increases as the load resistance R is increasing.

2.2.7 Enhanced Self-Lift P/O Luo-Converter

Enhanced self-lift positive output Luo-converter circuits and the equivalent circuits during switch-on and switch-off periods are shown in Figure 2.9. They are derived from the self-lift positive output Luo-converter in Figure 2.4 with swapping the positions of switch S and inductor L.

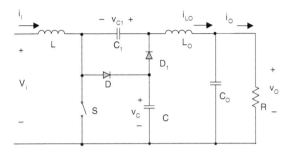

FIGURE 2.9
Enhanced self-lift P/O Luo-converter.

During switch-on period, S and D_1 are on, and D is off. Obtain:

$$V_C = V_{C1}$$

and

$$\Delta i_L = \frac{V_I}{L} kT$$

During switch-off period, D is on, and S and D_1 are off.

$$\Delta i_L = \frac{V_C - V_I}{L}(1-k)T$$

So that

$$V_C = \frac{1}{1-k} V_I$$

The output voltage and current and the voltage transfer gain are

$$V_O = V_I + V_{C1} = (1+\frac{1}{1-k})V_I \qquad (2.62)$$

$$I_O = \frac{1-k}{2-k} I_I \qquad (2.64)$$

$$M = 1+\frac{1}{1-k} = \frac{2-k}{1-k} \qquad (2.65)$$

Average voltages:

$$V_C = \frac{1}{1-k} V_I \qquad (2.66)$$

$$V_{C1} = \frac{1}{1-k} V_I \qquad (2.67)$$

Average currents:

$$I_{LO} = I_O \qquad (2.68)$$

$$I_L = \frac{2-k}{1-k} I_O = I_I \qquad (2.69)$$

Therefore,

$$\frac{V_O}{V_I} = \frac{1}{1-k} + 1 = \frac{2-k}{1-k} \qquad (2.70)$$

2.3   Positive Output Luo-Converters

Positive output Luo-converters perform the voltage conversion from positive to positive voltages using VL technique. They work in the first quadrant with large voltage amplification. Five circuits have been introduced in the literature:

- Elementary circuit
- Self-lift circuit
- Re-lift circuit
- Triple-lift circuit
- Quadruple-lift circuit

The elementary circuit can perform step-down and step-up DC-DC conversion, which was introduced in previous section. Other positive output Luo-converters are derived from this elementary circuit, they are the self-lift circuit, re-lift circuit, and multiple-lift circuits (e.g., triple-lift and quadruple-lift circuits) shown in the corresponding figures. Switch S in these diagrams is a P-channel power MOSFET device (PMOS), and S_1 is an N-channel power MOSFET device (NMOS). They are driven by a PWM switch signal with repeating frequency f and conduction duty k. The switch repeating period is $T = 1/f$, so that the switch-on period is kT and switch-off period is $(1 - k)T$. For all circuits, the load is usually resistive, $R = V_O/I_O$; the combined inductor $L = L_1L_2/(L_1 + L_2)$; the normalized load is $z_N = R/fL$. Each converter consists of a positive Luo-pump and a low-pass filter L_2-C_O, and lift circuit (introduced in the following sections). The pump inductor L_1 transfers the energy from source to capacitor C during switch-off, and then the stored energy on capacitor C is delivered to load R during switch-on. Therefore, if the voltage V_C is higher the output voltage V_O should be higher.

When the switch S is turned off, the current i_D flows through the freewheeling diode D. This current descends in whole switch-off period $(1 - k)T$. If current i_D does not become zero before switch S turned on again, this working state is defined as the continuous conduction mode (CCM). If current i_D becomes zero before switch S turned on again, this working state is defined as the discontinuous conduction mode (DCM).

Assuming that the output power is equal to the input power,

$$P_O = P_{IN} \quad \text{or} \quad V_OI_O = V_II_I$$

The voltage transfer gain in continuous mode is

$$M = \frac{V_O}{V_I} = \frac{I_I}{I_O}$$

Variation ratio of current i_{L1}:

$$\xi_1 = \frac{\Delta i_{L1}/2}{I_{L1}}$$

Variation ratio of current i_{L2}:

$$\xi_2 = \frac{\Delta i_{L2}/2}{I_{L2}}$$

Variation ratio of current i_D:

$$\zeta = \frac{\Delta i_D/2}{I_{L1} + I_{L2}}$$

Variation ratio of current i_{L2+j} is

$$\chi_j = \frac{\Delta i_{L2+j}/2}{I_{L2+j}} \quad j = 1, 2, 3, \dots$$

Variation ratio of voltage v_C:

$$\rho = \frac{\Delta v_C / 2}{V_C}$$

Variation ratio of voltage v_{Cj}:

$$\sigma_j = \frac{\Delta v_{Cj} / 2}{V_{Cj}} \quad j = 1, 2, 3, 4, \ldots$$

Variation ratio of output voltage v_{CO}:

$$\varepsilon = \frac{\Delta v_O / 2}{V_O}$$

2.3.1 Elementary Circuit

Elementary circuit and its switch-on and -off equivalent circuits are shown in Figure 2.10. Capacitor C acts as the primary means of storing and transferring energy from the input source to the output load via the pump inductor L_1. Assuming capacitor C to be sufficiently large, the variation of the voltage across capacitor C from its average value V_C can be neglected in steady state, i.e., $v_C(t) \approx V_C$, even though it stores and transfers energy from the input to the output.

2.3.1.1 *Circuit Description*

When switch S is on, the source current $i_I = i_{L1} + i_{L2}$. Inductor L_1 absorbs energy from the source. In the mean time inductor L_2 absorbs energy from source and capacitor C, both currents i_{L1} and i_{L2} increase. When switch S is off, source current $i_I = 0$. Current i_{L1} flows through the free-wheeling diode D to charge capacitor C. Inductor L_1 transfers its stored energy to capacitor C. In the mean time current i_{L2} flows through the $(C_O - R)$ circuit and free-wheeling diode D to keep itself continuous. Both currents i_{L1} and i_{L2} decrease. In order to analyze the circuit working procession, the equivalent circuits in switch-on and -off states are shown in Figures 2.10b, c, and d.

Actually, the variations of currents i_{L1} and i_{L2} are small so that $i_{L1} \approx I_{L1}$ and $i_{L2} \approx I_{L2}$.

The charge on capacitor C increases during switch off:

$$Q+ = (1 - k)T \, I_{L1}.$$

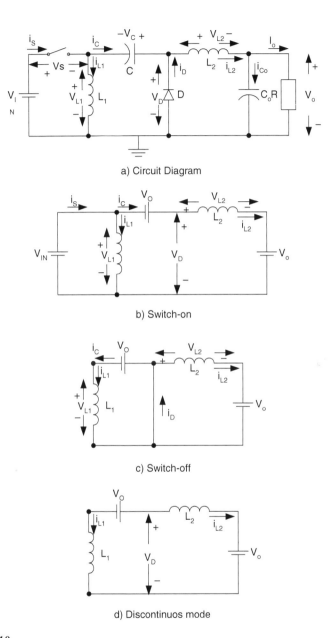

a) Circuit Diagram

b) Switch-on

c) Switch-off

d) Discontinuos mode

FIGURE 2.10
Elementary circuit of positive output Luo-converter(a) Circuit diagram. (b) Switch-on. (c) Switch-off. (d) Discontinuous mode.

It decreases during switch-on:

$$Q- = kTI_{L2}$$

In a whole period investigation, $Q+ = Q-$. Thus,

$$I_{L2} = \frac{1-k}{k} I_{L1} \tag{2.71}$$

Since capacitor C_O performs as a low-pass filter, the output current

$$I_{L2} = I_O \tag{2.72}$$

These two Equations (2.71) and (2.72) are available for all positive output Luo-converters.

The source current is $i_I = i_{L1} + i_{L2}$ during switch-on period, and $i_I = 0$ during switch-off. Thus, the average source current I_I is

$$I_I = k \times i_I = k(i_{L1} + i_{L2}) = k(1 + \frac{1-k}{k})I_{L1} = I_{L1} \tag{2.73}$$

Therefore, the output current is

$$I_O = \frac{1-k}{k} I_I \tag{2.74}$$

Hence, output voltage is

$$V_O = \frac{k}{1-k} V_I \tag{2.75}$$

The voltage transfer gain in continuous mode is

$$M_E = \frac{V_O}{V_I} = \frac{k}{1-k} \tag{2.76}$$

The curve of M_E vs. k is shown in Figure 2.11.

Current i_{L1} increases and is supplied by V_I during switch-on. It decreases and is inversely biased by $-V_C$ during switch-off. Therefore,

$$kTV_I = (1-k)TV_C$$

The average voltage across capacitor C is

$$V_C = \frac{k}{1-k} V_I = V_O \tag{2.77}$$

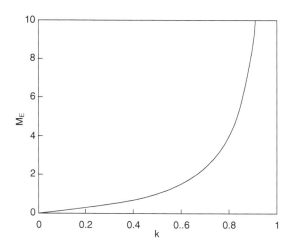

FIGURE 2.11
Voltage transfer gain M_E vs. k

2.3.1.2 Variations of Currents and Voltages

To analyze the variations of currents and voltages, some voltage and current waveforms are shown in Figure 2.12.

Current i_{L1} increases and is supplied by V_I during switch-on. It decreases and is inversely biased by $-V_C$ during switch-off. Therefore, its peak-to-peak variation is

$$\Delta i_{L1} = \frac{kTV_I}{L_1}$$

Considering Equation (2.73), the variation ratio of the current i_{L1} is

$$\xi_1 = \frac{\Delta i_{L1}/2}{I_{L1}} = \frac{kTV_I}{2L_1 I_I} = \frac{1-k}{2M_E}\frac{R}{fL_1} \tag{2.78}$$

Current i_{L2} increases and is supplied by the voltage $(V_I + V_C - V_O) = V_I$ during switch-on. It decreases and is inversely biased by $-V_O$ during switch-off. Therefore its peak-to-peak variation is

$$\Delta i_{L2} = \frac{kTV_I}{L_2} \tag{2.79}$$

Considering Equation (2.72), the variation ratio of current i_{L2} is

$$\xi_2 = \frac{\Delta i_{L2}/2}{I_{L2}} = \frac{kTV_I}{2L_2 I_O} = \frac{k}{2M_E}\frac{R}{fL_2} \tag{2.80}$$

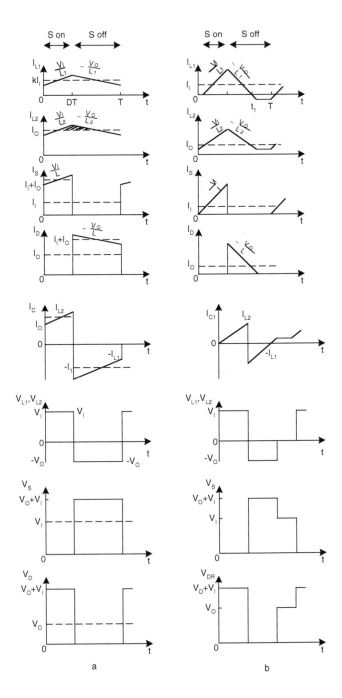

FIGURE 2.12
Some voltage and current waveforms of elementary circuit.

When switch is off, the free-wheeling diode current is $i_D = i_{L1} + i_{L2}$ and

$$\Delta i_D = \Delta i_{L1} + \Delta i_{L2} = \frac{kTV_I}{L_1} + \frac{kTV_I}{L_2} = \frac{kTV_I}{L} = \frac{(1-k)TV_O}{L} \tag{2.81}$$

Considering Equation (2.71) and Equation (2.72), the average current in switch-off period is

$$I_D = I_{L1} + I_{L2} = \frac{I_O}{1-k}$$

The variation ratio of current i_D is

$$\zeta = \frac{\Delta i_D / 2}{I_D} = \frac{(1-k)^2 TV_O}{2LI_O} = \frac{k(1-k)R}{2M_E fL} = \frac{k^2}{M_E^2} \frac{R}{2fL} \tag{2.82}$$

The peak-to-peak variation of v_C is

$$\Delta v_C = \frac{Q+}{C} = \frac{1-k}{C} TI_I$$

Considering Equation (2.77), the variation ratio of v_C is

$$\rho = \frac{\Delta v_C / 2}{V_C} = \frac{(1-k)TI_I}{2CV_O} = \frac{k}{2} \frac{1}{fCR} \tag{2.83}$$

If $L_1 = L_2 = 1$ mH, $C = C_O = 20$ μF, $R = 10$ Ω, $f = 50$ kHz and $k = 0.5$, we get $\xi_1 = \xi_2 = 0.05$, $\zeta = 0.025$ and $\rho = 0.025$. Therefore, the variations of i_{L1}, i_{L2}, and v_C are small.

In order to investigate the variation of output voltage v_O, we have to calculate the charge variation on the output capacitor C_O, because

$$Q = C_O V_O$$

and

$$\Delta Q = C_O \Delta v_O$$

ΔQ is caused by Δi_{L2} and corresponds to the **area** of the triangle with the **height** of half of Δi_{L2} and the **width** of half of the repeating period $T/2$, which is shown in Figure 2.12. Considering Equation (2.79),

$$\Delta Q = \frac{1}{2} \frac{\Delta i_{L2}}{2} \frac{T}{2} = \frac{T}{8} \frac{kTV_I}{L_2}$$

Thus, the half peak-to-peak variation of output voltage v_O and v_{CO} is

$$\frac{\Delta v_O}{2} = \frac{\Delta Q}{C_O} = \frac{kT^2 V_I}{8C_O L_2}$$

The variation ratio of output voltage v_O is

$$\varepsilon = \frac{\Delta v_O / 2}{V_O} = \frac{kT^2}{8C_O L_2} \frac{V_I}{V_O} = \frac{k}{8M_E} \frac{1}{f^2 C_O L_2} \tag{2.84}$$

If $L_2 = 1$ mH, $C_O = 20$ μF, $f = 50$ kHz and $k = 0.5$, we obtain that $\varepsilon = 0.00125$. Therefore, the output voltage V_O is almost a real DC voltage with very small ripple. Because of the resistive load, the output current $i_O(t)$ is almost a real DC waveform with very small ripple as well, and $I_O = V_O/R$.

2.3.1.3   *Instantaneous Values of Currents and Voltages*

Referring to Figure 2.12, the instantaneous values of the currents and voltages are listed below:

$$v_S = \begin{cases} 0 & for \quad 0 < t \le kT \\ V_O + V_I & for \quad kT < t \le T \end{cases} \tag{2.85}$$

$$v_D = \begin{cases} V_I + V_O & for \quad 0 < t \le kT \\ 0 & for \quad kT < t \le T \end{cases} \tag{2.86}$$

$$v_{L1} = \begin{cases} V_I & for \quad 0 < t \le kT \\ -V_O & for \quad kT < t \le T \end{cases} \tag{2.87}$$

$$v_{L2} = \begin{cases} V_I & for \quad 0 < t \le kT \\ -V_O & for \quad kT < t \le T \end{cases} \tag{2.88}$$

$$i_I = i_S = \begin{cases} i_{L1}(0) + i_{L2}(0) + \dfrac{V_I}{L} t & for \quad 0 < t \le kT \\ 0 & for \quad kT < t \le T \end{cases} \tag{2.89}$$

$$i_{L1} = \begin{cases} i_{L1}(0) + \dfrac{V_I}{L_1} t & for \quad 0 < t \le kT \\ i_{L1}(kT) - \dfrac{V_O}{L_1}(t - kT) & for \quad kT < t \le T \end{cases} \tag{2.90}$$

$$i_{L2} = \begin{cases} i_{L2}(0) + \dfrac{V_I}{L_2} t & \text{for} \quad 0 < t \le kT \\ i_{L2}(kT) - \dfrac{V_O}{L_2}(t - kT) & \text{for} \quad kT < t \le T \end{cases} \tag{2.91}$$

$$i_D = \begin{cases} 0 & \text{for} \quad 0 < t \le kT \\ i_{L1}(kT) + i_{L2}(kT) - \dfrac{V_O}{L}(t - kT) & \text{for} \quad kT < t \le T \end{cases} \tag{2.92}$$

$$i_C \approx \begin{cases} i_{L2}(0) + \dfrac{V_I}{L_2} t & \text{for} \quad 0 < t \le kT \\ -i_{L1}(kT) + \dfrac{V_O}{L_1}(t - kT) & \text{for} \quad kT < t \le T \end{cases} \tag{2.93}$$

$$i_{CO} \approx \begin{cases} i_{L2}(0) + \dfrac{V_I}{L_2} t - I_O & \text{for} \quad 0 < t \le kT \\ -i_{L1}(kT) + \dfrac{V_O}{L_1}(t - kT) - I_O & \text{for} \quad kT < t \le T \end{cases} \tag{2.94}$$

where

$$i_{L1}(0) = \frac{kI_O}{1-k} - \frac{(1-k)V_O}{2fL_1}$$

$$i_{L1}(kT) = \frac{kI_O}{1-k} + \frac{(1-k)V_O}{2fL_1}$$

$$i_{L2}(0) = I_O - \frac{(1-k)V_O}{2fL_2}$$

$$i_{L2}(kT) = I_O + \frac{(1-k)V_O}{2fL_2}$$

2.3.1.4 Discontinuous Mode

Referring to Figure 2.10d, we can see that the diode current i_D becomes zero during switch off before next period switch on. The condition for discontinuous mode is

$$\zeta \ge 1$$

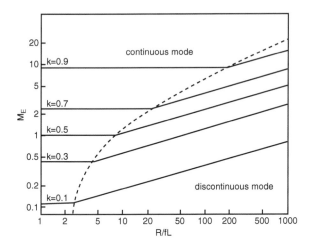

FIGURE 2.13
The boundary between continuous and discontinuous modes and the output voltage vs. the normalized load $z_N = R/fL$.

i.e.,
$$\frac{k^2}{M_E^2}\frac{R}{2fL} \geq 1$$

or
$$M_E \leq k\sqrt{\frac{R}{2fL}} = k\sqrt{\frac{z_N}{2}} \qquad (2.95)$$

The graph of the boundary curve vs. the normalized load $z_N = R/fL$ is shown in Figure 2.13. It can be seen that the boundary curve is a monorising function of the parameter k.

In this case the current i_D exists in the period between kT and $t_1 = [k + (1 - k)m_E]T$, where m_E is the **filling efficiency** and it is defined as:

$$m_E = \frac{1}{\varsigma} = \frac{M_E^2}{k^2\dfrac{R}{2fL}} \qquad (2.96)$$

Considering Equation (2.95), therefore $0 < m_E < 1$. Since the diode current i_D becomes zero at $t = kT + (1 - k)m_E T$, for the current i_L, then

$$kTV_I = (1 - k)m_E TV_C$$

or

$$V_C = \frac{k}{(1-k)m_E}V_I = k(1-k)\frac{R}{2fL}V_I$$

with

$$\sqrt{\frac{R}{2fL}} \geq \frac{1}{1-k}$$

and for the current i_{LO}

$$kT(V_I + V_C - V_O) = (1-k)m_E TV_O$$

Therefore, output voltage in discontinuous mode is

$$V_O = \frac{k}{(1-k)m_E} V_I = k(1-k)\frac{R}{2fL} V_I$$

with

$$\sqrt{\frac{R}{2fL}} \geq \frac{1}{1-k} \tag{2.97}$$

i.e., the output voltage will linearly increase during load resistance R increasing. The output voltage vs. the normalized load $z_N = R/fL$ is shown in Figure 2.13. It can be seen that larger load resistance R may cause higher output voltage in discontinuous conduction mode.

2.3.1.5 Stability Analysis

Stability analysis is of vital importance for any converter circuit. Considering the various methods including the Bode plot, the root-locus method in s-plane is used for this analysis. According to the circuit network and control system theory, the transfer function in s-domain for switch-on and -off are obtained:

$$G_{on} = \{\frac{\delta V_O(s)}{\delta V_I(s)}\}_{on} = \frac{sCR}{s^3 CC_O L_2 R + s^2 CL_2 + s(C + C_O)R + 1} \tag{2.98}$$

$$G_{off} = \{\frac{\delta V_O(s)}{\delta V_I(s)}\}_{off} = \frac{sCR}{s^3 CC_O L_2 R + s^2 CL_2 + s(C + C_O)R + 1} \tag{2.99}$$

where s is the Laplace operator. From Equation (2.98) and Equation (2.99) in Laplace transform it can be seen that the elementary converter is a third order control circuit. The zero is determined by the equations where the numerator is equal to zero, and the poles are determined by the equation where the denominator is equal to zero. There is a zero at original point (0,

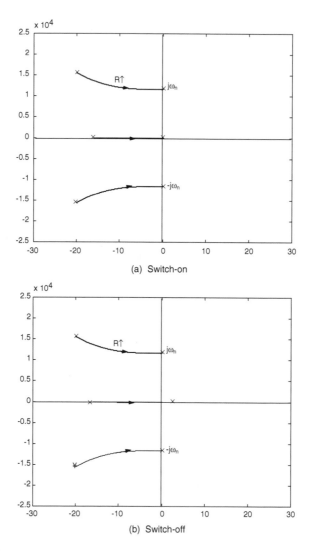

FIGURE 2.14
Stability analysis of elementary circuit. (a) Switch-on. (b) Switch-off.

0) and three poles located in the left-hand half plane in Figure 2.14, so that this converter is stable. Since the equations to determine the poles are the equations with all positive real coefficients, according to the **Gauss theorem**, the three poles are one negative real pole and a pair of conjugate complex poles with negative real part. When the load resistance R increases and tends toward infinity, the three poles move. The real pole goes to the original point and eliminates with the zero. The pair of conjugate complex poles becomes a pair of imaginary poles located on the image axis. Assuming $C = C_O$ and $L_1 = L_2$ {$L = L_1 L_2/(L_1 + L_2)$ **or** $L_2 = 2L$}, the pair of imaginary poles are

$$s = \pm j \sqrt{\frac{C + C_0}{CC_0 L_2}} = \pm j \sqrt{\frac{1}{CL}} = \pm j\omega_n \quad \text{for switch on} \qquad (2.100)$$

$$s = \pm j \sqrt{\frac{C + C_0}{CC_0 L_2}} = \pm j \sqrt{\frac{1}{CL}} = \pm j\omega_n \quad \text{for switch off} \qquad (2.101)$$

where $\omega_n = \sqrt{1/CL}$ is the converter normal angular frequency. They are locating on the stability boundary. Therefore, the circuit works in the critical state. This fact is verified by experiment and computer simulation. When $R = \infty$, the output voltage v_O intends to be very high value. The output voltage V_O cannot be infinity because of the leakage current penetrating the capacitor C_O.

2.3.2 Self-Lift Circuit

Self-lift circuit and its switch-on and -off equivalent circuits are shown in Figure 2.15, which is derived from the elementary circuit. Comparing to Figure 2.10 and Figure 2.15, it can be seen that the pump circuit and filter are retained and there is only one capacitor C_1 and one diode D_1 more, as a lift circuit is added into the circuit. Capacitor C_1 functions to lift the capacitor voltage V_C by a source voltage V_{in}. Current $i_{C1}(t) = \delta(t)$ is an exponential function. It has a large value at the power on moment, but it is small in the steady state because $V_{C1} = V_{in}$.

2.3.2.1 *Circuit Description*

When switch S is on, the instantaneous source current is $i_I = i_{L1} + i_{L2} + i_{C1}$. Inductor L_1 absorbs energy from the source. In the mean time inductor L_2 absorbs energy from source and capacitor C. Both currents i_{L1} and i_{L2} increase, and C_1 is charged to $v_{C1} = V_I$. When switch S is off, the instantaneous source current is $i_I = 0$. Current i_{L1} flows through capacitor C_1 and diode D to charge capacitor C. Inductor L_1 transfers its stored energy to capacitor C. In the mean time, current i_{L2} flows through the $(C_O - R)$ circuit, capacitor C_1 and diode D, to keep itself continuous. Both currents i_{L1} and i_{L2} decrease. In order to analyze the circuit working procession, the equivalent circuits in switch-on and -off states are shown in Figures 2.15b, c and d. Assuming that capacitor C_1 is sufficiently large, voltage V_{C1} is equal to V_I in steady state.

Current i_{L1} increases in switch-on period kT, and decreases in switch-off period $(1 - k)T$. The corresponding voltages applied across L_1 are V_I and $-(V_C - V_I)$ respectively. Therefore,

$$kTV_I = (1 - k)T(V_C - V_I)$$

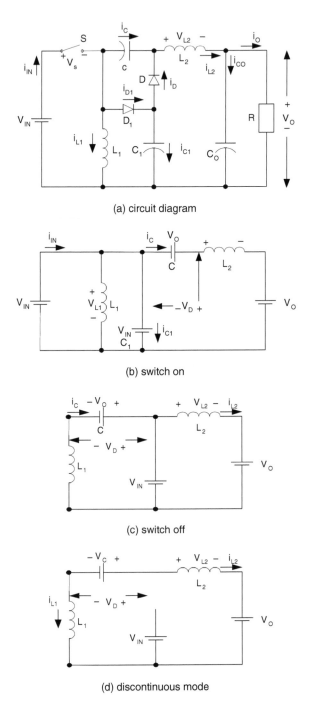

FIGURE 2.15
Self-lift circuit. (a) Circuit diagram. (b) Switch on. (c) Switch off. (d) Discontinuous mode.

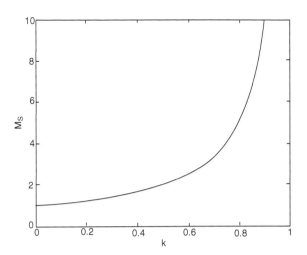

FIGURE 2.16
Voltage transfer gain M_S vs. k.

Hence,
$$V_C = \frac{1}{1-k}V_I \qquad (2.102)$$

Current i_{L2} increases in switch-on period kT, and decreases in switch-off period $(1-k)T$. The corresponding voltages applied across L_2 are $(V_I + V_C - V_O)$ and $-(V_O - V_I)$. Therefore,

$$kT(V_C + V_I - V_O) = (1-k)T(V_O - V_I)$$

Hence,
$$V_O = \frac{1}{1-k}V_I \qquad (2.103)$$

and the output current is

$$I_O = (1-k)I_I \qquad (2.104)$$

Therefore, the voltage transfer gain in continuous mode is

$$M_S = \frac{V_O}{V_I} = \frac{1}{1-k} \qquad (2.105)$$

The curve of M_S vs. k is shown in Figure 2.16.

2.3.2.2   *Average Current I_{C1} and Source Current I_S*

During switch-off period $(1-k)T$, current i_{C1} is equal to $(i_{L1} + i_{L2})$, and the charge on capacitor C_1 decreases. During switch-on period kT, the charge increases, so its average current in switch-on period is

$$I_{C1} = \frac{1-k}{k}(i_{L1} + i_{L2}) = \frac{1-k}{k}(I_{L1} + I_{L2}) = \frac{I_O}{k} \tag{2.106}$$

During switch-off period $(1-k)T$ the source current i_I is 0, and in the switch-on period kT,

$$i_I = i_{L1} + i_{L2} + i_{C1}$$

Hence, $$I_I = k(i_{L1} + i_{L2} + i_{C1}) = k(I_{L1} + I_{L2} + I_{C1})$$

$$= k(I_{L1} + I_{L2})(1 + \frac{1-k}{k}) = k\frac{I_{L2}}{1-k}\frac{1}{k} = \frac{I_O}{1-k} \tag{2.107}$$

2.3.2.3   *Variations of Currents and Voltages*

To analyze the variations of currents and voltages, some voltage and current waveforms are shown in Figure 2.17. Current i_{L1} increases and is supplied by V_I during switch-on period kT. It decreases and is reversely biased by $-(V_C - V_I)$ during switch-off. Therefore, its peak-to-peak variation is

$$\Delta i_{L1} = \frac{kTV_I}{L_1}$$

Hence, the variation ratio of current i_{L1} is

$$\xi_1 = \frac{\Delta i_{L1}/2}{I_{L1}} = \frac{kV_I T}{2kL_1 I_I} = \frac{1-k}{2M_S}\frac{R}{fL_1} \tag{2.108}$$

Current i_{L2} increases and is supplied by the voltage $(V_I + V_C - V_O) = V_I$ in switch-on period kT. It decreases and is inversely biased by $-(V_C - V_I)$ during switch-off. Therefore its peak-to-peak variation is

$$\Delta i_{L2} = \frac{kTV_I}{L_2}$$

Thus, the variation ratio of current i_{L2} is

$$\xi_2 = \frac{\Delta i_{L2}/2}{I_{L2}} = \frac{kV_I T}{2L_2 I_O} = \frac{k}{2M_S}\frac{R}{fL_2} \tag{2.109}$$

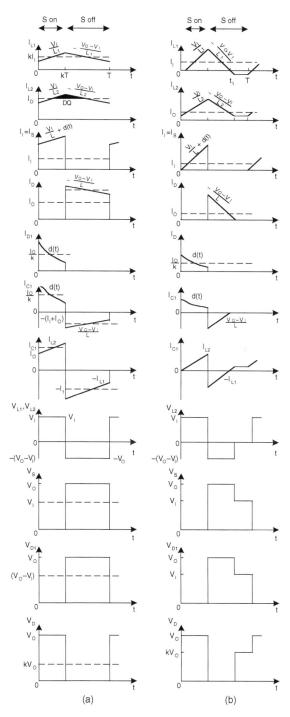

FIGURE 2.17
Some voltage and current waveforms of self-lift circuit.

When switch is off, the free-wheeling diode current is $i_D = i_{L1} + i_{L2}$ and

$$\Delta i_D = \Delta i_{L1} + \Delta i_{L2} = \frac{kTV_I}{L} = \frac{k(1-k)V_O}{L}T \qquad (2.110)$$

Considering Equation (2.71) and Equation (2.72),

$$I_D = I_{L1} + I_{L2} = \frac{I_O}{1-k}$$

The variation ratio of current i_D is

$$\zeta = \frac{\Delta i_D / 2}{I_D} = \frac{k(1-k)^2 TV_O}{2LI_O} = \frac{k}{M_S^2}\frac{R}{2fL} \qquad (2.111)$$

The peak-to-peak variation of voltage v_C is

$$\Delta v_C = \frac{Q+}{C} = \frac{(1-k)TI_{L1}}{C} = \frac{1-k}{C}kTI_I$$

Hence, its variation ratio is

$$\rho = \frac{\Delta v_C / 2}{V_C} = \frac{(1-k)^2 kI_I T}{2CV_I} = \frac{k}{2fCR} \qquad (2.112)$$

The charge on capacitor C_1 increases during switch-on, and decreases during switch-off period $(1-k)T$ by the current $(I_{L1} + I_{L2})$. Therefore, its peak-to-peak variation is

$$\Delta v_{C1} = \frac{(1-k)T(I_{L1} + I_{L2})}{C_1} = \frac{I_O}{fC_1}$$

Considering $V_{C1} = V_I$, the variation ratio of voltage v_{C1} is

$$\sigma = \frac{\Delta v_{C1} / 2}{V_{C1}} = \frac{I_O}{2fC_1 V_I} = \frac{M_S}{2fC_1 R} \qquad (2.113)$$

If $L_1 = L_2 = 1\text{mH}$, $C = C_1 = C_O = 20\ \mu\text{F}$, $R = 40\ \Omega$, $f = 50\ \text{kHz}$ and $k = 0.5$, we obtained that $\xi_1 = 0.1$, $\xi_2 = 0.1$, $\zeta = 0.1$, $\rho = 0.006$ and $\sigma = 0.025$. Therefore, the variations of i_{L1}, i_{L2}, v_{C1} and v_C are small.

Considering Equation (2.84) and Equation (2.105), the variation ratio of output voltage v_O is

$$\varepsilon = \frac{\Delta v_O / 2}{V_O} = \frac{kT^2}{8C_O L_2} \frac{V_I}{V_O} = \frac{k}{8M_S} \frac{1}{f^2 C_O L_2} \tag{2.114}$$

If $L_2 = 1$ mH, $C_0 = 20$ μF, $f = 50$ kHz and $k = 0.5$, $\varepsilon = 0.0006$. Therefore, the output voltage V_O is almost a real DC voltage with very small ripple. Because of the resistive load, the output current $i_O(t)$ is almost a real DC waveform with very small ripple as well, and $I_O = V_O / R$.

2.3.2.4   Instantaneous Value of the Currents and Voltages

Referring to Figure 2.17, the instantaneous values of the currents and voltages are listed below:

$$v_S = \begin{cases} 0 & \text{for} \quad 0 < t \le kT \\ V_O & \text{for} \quad kT < t \le T \end{cases} \tag{2.115}$$

$$v_D = \begin{cases} V_O & \text{for} \quad 0 < t \le kT \\ 0 & \text{for} \quad kT < t \le T \end{cases} \tag{2.116}$$

$$v_{D1} = \begin{cases} 0 & \text{for} \quad 0 < t \le kT \\ V_O & \text{for} \quad kT < t \le T \end{cases} \tag{2.117}$$

$$v_{L1} = v_{L2} = \begin{cases} V_I & \text{for} \quad 0 < t \le kT \\ -(V_O - V_I) & \text{for} \quad kT < t \le T \end{cases} \tag{2.118}$$

$$i_I = i_S = \begin{cases} i_{L1}(0) + i_{L2}(0) + \delta(t) + \dfrac{V_I}{L} t & \text{for} \quad 0 < t \le kT \\ 0 & \text{for} \quad kT < t \le T \end{cases} \tag{2.119}$$

$$i_{L1} = \begin{cases} i_{L1}(0) + \dfrac{V_I}{L_1} t & \text{for} \quad 0 < t \le kT \\ i_{L1}(kT) - \dfrac{V_O - V_I}{L_1}(t - kT) & \text{for} \quad kT < t \le T \end{cases} \tag{2.120}$$

$$i_{L2} = \begin{cases} i_{L2}(0) + \dfrac{V_I}{L_2} t & \text{for} \quad 0 < t \le kT \\ i_{L2}(kT) - \dfrac{V_O - V_I}{L_2}(t - kT) & \text{for} \quad kT < t \le T \end{cases} \tag{2.121}$$

$$i_D = \begin{cases} 0 & \text{for} \quad 0 < t \le kT \\ i_{L1}(kT) + i_{L2}(kT) - \dfrac{V_O - V_I}{L}(t - kT) & \text{for} \quad kT < t \le T \end{cases} \tag{2.122}$$

$$i_{D1} = \begin{cases} \delta(t) & \text{for} \quad 0 < t \le kT \\ 0 & \text{for} \quad kT < t \le T \end{cases} \tag{2.123}$$

$$i_C = \begin{cases} i_{L2}(0) + \dfrac{V_I}{L_2}t & \text{for} \quad 0 < t \le kT \\ i_{L1}(kT) - \dfrac{V_O - V_I}{L_1}(t - kT) & \text{for} \quad kT < t \le T \end{cases} \tag{2.124}$$

$$i_{C1} = \begin{cases} \delta(t) & \text{for} \quad 0 < t \le kT \\ -i_{L1}(kT) - i_{L2}(kT) + \dfrac{V_O - V_I}{L}(t - kT) & \text{for} \quad kT < t \le T \end{cases} \tag{2.125}$$

$$i_{CO} = \begin{cases} i_{L2}(0) + \dfrac{V_I}{L_2}t - I_O & \text{for} \quad 0 < t \le kT \\ i_{L2}(kT) - \dfrac{V_O - V_I}{L_2}(t - kT) - I_O & \text{for} \quad kT < t \le T \end{cases} \tag{2.126}$$

where

$$i_{L1}(0) = k\,I_I - k\,V_I/2f\,L_1$$

$$i_{L1}(kT) = k\,I_I + k\,V_I/2f\,L_1$$

and

$$i_{L2}(0) = I_O - k\,V_O/2f\,M\,L_2$$

$$i_{L2}(kT) = I_O + k\,V_O/2f\,M\,L_2$$

2.3.2.5 Discontinuous Mode

Referring to Figure 2.15d, we can see that the diode current i_D becomes zero during switch off before next period switch on. The condition for discontinuous mode is $\zeta \ge 1$,

i.e.,

$$\frac{k}{M_S^{\,2}} \frac{R}{2fL} \ge 1$$

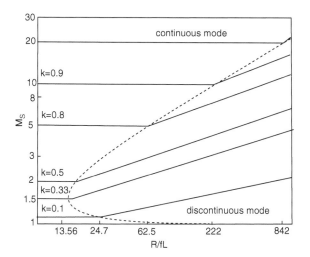

FIGURE 2.18
The boundary between CCM and DCM and the output voltage vs. the normalized load $z_N = R/fL$.

or

$$M_S \leq \sqrt{k}\sqrt{\frac{R}{2fL}} = \sqrt{k}\sqrt{\frac{z_N}{2}} \tag{2.127}$$

The graph of the boundary curve vs. the normalized load $z_N = R/fL$ is shown in Figure 2.18. It can be seen that the boundary curve has a minimum value of 1.5 at $k = \frac{1}{3}$.

In this case the current i_D exists in the period between kT and $t_1 = [k + (1-k)m_S]T$, where m_S is the **filling efficiency** and it is defined as:

$$m_S = \frac{1}{\zeta} = \frac{M_S^2}{k\dfrac{R}{2fL}} \tag{2.128}$$

Considering Equation (2.127), therefore $0 < m_S < 1$. Since the diode current i_D becomes zero at $t = kT + (1-k)m_S T$, for the current i_L

$$kTV_I = (1-k)m_S T(V_C - V_I)$$

or

$$V_C = [1 + \frac{k}{(1-k)m_S}]V_I = [1 + k^2(1-k)\frac{R}{2fL}]V_I$$

with

$$\sqrt{k}\sqrt{\frac{R}{2fL}} \geq \frac{1}{1-k}$$

and for the current i_{LO}

$$kT(V_I + V_C - V_O) = (1-k)m_S T(V_O - V_I)$$

Therefore, output voltage in discontinuous mode is

$$V_O = [1 + \frac{k}{(1-k)m_S}]V_I = [1 + k^2(1-k)\frac{R}{2fL}]V_I \quad \text{with} \quad \sqrt{k}\sqrt{\frac{R}{2fL}} \geq \frac{1}{1-k} \quad (2.129)$$

i.e., the output voltage will linearly increase while load resistance R increases. The output voltage V_O vs. the normalized load $z_N = R/fL$ is shown in Figure 2.18. Larger load resistance R causes higher output voltage in discontinuous conduction mode.

2.3.2.6 Stability Analysis

Taking the root-locus method in s-plane for stability analysis the transfer functions in s-domain for switch-on and -off are obtained:

$$G_{on} = \{\frac{\delta V_O(s)}{\delta V_I(s)}\}_{on} = \frac{sCR}{s^3 CC_O L_2 R + s^2 CL_2 + s(C + C_O)R + 1} \quad (2.130)$$

$$G_{off} = \{\frac{\delta V_O(s)}{\delta V_I(s)}\}_{off} = \frac{sCR}{s^3 (C + C_1)C_O L_2 R + s^2 (C + C_1)L_2 + s(C + C_1 + C_O)R + 1} \quad (2.131)$$

where s is the Laplace operator. From Equations (2.130) and (2.131) in Laplace transform it can be seen that the self-lift converter is a third order control circuit. The zero is determined by the equation when the numerator is equal to zero, and the poles are determined by the equation when the denominator is equal to zero. There is a zero at origin point (0, 0) and three poles located in the left-hand half plane in Figure 2.19, so that the self-lift converter is stable. Since the equations to determine the poles are the equations with all positive real coefficients, according to the **Gauss theorem**, the three poles are one negative real pole and a pair of conjugate complex poles with negative real part. When the load resistance R increases and tends towards infinity, the three poles move. The real pole goes to the origin point and eliminates with the zero. The pair of conjugate complex poles becomes a pair of imaginary poles locating on the image axis. Assuming $C = C_1 = C_O$ and $L_1 = L_2$ {$L = L_1 L_2/(L_1 + L_2)$ **or** $L_2 = 2L$}, the pair of imaginary poles are

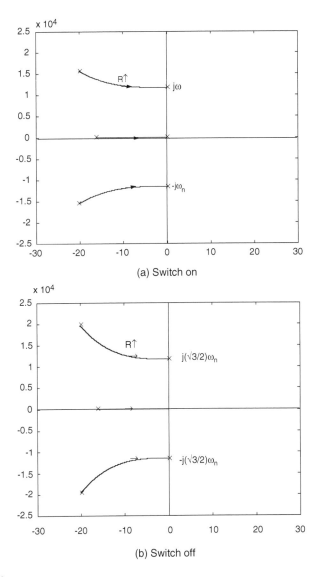

FIGURE 2.19
Stability analysis of self-lift circuit. (a) Switch-on. (b) Switch-off.

$$s = \pm j \sqrt{\frac{C + C_0}{CC_0 L_2}} = \pm j \sqrt{\frac{1}{CL}} = \pm j \omega_n \quad \text{for switch on} \qquad (2.132)$$

$$s = \pm j \sqrt{\frac{C + C_1 + C_0}{(C + C_1)C_0 L_2}} = \pm j \sqrt{\frac{3}{4CL}} = \pm j \frac{\sqrt{3}}{2} \omega_n \quad \text{for switch off} \qquad (2.133)$$

where $\omega_n = \sqrt{1/CL}$ is the self-lift converter normal angular frequency. They are locating on the stability boundary. Therefore, the circuit works in the critical state. This fact is verified by experiment and computer simulation. When $R = 8$, the output voltage v_O intends to be a very high value. The output voltage V_O cannot be infinity because of the leakage current penetrating the capacitor C_O.

2.3.3   Re-Lift Circuit

Re-lift circuit, and its switch-on and -off equivalent circuits are shown in Figure 2.20, which is derived from the self-lift circuit. It consists of two static switches S and S_1; three diodes D, D_1, and D_2; three inductors L_1, L_2 and L_3; four capacitors C, C_1, C_2, and C_O. From Figure 2.10, Figure 2.15, and Figure 2.20, it can be seen that the pump circuit and filter are retained and there are one capacitor C_2, one inductor L_3 and one diode D_2 added into the re-lift circuit. The lift circuit consists of D_1-C_1-L_3D_2-S_1-C_2. Capacitors C_1 and C_2 perform characteristics to lift the capacitor voltage V_C by twice the source voltage V_I. L_3 performs the function as a ladder joint to link the two capacitors C_1 and C_2 and lift the capacitor voltage V_C up. Current $i_{C1}(t) = \delta_1(t)$ and $i_{C2}(t) = \delta_2(t)$ are exponential functions. They have large values at the moment of power on, but they are small because $v_{C1} = v_{C2} = V_I$ in steady state.

2.3.3.1   *Circuit Description*

When switches S and S_1 turn on, the source instantaneous current $i_I = i_{L1} + i_{L2} + i_{C1} + i_{L3} + i_{C2}$. Inductors L_1 and L_3 absorb energy from the source. In the mean time inductor L_2 absorbs energy from source and capacitor C. Three currents i_{L1}, i_{L3} and i_{L2} increase. When switches S and S_1 turn off, source current $i_I = 0$. Current i_{L1} flows through capacitor C_1, inductor L_3, capacitor C_2 and diode D to charge capacitor C. Inductor L_1 transfers its stored energy to capacitor C. In the mean time, current i_{L2} flows through the $(C_O - R)$ circuit, capacitor C_1, inductor L_3, capacitor C_2 and diode D to keep itself continuous. Both currents i_{L1} and i_{L2} decrease. In order to analyze the circuit working procession, the equivalent circuits in switch-on and -off states are shown in Figure 2.20b, c, and d. Assuming capacitor C_1 and C_2 are sufficiently large, and the voltages V_{C1} and V_{C2} across them are equal to V_I in steady state.

Voltage v_{L3} is equal to V_I during switch-on. The peak-to-peak variation of current i_{L3} is

$$\Delta i_{L3} = \frac{V_I kT}{L_3} \tag{2.134}$$

This variation is equal to the current reduction when it is switch-off. Suppose its voltage is $-V_{L3-off}$, so

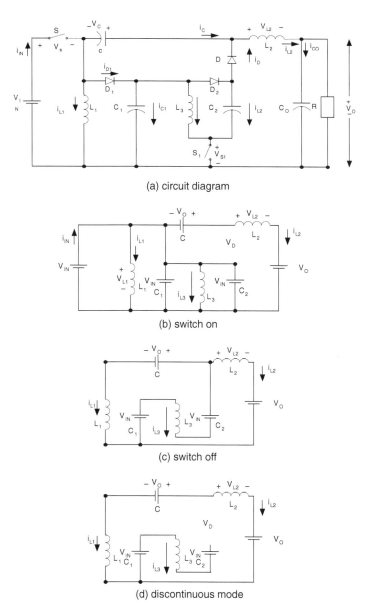

FIGURE 2.20
Re-lift circuit: (a) circuit diagram; (b) switch on; (c) switch off; (d) discontinuous mode.

$$\Delta i_{L3} = \frac{V_{L3-\textit{off}}(1-k)T}{L_3}$$

Thus, during switch-off the voltage drop across inductor L_3 is

$$V_{L3-off} = \frac{k}{1-k} V_I \tag{2.135}$$

Current i_{L1} increases in switch-on period kT, and decreases in switch-off period $(1-k)T$. The corresponding voltages applied across L_1 are V_I and $-(V_C - 2V_I - V_{L3-off})$. Therefore,

$$kTV_I = (1-k)T(V_C - 2V_I - V_{L3-off})$$

Hence,
$$V_C = \frac{2}{1-k} V_I \tag{2.136}$$

Current i_{L2} increases in switch-on period kT, and it decreases in switch-off period $(1-k)T$. The corresponding voltages applied across L_2 are $(V_I + V_C - V_O)$ and $-(V_O - 2V_I - V_{L3-off})$. Therefore,

$$kT(V_C + V_I - V_O) = (1-k)T(V_O - 2V_I - V_{L3-off})$$

Hence,
$$V_O = \frac{2}{1-k} V_I \tag{2.137}$$

and the output current is

$$I_O = \frac{1-k}{2} I_I \tag{2.138}$$

The voltage transfer gain in continuous mode is

$$M_R = \frac{V_O}{V_I} = \frac{2}{1-k} \tag{2.139}$$

The curve of M_R vs. k is shown in Figure 2.21.

2.3.3.2   *Other Average Currents*
Considering Equation (2.71),

$$I_{L1} = \frac{k}{1-k} I_O = \frac{k}{2} I_I \tag{2.140}$$

and
$$I_{L3} = I_{L1} + I_{L2} = \frac{1}{1-k} I_O \tag{2.141}$$

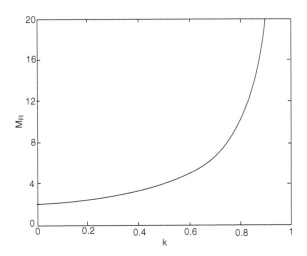

FIGURE 2.21
Voltage transfer gain M_R vs. k.

Currents i_{C1} and i_{C2} equal to $(i_{L1} + i_{L2})$ during **switch-off** period $(1 - k)T$, and the charges on capacitors C_1 and C_2 decrease, i.e.,

$$i_{C1} = i_{C2} = (i_{L1} + i_{L2}) = \frac{1}{1-k} I_O$$

The charges increase during **switch-on** period kT, so their average currents are

$$I_{C1} = I_{C2} = \frac{1-k}{k}(I_{L1} + I_{L2}) = \frac{1-k}{k}(\frac{k}{1-k} + 1)I_O = \frac{I_O}{k} \tag{2.142}$$

During switch-off the source current i_I is 0, and in the switch-on period kT, it is

$$i_I = i_{L1} + i_{L2} + i_{C1} + i_{L3} + i_{C2}$$

Hence,

$$I_I = ki_I = k(I_{L1} + I_{L2} + I_{C1} + I_{L3} + I_{C2}) = k[2(I_{L1} + I_{L2}) + 2I_{C1}]$$

$$= 2k(I_{L1} + I_{L2})(1 + \frac{1-k}{k}) = 2k\frac{I_{L2}}{1-k}\frac{1}{k} = \frac{2}{1-k}I_O \tag{2.143}$$

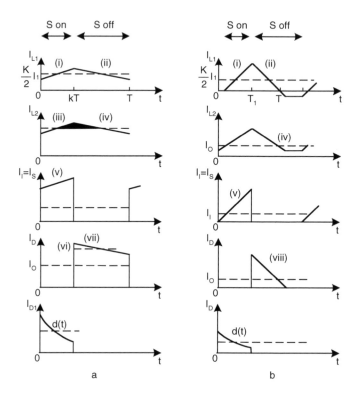

FIGURE 2.22
Some voltage and current waveforms of re-lift circuit.

2.3.3.3 Variations of Currents and Voltages

To analyze the variations of currents and voltages, some voltage and current waveforms are shown in Figure 2.22. Current i_{L1} increases and is supplied by V_I during switch-on period kT. It decreases and is reversely biased by $-(V_C - 2V_I - V_{L3})$ during switch-off period $(1 - k)T$. Therefore, its peak-to-peak variation is

$$\Delta i_{L1} = \frac{kTV_I}{L_1}$$

Considering Equation (2.140), the variation ratio of current i_{L1} is

$$\xi_1 = \frac{\Delta i_{L1} / 2}{I_{L1}} = \frac{kV_I T}{kL_1 I_I} = \frac{1-k}{2M_R} \frac{R}{fL_1} \tag{2.144}$$

Current i_{L2} increases and is supplied by the voltage $(V_I + V_C - V_O) = V_I$ during switch-on period kT. It decreases and is reversely biased by $-(V_O - 2V_I - V_{L3})$ during switch-off. Therefore, its peak-to-peak variation is

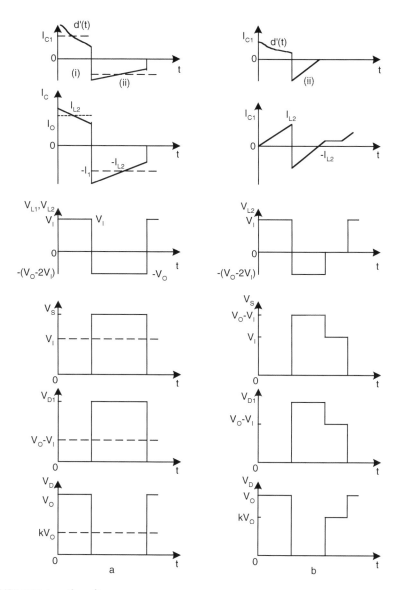

FIGURE 2.22 (continued)

$$\Delta i_{L2} = \frac{kTV_I}{L_2}$$

Considering Equation (2.72), the variation ratio of current i_{L2} is

$$\xi_2 = \frac{\Delta i_{L2}/2}{I_{L2}} = \frac{kTV_I}{2L_2 I_O} = \frac{k}{2M_R}\frac{R}{fL_2} \qquad (2.145)$$

When switch is off, the free-wheeling diode current is $i_D = i_{L1} + i_{L2}$ and

$$\Delta i_D = \Delta i_{L1} + \Delta i_{L2} = \frac{kTV_I}{L} = \frac{k(1-k)V_O}{2L}T \qquad (2.146)$$

Considering Equation (2.71) and Equation (2.72),

$$I_D = I_{L1} + I_{L2} = \frac{I_O}{1-k}$$

The variation ratio of current i_D is

$$\zeta = \frac{\Delta i_D/2}{I_D} = \frac{k(1-k)^2 TV_O}{4LI_O} = \frac{k(1-k)R}{2M_R fL} = \frac{k}{M_R^2}\frac{R}{fL} \qquad (2.147)$$

Considering Equation (2.134) and Equation (2.141), the variation ratio of current i_{L3} is

$$\chi_1 = \frac{\Delta i_{L3}/2}{I_{L3}} = \frac{kV_I T}{2L_3 \frac{1}{1-k}I_O} = \frac{k}{M_R^2}\frac{R}{fL_3} \qquad (2.148)$$

If $L_1 = L_2 = 1$ mH, $L_3 = 0.5$ mH, $R = 160\ \Omega$, $f = 50$ kHz and $k = 0.5$, we obtained that $\xi_1 = 0.2$, $\xi_2 = 0.2$, $\zeta = 0.1$ and $\chi_1 = 0.2$. Therefore, the variations of i_{L1}, i_{L2} and i_{L3} are small.

The peak-to-peak variation of v_C is

$$\Delta v_C = \frac{Q+}{C} = \frac{1-k}{C}TI_{L1} = \frac{k(1-k)}{2C}TI_I$$

Considering Equation (2.136), the variation ratio is

$$\rho = \frac{\Delta v_C/2}{V_C} = \frac{k(1-k)TI_I}{4CV_O} = \frac{k}{2fCR} \qquad (2.149)$$

The charges on capacitors C_1 and C_2 increase during switch-on period kT, and decrease during switch-off period $(1-k)T$ by the current $(I_{L1} + I_{L2})$. Therefore their peak-to-peak variations are

$$\Delta v_{C1} = \frac{(1-k)T(I_{L1} + I_{L2})}{C_1} = \frac{(1-k)I_I}{2C_1 f}$$

$$\Delta v_{C2} = \frac{(1-k)T(I_{L1} + I_{L2})}{C_2} = \frac{(1-k)I_I}{2C_2 f}$$

Considering $V_{C1} = V_{C2} = V_I$, the variation ratios of voltages v_{C1} and v_{C2} are

$$\sigma_1 = \frac{\Delta v_{C1}/2}{V_{C1}} = \frac{(1-k)I_I}{4fC_1V_I} = \frac{M_R}{2fC_1R} \tag{2.150}$$

$$\sigma_2 = \frac{\Delta v_{C2}/2}{V_{C2}} = \frac{(1-k)I_I}{4V_IC_2f} = \frac{M_R}{2fC_2R} \tag{2.151}$$

Considering Equation (2.84), the variation ratio of output voltage v_O is

$$\varepsilon = \frac{\Delta v_O/2}{V_O} = \frac{kT^2}{8C_OL_2}\frac{V_I}{V_O} = \frac{k}{8M_R}\frac{1}{f^2C_OL_2} \tag{2.152}$$

If $C = C_1 = C_2 = C_O = 20\ \mu F$, $L_2 = 1$ mH, $R = 160\ \Omega$, $f = 50$ kHz and $k = 0.5$, we obtained that $\rho = 0.0016$, $\sigma_1 = \sigma_2 = 0.0125$, and $\varepsilon = 0.0003$. The ripples of v_C, v_{C1}, v_{C2} and v_{CO} are small. Therefore, the output voltage v_O is almost a real DC voltage with very small ripple. Because of the resistive load, the output current $i_O(t)$ is almost a real DC waveform with very small ripple as well, and $I_O = V_O/R$.

2.3.3.4 Instantaneous Value of the Currents and Voltages

Referring to Figure 2.22, the instantaneous current and voltage values are listed below:

$$v_S = \begin{cases} 0 & \text{for} \quad 0 < t \le kT \\ V_O & \text{for} \quad kT < t \le T \end{cases} \tag{2.153}$$

$$v_D = \begin{cases} V_O & \text{for} \quad 0 < t \le kT \\ 0 & \text{for} \quad kT < t \le T \end{cases} \tag{2.154}$$

$$v_{D1} + v_{D2} = \begin{cases} 0 & \text{for} \quad 0 < t \le kT \\ V_O & \text{for} \quad kT < t \le T \end{cases} \tag{2.155}$$

$$v_{L3} = \begin{cases} V_I & \text{for} \quad 0 < t \le kT \\ -\dfrac{k}{1-k}V_I & \text{for} \quad kT < t \le T \end{cases} \tag{2.156}$$

$$v_{L1} = v_{L2} = \begin{cases} V_I & \text{for} \quad 0 < t \le kT \\ -[V_O - (2 - \dfrac{k}{1-k})V_I] & \text{for} \quad kT < t \le T \end{cases} \tag{2.157}$$

$$i_I = i_S = \begin{cases} i_{L1}(0) + i_{L2}(0) + i_{L3}(0) + \delta_1(t) + \delta_2(t) + \dfrac{V_I}{L}t + \dfrac{V_I}{L_3}t & \text{for } 0 < t \le kT \\ 0 & \text{for } kT < t \le T \end{cases} \tag{2.158}$$

$$i_{L1} = \begin{cases} i_{L1}(0) + \dfrac{V_I}{L_1}t & \text{for} \quad 0 < t \le kT \\ i_{L1}(kT) - \dfrac{V_O - (2 - \frac{k}{1-k})V_I}{L_1}(t - kT) & \text{for} \quad kT < t \le T \end{cases} \tag{2.159}$$

$$i_{L2} = \begin{cases} i_{L2}(0) + \dfrac{V_I}{L_2}t & \text{for} \quad 0 < t \le kT \\ i_{L2}(kT) - \dfrac{V_O - (2 - \frac{k}{1-k})V_I}{L_2}(t - kT) & \text{for} \quad kT < t \le T \end{cases} \tag{2.160}$$

$$i_{L3} = \begin{cases} i_{L3}(0) + \dfrac{V_I}{L_3}t & \text{for} \quad 0 < t \le kT \\ i_{L3}(kT) - \dfrac{\frac{k}{1-k}V_I}{L_3}(t - kT) & \text{for} \quad kT < t \le T \end{cases} \tag{2.161}$$

$$i_D = \begin{cases} 0 & \text{for } 0 < t \le kT \\ i_{L1}(kT) + i_{L2}(kT) - \dfrac{V_O - (2 - \frac{k}{1-k})V_I}{L}(t - kT) & \text{for } kT < t \le T \end{cases} \tag{2.162}$$

$$i_{D1} = \begin{cases} \delta_1(t) + \delta_2(t) & \text{for} \quad 0 < t \le kT \\ 0 & \text{for} \quad kT < t \le T \end{cases} \tag{2.163}$$

$$i_{D2} = \begin{cases} \delta_2(t) & \text{for} \quad 0 < t \le kT \\ 0 & \text{for} \quad kT < t \le T \end{cases} \tag{2.164}$$

$$i_C = \begin{cases} i_{L2}(0) + \dfrac{V_I}{L_2}t & \text{for} \quad 0 < t \le kT \\ i_{L1}(kT) - \dfrac{V_O - V_I}{L_1}(t - kT) & \text{for} \quad kT < t \le T \end{cases} \tag{2.165}$$

$$
i_{C1} = \begin{cases} \delta_1(t) & for \quad 0 < t \le kT \\ -i_{L1}(kT) - i_{L2}(kT) + \dfrac{V_O - (2 + \dfrac{k}{1-k})V_I}{L}(t - kT) & for \quad kT < t \le T \end{cases} \tag{2.166}
$$

$$
i_{C2} = \begin{cases} \delta_2(t) & for \quad 0 < t \le kT \\ -i_{L1}(kT) - i_{L2}(kT) + \dfrac{V_O - (2 + \dfrac{k}{1-k})V_I}{L}(t - kT) & for \quad kT < t \le T \end{cases} \tag{2.167}
$$

$$
i_{CO} = \begin{cases} i_{L2}(0) + \dfrac{V_I}{L_2}t - I_O & for \quad 0 < t \le kT \\ i_{L2}(kT) - \dfrac{V_O - (2 + \dfrac{k}{1-k})V_I}{L_2}(t - kT) - I_O & for \quad kT < t \le T \end{cases} \tag{2.168}
$$

where $i_{L1}(0) = k\, I_I/2 - k\, V_I/2\, f\, L_1$, $i_{L1}(kT) = k\, I_I/2 + k\, V_I/2\, f\, L_1$, and $i_{L2}(0) = I_O - k\, V_I/2\, f\, L_2$, $i_{L2}(kT) = I_O + k\, V_I/2\, f\, L_2$, and $i_{L3}(0) = I_O + k\, I_I/2 - k\, V_I/2\, f\, L_3$, $i_{L3}(kT) = I_O + k\, I_I/2 + k\, V_I/2\, f\, L_3$.

2.3.3.5   *Discontinuous Mode*

Referring to Figure 2.20d, we can see that the diode current i_D becomes zero during switch off before next period switch on. The condition for discontinuous mode is

$$
\zeta \ge 1
$$

i.e.,

$$
\frac{k}{M_R{}^2}\frac{R}{fL} \ge 1
$$

or

$$
M_R \le \sqrt{k}\sqrt{\frac{R}{fL}} = \sqrt{k}\sqrt{z_N} \tag{2.169}
$$

The graph of the boundary curve vs. the normalized load $z_N = R/fL$ is shown in Figure 2.23. It can be seen that the boundary curve has a minimum value of 3.0 at $k = \frac{1}{3}$.

In this case the current i_D exists in the period between kT and $t_1 = [k + (1-k)m_R]T$, where m_R is the **filling efficiency** and it is defined as:

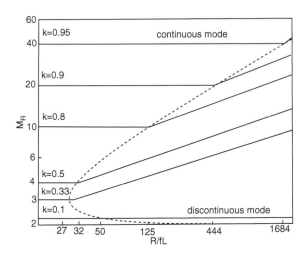

FIGURE 2.23

The boundary between continuous and discontinuous modes and the output voltage vs. the normalized load $z_N = R/fL$.

$$m_R = \frac{1}{\zeta} = \frac{M_R^2}{k\dfrac{R}{fL}} \tag{2.170}$$

Considering Equation (2.169), therefore $0 < m_R < 1$. Since the diode current i_D becomes zero at $t = kT + (1-k)m_RT$, for the current i_L

$$kTV_I = (1-k)m_RT(V_C - 2V_I - V_{L3-off})$$

or

$$V_C = [2 + \frac{k}{1-k} + \frac{k}{(1-k)m_R}]V_I = [2 + \frac{k}{1-k} + k^2(1-k)\frac{R}{4fL}]V_I$$

with

$$\sqrt{k}\sqrt{\frac{R}{fL}} \geq \frac{2}{1-k}$$

and for the current i_{LO} $kT(V_I + V_C - V_O) = (1-k)m_RT(V_O - 2V_I - V_{L3-off})$
  Therefore, output voltage in discontinuous mode is

$$V_O = [2 + \frac{k}{1-k} + \frac{k}{(1-k)m_R}]V_I = [2 + \frac{k}{1-k} + k^2(1-k)\frac{R}{4fL}]V_I$$

with
$$\sqrt{k}\sqrt{\frac{R}{fL}} \geq \frac{2}{1-k} \qquad (2.171)$$

i.e., the output voltage will linearly increase during load resistance R increasing. The output voltage vs. the normalized load $z_N = R/fL$ is shown in Figure 2.23. Larger load resistance R may cause higher output voltage in discontinuous mode.

2.3.3.6 Stability Analysis

Stability analysis is of vital importance for any converter circuit. According to the circuit network and control systems theory, the transfer functions in s-domain for switch-on and -off states are obtained:

$$G_{on} = \{\frac{\delta V_O(s)}{\delta V_I(s)}\}_{on} = \frac{sCR}{s^3 L_2 CC_O R + s^2 L_2 C + s(C + C_O)R + 1} \qquad (2.172)$$

$$G_{off} = \{\frac{\delta V_O(s)}{\delta V_I(s)}\}_{off}$$

$$= \frac{\dfrac{R}{1+sC_O R} \; sC \; \dfrac{s(C_1 + C_2) + s^3 L_3 C_1 C_2}{s^2 C_1 C_2}}{\left(\begin{array}{c} sC \dfrac{s(C_1 + C_2) + s^3 L_3 C_1 C_2}{s^2 C_1 C_2} \dfrac{R + sL_2 + s^2 L_2 C_O R}{1 + sC_O R} \\ + \dfrac{s(C_1 + C_2) + s^3 L_3 C_1 C_2}{s^2 C_1 C_2} + \dfrac{R + sL_2 + s^2 L_2 C_O R}{1 + sC_O R} \end{array} \right)}$$

$$= \frac{sCR[(C_1 + C_2) + s^2 L_3 C_1 C_2]}{\left(\begin{array}{c} sC[(C_1 + C_2) + s^2 L_3 C_1 C_2][R + sL_2 + s^2 L_2 C_O R] + (1 + sC_O R)[(C_1 + C_2) \\ + s^2 L_3 C_1 C_2] + sC_1 C_2[R + sL_2 + s^2 L_2 C_O R] \end{array} \right)} \qquad (2.173)$$

where s is the Laplace operator. From Equation (2.172) and Equation (2.173) in Laplace transform we can see that the re-lift converter is a third order control circuit for switch-on state and a fifth order control circuit for switch-off state.

For the switch-on state, the zeros are determined by the equation when the numerator of Equation (2.172) is equal to zero, and the poles are determined by the equation when the denominator of Equation (2.172) is equal to zero. There is a zero at the origin point (0, 0). Since the equation to determine the poles is the equation with all positive real coefficients, according to the **Gauss theorem**, the three poles are: one negative real pole (p_3)

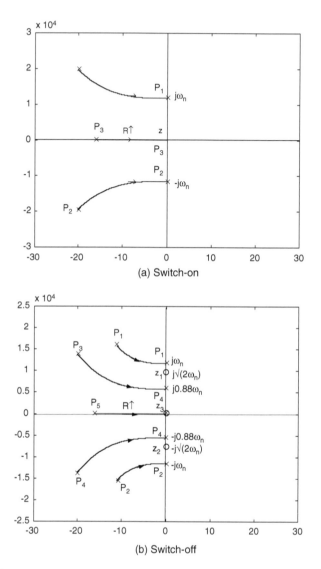

FIGURE 2.24
Stability analysis of re-lift circuit. (a) Switch-on. (b) Switch-off.

and a pair of conjugate complex poles with negative real part ($p_{1,2}$). The three poles are located in the left half plane in Figure 2.24, so that the re-lift converter is stable. When the load resistance R increases and intends towards infinity, the three poles move. The real pole goes to the origin point and eliminates with the zero. The pair of conjugate complex poles becomes a pair of imaginary poles locating on the imaginary axis. Assuming that all capacitors have same capacitance C, and $L_1 = L_2$ {$L = L_1 L_2/(L_1 + L_2)$ or $L_2 = 2L$} and $L_3 = L$, Equation (2.172) becomes:

$$G_{on} = \{\frac{\delta V_O(s)}{\delta V_I(s)}\}_{on} = \frac{1}{s^2 L_2 C_O + \dfrac{C + C_O}{C}} = \frac{1}{2s^2 LC + 2} \quad (2.174)$$

and the pair of imaginary poles is

$$p_{1,2} = \pm j\sqrt{\frac{C + C_O}{L_2 CC_O}} = \pm j\sqrt{\frac{1}{LC}} = \pm j\omega_n \quad \text{poles for switch on} \quad (2.175)$$

where $\omega_n = (LC)^{-1/2}$ is the re-lift converter normal angular frequency.

For the switch-off state, the zeros are determined by the equation when the numerator of Equation (2.173) is equal to zero, and the poles are determined by the equation when the denominator of Equation (2.173) is equal to zero. There are three zeros: one (z_3) at the original point (0, 0) and two zeros ($z_{1,2}$) on the imaginary axis which are

$$z_{1,2} = \pm j\sqrt{\frac{C_1 + C_2}{L_3 C_1 C_2}} = \pm j\sqrt{\frac{2}{LC}} = \pm j\sqrt{2}\omega_n \quad \text{zeros for switch off} \quad (2.176)$$

Since the equation to determine the poles is the equation with all positive real coefficients, according to the **Gauss theorem**, the five poles are one negative real pole (p_5) and two pairs of conjugate complex poles with negative real parts ($p_{1,2}$ and $p_{3,4}$). There are five poles located in the left-hand half plane in Figure 2.24, so that the re-lift converter is stable. When the load resistance R increases and intends towards infinity, the five poles move. The real pole goes to the origin point and eliminates with the zero. The two pairs of conjugate complex poles become two pairs of imaginary poles locating on the imaginary axis. Assuming that all capacitors have same capacitance C, and $L_1 = L_2$ {$L = L_1 L_2/(L_1 + L_2)$ or $L_2 = 2L$} and $L_3 = L$, Equation (2.173) becomes:

$$G_{off} = \{\frac{\delta V_O(s)}{\delta V_I(s)}\}_{off} = \frac{C(C_1 + C_2) + s^2 L_3 CC_1 C_2}{\left(\begin{array}{c}(CC_1 + CC_2 + C_1 C_2 + s^2 L_3 CC_1 C_2)(1 + s^2 L_2 C_O) \\ + (C_O C_1 + C_O C_2 + s^2 L_3 C_O C_1 C_2)\end{array}\right)}$$

$$= \frac{2C^2 + s^2 LC^3}{(3C^2 + s^2 LC^3)(1 + 2s^2 LC) + (2C^2 + s^2 LC^3)} = \frac{2 + s^2 LC}{2s^4 L^2 C^2 + 8s^2 LC + 5} \quad (2.177)$$

and the two pairs of imaginary poles are

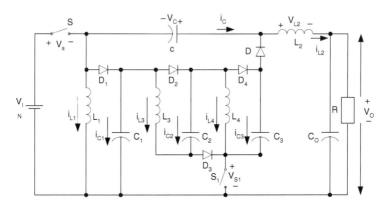

FIGURE 2.25
Triple-lift circuit.

$$s^2LC = \frac{-8 \pm \sqrt{64-40}}{4} = -2 \pm \frac{\sqrt{6}}{2} = \begin{cases} -3.225 \\ -0.775 \end{cases}$$

poles for switch off, so that

$$p_{1,2} = \pm j\,1.8\,\omega_n$$

and
$$p_{3,4} = \pm j\,0.88\,\omega_n \tag{2.178}$$

For both states when R tends to infinity all poles are locating on the stability boundary. Therefore, the circuit works in the critical state. From Equation (2.171) the output voltage will be infinity. This fact is verified by the experimental results and computer simulation results. When $R = \infty$, the output voltage v_O tends to be a very high value. In this particular circuit since there is some leakage current across the capacitor C_O, the output voltage v_O can not be infinity.

2.3.4 Multiple-Lift Circuits

Referring to Figure 2.20a, it is possible to build multiple-lift circuits using the parts $(L_3\text{-}C_2\text{-}S_1\text{-}D_2)$ multiple times. For example in Figure 2.25 the parts $(L_4\text{-}C_3D_3\text{-}D_4)$ were added in the triple-lift circuit. Because the voltage at the point of the joint $(L_4\text{-}C_3)$ is positive value and higher than that at the point of the joint $(L_3\text{-}C_2)$, so that we can use a diode D_3 to replace the switch (S_2). For multiple-lift circuits all further switches can be replaced by diodes. According to this principle, triple-lift circuits and quadruple-lift circuits were built as shown in Figure 2.25 and Figure 2.28. In this book it is not necessary to introduce the particular analysis and calculations one by one to readers. However, their formulas are shown in this section.

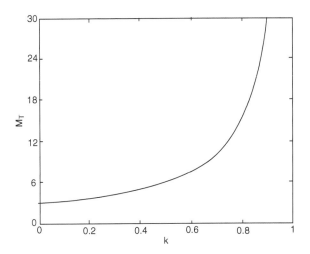

FIGURE 2.26
Voltage transfer gain M_T vs. k.

2.3.4.1 Triple-Lift Circuit

A triple-lift circuit is shown in Figure 2.25, and it consists of two static switches S and S_1; four inductors L_1, L_2, L_3, and L_4; and five capacitors C, C_1, C_2, C_3, and C_O; and five diodes. Capacitors C_1, C_2, and C_3 perform characteristics to lift the capacitor voltage V_C by three times the source voltage V_I. L_3 and L_4 perform the function as ladder joints to link the three capacitors C_1, C_2, and C_3 and lift the capacitor voltage V_C up. Current $i_{C1}(t)$, $i_{C2}(t)$, and $i_{C3}(t)$ are exponential functions. They have large values at the moment of power on, but they are small because $v_{C1} = v_{C2} = v_{C3} = V_I$ in steady state.

The output voltage and current are

$$V_O = \frac{3}{1-k} V_I \tag{2.179}$$

and

$$I_O = \frac{1-k}{3} I_I \tag{2.180}$$

The voltage transfer gain in continuous mode is

$$M_T = \frac{V_O}{V_I} = \frac{3}{1-k} \tag{2.181}$$

The curve of M_T vs. k is shown in Figure 2.26.

Other average voltages:

$$V_C = V_O \quad V_{C1} = V_{C2} = V_{C3} = V_I$$

Other average currents:

$$I_{L2} = I_O \quad I_{L1} = \frac{k}{1-k} I_O$$

$$I_{L3} = I_{L4} = I_{L1} + I_{L2} = \frac{1}{1-k} I_O$$

Current variations:

$$\xi_1 = \frac{1-k}{2M_T} \frac{R}{fL_1} \quad \xi_2 = \frac{k}{2M_T} \frac{R}{fL_2} \quad \zeta = \frac{k(1-k)R}{2M_T fL} = \frac{k}{M_T^2} \frac{3R}{2fL}$$

$$\chi_1 = \frac{k}{M_T^2} \frac{R}{fL_3} \quad \chi_2 = \frac{k}{M_T^2} \frac{R}{fL_4}$$

Voltage variations:

$$\rho = \frac{k}{2fCR} \quad \sigma_1 = \frac{M_T}{2fC_1R}$$

$$\sigma_2 = \frac{M_T}{2fC_2R} \quad \sigma_3 = \frac{M_T}{2fC_3R}$$

The variation ratio of output voltage v_C is

$$\varepsilon = \frac{k}{8M_T} \frac{1}{f^2 C_O L_2} \tag{2.182}$$

The output voltage ripple is very small. The boundary between continuous and discontinuous conduction modes is

$$M_T \leq \sqrt{k} \sqrt{\frac{3R}{2fL}} = \sqrt{\frac{3kz_N}{2}} \tag{2.183}$$

This boundary curve is shown in Figure 3.27. Comparing with Equations (2.95), (2.165) (2.169), and (2.183), it can be seen that the boundary curve has a minimum value of M_T that is equal to 4.5, corresponding to $k = 1/3$.

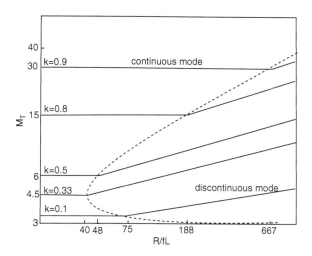

FIGURE 2.27
The boundary between continuous and discontinuous modes and the output voltage vs. the normalized load $z_N = R/fL$.

In discontinuous mode the current i_D exists in the period between kT and $[k + (1 - k)m_T]T$, where m_T is the filling efficiency that is

$$m_T = \frac{1}{\zeta} = \frac{M_T{}^2}{k\dfrac{3R}{2fL}} \tag{2.184}$$

Considering Equation (2.183), therefore, $0 < m_T < 1$. Since the diode current i_D becomes zero at $t = kT + (1 - k)m_T T$, for the current i_{L1}

$$kTV_I = (1 - k)m_T T(V_C - 3V_I - V_{L3\text{-}off} - V_{L4\text{-}off})$$

or

$$V_C = [3 + \frac{2k}{1-k} + \frac{k}{(1-k)m_T}]V_I = [3 + \frac{2k}{1-k} + k^2(1-k)\frac{R}{6fL}]V_I$$

with $\sqrt{k}\sqrt{\dfrac{3R}{2fL}} \geq \dfrac{3}{1-k}$

and for the current i_{L2} $kT(V_I + V_C - V_O) = (1 - k)m_T T(V_O - 2V_I - V_{L3\text{-}off} - V_{L4\text{-}off})$
Therefore, output voltage in discontinuous mode is

$$V_O = [3 + \frac{2k}{1-k} + \frac{k}{(1-k)m_T}]V_I = [3 + \frac{2k}{1-k} + k^2(1-k)\frac{R}{6fL}]V_I$$

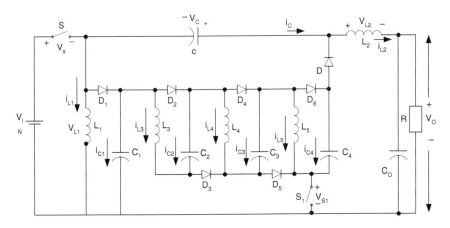

FIGURE 2.28
Quadruple-lift circuit.

with
$$\sqrt{k}\sqrt{\frac{3R}{2fL}} \geq \frac{3}{1-k} \qquad (2.185)$$

i.e., the output voltage will linearly increase during load resistance R increasing, as shown in Figure 2.27.

2.3.4.2 Quadruple-Lift Circuit

Quadruple-lift circuit shown in Figure 2.28 consists of two static switches S and S_1; five inductors L_1, L_2, L_3, L_4, and L_5; and six capacitors C, C_1, C_2, C_3, C_4, and C_O; and seven diodes. Capacitors C_1, C_2, C_3, and C_4 perform characteristics to lift the capacitor voltage V_C by four times the source voltage V_I. L_3, L_4, and L_5 perform the function as ladder joints to link the four capacitors C_1, C_2, C_3, and C_4 and lift the output capacitor voltage V_C up. Current $i_{C1}(t)$, $i_{C2}(t)$, $i_{C3}(t)$, and $i_{C4}(t)$ are exponential functions. They have large values at the moment of power on, but they are small because $v_{C1} = v_{C2} = v_{C3} = v_{C4} = V_I$ in steady state.

The output voltage and current are

$$V_O = \frac{4}{1-k}V_I \qquad (2.186)$$

and

$$I_O = \frac{1-k}{4}I_I \qquad (2.187)$$

The voltage transfer gain in continuous mode is

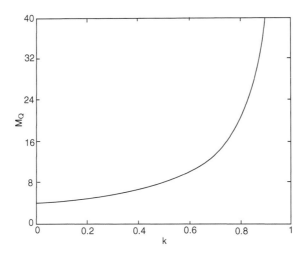

FIGURE 2.29
Voltage transfer gain M_Q vs. k.

$$M_Q = \frac{V_O}{V_I} = \frac{4}{1-k} \tag{2.188}$$

The curve of M_Q vs. k is shown in Figure 2.29. Other average voltages:

$$V_C = V_O; \quad V_{C1} = V_{C2} = V_{C3} = V_{C4} = V_I$$

Other average currents:

$$I_{L2} = I_O; \quad I_{L1} = \frac{k}{1-k} I_O$$

$$I_{L3} = I_{L4} = L_{L5} = I_{L1} + I_{L2} = \frac{1}{1-k} I_O$$

Current variations:

$$\xi_1 = \frac{1-k}{2M_Q} \frac{R}{fL_1} \quad \xi_2 = \frac{k}{2M_Q} \frac{R}{fL_2} \quad \zeta = \frac{k(1-k)R}{2M_QfL} = \frac{k}{M_Q{}^2} \frac{2R}{fL}$$

$$\chi_1 = \frac{k}{M_Q{}^2} \frac{R}{fL_3} \quad \chi_2 = \frac{k}{M_Q{}^2} \frac{R}{fL_4} \quad \chi_3 = \frac{k}{M_Q{}^2} \frac{R}{fL_5}$$

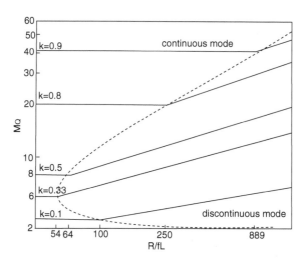

FIGURE 2.30
The boundary between continuous and discontinuous modes and the output voltage vs. the normalized load $z_N = R/fL$.

Voltage variations:

$$\rho = \frac{k}{2fCR} \qquad \sigma_1 = \frac{M_Q}{2fC_1R}$$

$$\sigma_2 = \frac{M_Q}{2fC_2R} \qquad \sigma_3 = \frac{M_Q}{2fC_3R} \qquad \sigma_4 = \frac{M_Q}{2fC_4R}$$

The variation ratio of output voltage V_C is

$$\varepsilon = \frac{k}{8M_Q} \frac{1}{f^2 C_O L_2} \tag{2.189}$$

The output voltage ripple is very small.

The boundary between continuous and discontinuous modes is

$$M_Q \le \sqrt{k} \sqrt{\frac{2R}{fL}} = \sqrt{2kz_N} \tag{2.190}$$

This boundary curve is shown in Figure 2.30. Comparing with Equations (2.95), (2.127), (2.169), (2.183), and (2.190), it can be seen that this boundary curve has a minimum value of M_Q that is equal to 6.0, corresponding to $k = \frac{1}{3}$.

In discontinuous mode the current i_D exists in the period between kT and $[k + (1 - k)m_Q]T$, where m_Q is the filling efficiency that is

$$m_Q = \frac{1}{\zeta} = \frac{M_Q{}^2}{k\frac{2R}{fL}} \tag{2.191}$$

Considering Equation (2.190), therefore $0 < m_Q < 1$. Since the current i_D becomes zero at $t = kT + (1 - k)m_Q T$, for the current i_{L1} we have

$$kTV_I = (1 - k)m_Q T(V_C - 4V_I - V_{L3-off} - V_{L4-off} - V_{L5-off})$$

or

$$V_C = [4 + \frac{3k}{1-k} + \frac{k}{(1-k)m_Q}]V_I = [4 + \frac{3k}{1-k} + k^2(1-k)\frac{R}{8fL}]V_I$$

with

$$\sqrt{k}\sqrt{\frac{2R}{fL}} \ge \frac{4}{1-k}$$

and for current i_{L2} we have

$$kT(V_I + V_C - V_O) = (1 - k)m_Q T(V_O - 2V_I - V_{L3-off} - V_{L4-off} - V_{L5-off})$$

Therefore, output voltage in discontinuous mode is

$$V_O = [4 + \frac{3k}{1-k} + \frac{k}{(1-k)m_Q}]V_I = [4 + \frac{3k}{1-k} + k^2(1-k)\frac{R}{8fL}]V_I$$

with

$$\sqrt{k}\sqrt{\frac{2R}{fL}} \ge \frac{4}{1-k} \tag{2.192}$$

i.e., the output voltage will linearly increase during load resistance R increasing, as shown in Figure 2.30.

2.3.5 Summary

From the analysis and calculation in previous sections, the common formulas for all circuits can be obtained:

$$M = \frac{V_O}{V_I} = \frac{I_I}{I_O} \quad L = \frac{L_1 L_2}{L_1 + L_2} \quad z_N = \frac{R}{fL} \quad R = \frac{V_O}{I_O}$$

Current variations:

$$\xi_1 = \frac{1-k}{2M}\frac{R}{fL_1} \quad \xi_2 = \frac{k}{2M}\frac{R}{fL_2} \quad \chi_j = \frac{k}{M^2}\frac{R}{fL_{j+2}} \quad (j = 1, 2, 3, \ldots)$$

Voltage variations:

$$\rho = \frac{k}{2fCR} \quad \varepsilon = \frac{k}{8M}\frac{1}{f^2 C_O L_2} \quad \sigma_j = \frac{M}{2fC_j R} \quad (j = 1, 2, 3, 4, \ldots)$$

In order to write common formulas for the boundaries between continuous and discontinuous modes and output voltage for all circuits, the circuits can be numbered. The definition is that subscript 0 means the elementary circuit, subscript 1 means the self-lift circuit, subscript 2 means the re-lift circuit, subscript 3 means the triple-lift circuit, subscript 4 means the quadruple-lift circuit, and so on.

The voltage transfer gain is

$$M_j = \frac{k^{h(j)}[j + h(j)]}{1-k} \quad j = 0, 1, 2, 3, 4, \ldots \tag{2.193}$$

The free-wheeling diode current i_D's variation is

$$\zeta_j = \frac{k^{[1+h(j)]}}{M_j^{\,2}}\frac{j + h(j)}{2} z_N \tag{2.194}$$

The boundaries are determined by the condition:

$$\zeta_j \geq 1$$

or

$$\frac{k^{[1+h(j)]}}{M_j^{\,2}}\frac{j + h(j)}{2} z_N \geq 1 \quad j = 0, 1, 2, 3, 4, \ldots \tag{2.195}$$

Therefore, the boundaries between continuous and discontinuous modes for all circuits are

$$M_j = k^{\frac{1+h(j)}{2}} \sqrt{\frac{j+h(j)}{2} z_N} \qquad j = 0, 1, 2, 3, 4,\dots \qquad (2.196)$$

The filling efficiency is

$$m_j = \frac{1}{\varsigma_j} = \frac{M_j^{\,2}}{k^{[1+h(j)]}} \frac{2}{j+h(j)} \frac{1}{z_N} \qquad (2.197)$$

The output voltage in discontinuous mode for all circuits is

$$V_{O-j} = [j + \frac{j+h(j)-1}{1-k} + k^{[2-u(j)]} \frac{1-k}{2[j+h(j)]} z_N] V_I \qquad j = 0, 1, 2, 3, 4,\dots \qquad (2.198)$$

where

$$h(j) = \begin{cases} 0 & if \quad j \geq 1 \\ 1 & if \quad j = 0 \end{cases} \quad \text{is the \textbf{Hong Function}} \qquad (2.199)$$

Assuming that $f = 50$ kHz, $L_1 = L_2 = 1$ mH, $L_2 = L_3 = L_4 = L_5 = 0.5$ mH, $C = C_1 = C_2 = C_3 = C_4 = C_O = 20$ µF and the source voltage $V_I = 10$ V, the value of the output voltage V_O with various conduction duty k in continuous mode are shown in Figure 2.31. Typically, some values of the output voltage V_O and its ripples in conduction duty $k = 0.33, 0.5, 0.75$ and 0.9 are listed in Table 2.1. From these data it states the fact that the output voltage of all Luo-converters is almost a real DC voltage with very small ripple.

The boundaries between continuous and discontinuous modes of all circuits are shown in Figure 2.32. The curves of all M vs. z_N state that the continuous mode area increases from M_E via M_S, M_R, M_T to M_Q. The boundary of the elementary circuit is a monorising curve, but other curves are not monorising. There are minimum values of the boundaries of other circuits, which of M_S, M_R, M_T and M_Q correspond at $k = 1/3$.

2.3.6   Discussion

Some important points are vital for particular circuit design. They are discussed in the following sections.

2.3.6.1   *Discontinuous-Conduction Mode*

Usually, the industrial applications require the DC-DC converters to work in continuous mode. However, it is irresistible that DC-DC converter works

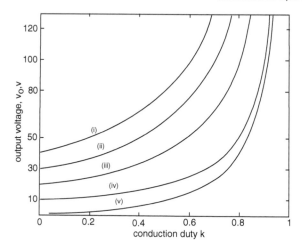

FIGURE 2.31
Output voltages of all positive output Luo-converters ($V_I = 10$ V).

TABLE 2.1

Comparison among Five Positive Output Luo-Converters

Positive Output Luo-Converters	I_O	V_O	V_O ($V_I = 10$ V)			
			$k = 0.33$	$k = 0.5$	$k = 0.75$	$k = 0.9$
Elementary Circuit	$I_O = \dfrac{1-k}{k} I_I$	$V_O = \dfrac{k}{1-k} V_I$	5 V	10 V	30 V	90 V
Self-Lift Circuit	$I_O = (1-k)I_I$	$V_O = \dfrac{1}{1-k} V_I$	15 V	20 V	40 V	100 V
Re-Lift Circuit	$I_O = \dfrac{1-k}{2} I_I$	$V_O = \dfrac{2}{1-k} V_I$	30 V	40 V	80 V	200 V
Triple-Lift Circuit	$I_O = \dfrac{1-k}{3} I_I$	$V_O = \dfrac{3}{1-k} V_I$	45 V	60 V	120 V	300 V
Quadruple-Lift Circuit	$I_O = \dfrac{1-k}{4} I_I$	$V_O = \dfrac{4}{1-k} V_I$	60 V	80 V	160 V	400 V

in discontinuous mode sometimes. The analysis in Section 2.3.2 through Section 2.3.5 shows that during switch-off if current i_D becomes zero before next period switch-on, the state is called discontinuous mode. The following factors affect the diode current i_D to become discontinuous:

1. Switch frequency f is too low
2. Conduction duty cycle k is too small
3. Combined inductor L is too small
4. Load resistance R is too big

Discontinuous mode means i_D is discontinuous during switch-off. The output current $i_O(t)$ is still continuous if L_2 and C_O are large enough.

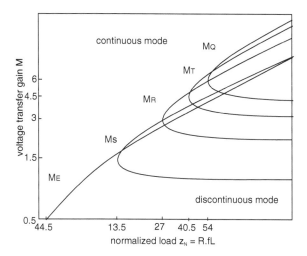

FIGURE 2.32
Boundaries between continuous and discontinuous modes of all positive output Luo-converters.

2.3.6.2 Output Voltage V_O vs. Conduction Duty k

Output voltage V_O is a positive value and is usually greater than the source voltage V_I when the conduction duty ratio is $k > 0.5$ for the elementary circuit, and any value in the range of $0 < k < 1$ for self-lift, re-lift, and multiple-lift circuits. Although small k results that the output voltage V_O of self-lift and re-lift circuits is greater than V_I and $2V_I$ and so on, when $k = 0$ it results in $V_O = 0$ because switch S is never turned on.

If k is close to the value of 1, the ideal output voltage V_O should be a very big value. Unfortunately, because of the effect of parasitic elements, output voltage V_O falls down very quickly. Finally, $k = 1$ results in $V_O = 0$, not infinity for all circuits. In this case the accident of i_{L1} toward infinity will happen. The recommended value range of the conduction duty k is

$$0 < k < 0.9$$

2.3.6.3 Switch Frequency f

In this paper the repeating frequency $f = 50$ kHz was selected. Actually, switch frequency f can be selected in the range between 10 kHz and 500 kHz. Usually, the higher the frequency, the lower the ripples.

2.4 Negative Output Luo-Converters

Negative output Luo-converters perform the voltage conversion from positive to negative voltages using VL technique. They work in the third quadrant with large voltage amplification. Five circuits have been introduced. They are

- Elementary circuit
- Self-lift circuit
- Re-lift circuit
- Triple-lift circuit
- Quadruple-lift circuit

As the positive output Luo-converters, the **negative output Luo-converters** are another series of DC-DC step-up converters, which were developed from prototypes using voltage lift technique. These converters perform positive to negative DC-DC voltage increasing conversion with high power density, high efficiency, and cheap topology in simple structure.

The elementary circuit can perform step-down and step-up DC-DC conversion. The other negative output Luo-converters are derived from this elementary circuit, they are the self-lift circuit, re-lift circuit, and multiple-lift circuits (e.g., triple-lift and quadruple-lift circuits) shown in the corresponding figures and introduced in the next sections respectively. Switch S in these diagrams is a P-channel power MOSFET device (PMOS). It is driven by a PWM switch signal with repeating frequency f and conduction duty **k**. In this book the switch repeating period is $T = 1/f$, so that the switch-on period is kT and switch-off period is $(1-k)T$. For all circuits, the load is usually resistive, i. e., $R = V_O/I_O$; the normalized load is $z_N = R/fL$. Each converter consists of a negative Luo-pump and a "Π"-type filter C-L_O-C_O, and a lift circuit (except elementary circuit). The pump inductor L absorbs the energy from source during switch-on and transfers the stored energy to capacitor C during switch-off. The energy on capacitor C is then delivered to load during switch-on. Therefore, if the voltage V_C is high the output voltage V_O is correspondingly high.

When the switch S is turned off the current i_D flows through the freewheeling diode D. This current descends in whole switch-off period $(1-k)T$. If current i_D does not become zero before switch S is turned on again, we define this working state to be continuous mode. If current i_D becomes zero before switch S is turned on again, we define this working state to be discontinuous mode.

The directions of all voltages and currents are indicated in the figures. All descriptions and calculations in the text are concentrated to the absolute values. In this paper for any component X, its instantaneous current and voltage values are expressed as i_X and v_X, or $i_X(t)$ and $v_X(t)$, and its average current and voltage values are expressed as I_X and V_X. For general description, the output voltage and current are V_O and I_O; the input voltage and current are V_I and I_I. Assuming the output power equals the input power,

$$P_O = P_{IN} \quad \text{or} \quad V_O I_O = V_I I_I$$

The following symbols are used in the text of this paper.

The voltage transfer gain is in **CCM**:

$$M = \frac{V_O}{V_I} = \frac{I_I}{I_O}$$

Variation ratio of current i_L:

$$\zeta = \frac{\Delta i_L / 2}{I_L}$$

Variation ratio of current i_{LO}:

$$\xi = \frac{\Delta i_{LO} / 2}{I_{LO}}$$

Variation ratio of current i_D:

$$\zeta = \frac{\Delta i_D / 2}{I_L} \quad \text{during switch-off, } i_D = i_L$$

Variation ratio of current i_{Lj} is

$$\chi_j = \frac{\Delta i_{Lj} / 2}{I_{Lj}} \quad j = 1, 2, 3, \ldots$$

Variation ratio of voltage v_C:

$$\rho = \frac{\Delta v_C / 2}{V_C}$$

Variation ratio of voltage v_{Cj}:

$$\sigma_j = \frac{\Delta v_{Cj} / 2}{V_{Cj}} \quad j = 1, 2, 3, 4, \ldots$$

Variation ratio of output voltage $v_O = v_{CO}$:

$$\varepsilon = \frac{\Delta v_O / 2}{V_O}$$

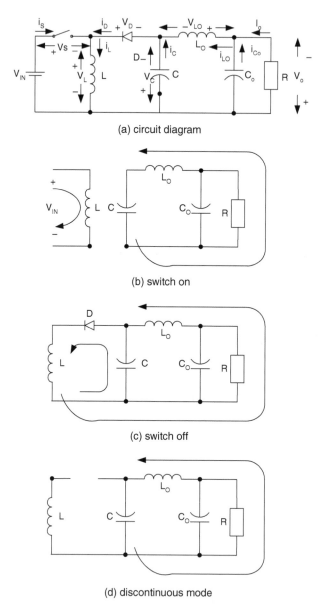

(a) circuit diagram

(b) switch on

(c) switch off

(d) discontinuous mode

FIGURE 2.33
Elementary circuit: (a) circuit diagram; (b) switch-on; (c) switch-off; (d) discontinuous mode.

2.4.1   Elementary Circuit

The elementary circuit, and its switch-on and -off equivalent circuits are shown in Figure 2.33. This circuit can be considered as a combination of an electronic pump S-L-D-C and a "Π"-type low-pass filter C-L_O-C_O. The electronic pump injects certain energy to the low-pass filter every cycle. Capacitor C in Figure 2.33 acts as the primary means of storing and transferring energy from the input source to the output load. Assuming capacitor C to be sufficiently large, the variation of the voltage across capacitor C from its average value V_C can be neglected in steady state, i.e., $v_C(t) \approx V_C$, even though it stores and transfers energy from the input to the output.

2.4.1.1 Circuit Description

When switch S is on, the equivalent circuit is shown in Figure 2.33b. In this case the source current $i_I = i_L$. Inductor L absorbs energy from the source, and current i_L linearly increases with slope V_I/L. In the mean time the diode D is blocked since it is inversely biased. Inductor L_O keeps the output current I_O continuous and transfers energy from capacitor C to the load R, i.e., $i_{C-on} = i_{LO}$. When switch S is off, the equivalent circuit is shown in Figure 2.33c. In this case the source current $i_I = 0$. Current i_L flows through the freewheeling diode D to charge capacitor C and enhances current i_{LO}. Inductor L transfers its stored energy to capacitor C and load R via inductor L_O, i.e., $i_L = i_{C-off} + i_{LO}$. Thus, currents i_L decrease.

2.4.1.2 Average Voltages and Currents

The output current $I_O = I_{LO}$ because the capacitor C_O does not consume any energy in the steady state. The average output current is

$$I_O = I_{LO} = I_{C-on} \tag{2.200}$$

The charge on the capacitor C increases during switch-off:

$$Q+ = (1 - k)\, T\, I_{C-off}$$

And it decreases during switch-on:

$$Q- = k\, T\, I_{C-on} \tag{2.201}$$

In a whole repeating period T,

$$Q+ = Q-, \quad I_{C-off} = \frac{k}{1-k} I_{C-on} = \frac{k}{1-k} I_O$$

Therefore, the inductor current I_L is

$$I_L = I_{C-off} + I_O = \frac{I_O}{1-k} \tag{2.202}$$

Equation (2.200) and Equation (2.202) are available for all circuit of negative output Luo-converters. The source current is $i_I = i_L$ during switch-on period. Therefore, its average source current I_I is

$$I_I = k \times i_I = ki_L = kI_L = \frac{k}{1-k} I_O$$

or

$$I_O = \frac{1-k}{k} I_I \tag{2.203}$$

and the output voltage is

$$V_O = \frac{k}{1-k} V_I \tag{2.204}$$

The voltage transfer gain in continuous mode is

$$M_E = \frac{V_O}{V_I} = \frac{I_I}{I_O} = \frac{k}{1-k} \tag{2.205}$$

The curve of M_E vs. k is shown in Figure 2.34. Current i_L increases and is supplied by V_I during switch-on. It decreases and is inversely biased by $-V_C$ during switch-off,

$$kTV_I = (1-k)TV_C \tag{2.206}$$

Therefore,

$$V_C = V_O = \frac{k}{1-k} V_I \tag{2.207}$$

2.4.1.3   *Variations of Currents and Voltages*

To analyze the variations of currents and voltages, some voltage and current waveforms are shown in Figure 2.35. Current i_L increases and is supplied by V_I during switch-on. Thus, its peak-to-peak variation is

$$\Delta i_L = \frac{kTV_I}{L}$$

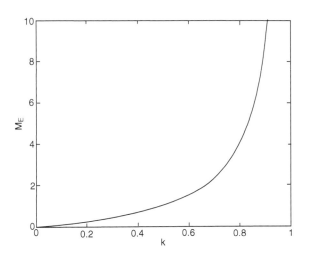

FIGURE 2.34
Voltage transfer gain M_E vs. k.

Considering Equation (2.202) and Equation (2.205), and $R = V_O/I_O$, the variation ratio of the current i_L is

$$\zeta = \frac{\Delta i_L / 2}{I_L} = \frac{k(1-k)V_I T}{2LI_O} = \frac{k(1-k)R}{2M_E fL} = \frac{k^2}{M_E^2} \frac{R}{2fL} \tag{2.208}$$

Considering Equation (2.201), the peak-to-peak variation of voltage v_C is

$$\Delta v_C = \frac{Q-}{C} = \frac{k}{C} T I_O \tag{2.209}$$

The variation ratio of voltage v_C is

$$\rho = \frac{\Delta v_C / 2}{V_C} = \frac{k I_O T}{2 C V_O} = \frac{k}{2} \frac{1}{fCR} \tag{2.210}$$

Since voltage V_O variation is very small, the peak-to-peak variation of current i_{LO} is calculated by the area (B) of the triangle with the width of $T/2$ and height $\Delta v_C/2$.

$$\Delta i_{LO} = \frac{B}{L_O} = \frac{1}{2} \frac{T}{2} \frac{k}{2CL_O} T I_O = \frac{k}{8f^2 CL_O} I_O \tag{2.211}$$

Considering Equation (2.200), the variation ratio of current i_{LO} is

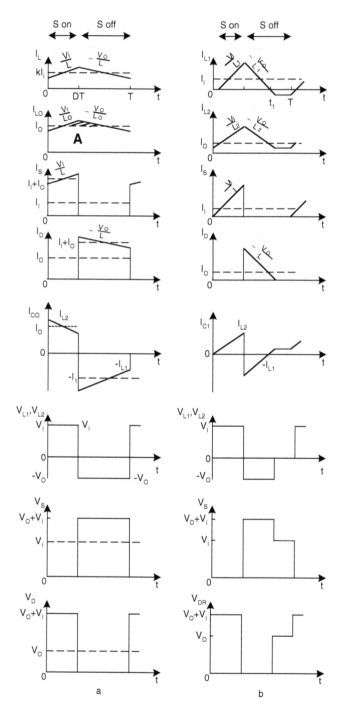

FIGURE 2.35
Some voltage and current waveforms of elementary circuit.

$$\xi = \frac{\Delta i_{LO}/2}{I_{LO}} = \frac{k}{16}\frac{1}{f^2 C L_O} \qquad (2.212)$$

Since the voltage v_C is a triangle waveform, the difference between v_C and output voltage V_O causes the ripple of current i_{LO}, and the difference between i_{LO} and output current I_O causes the ripple of output voltage v_O. The ripple waveform of current i_{LO} should be a partial parabola in Figure 2.35 because of the triangle waveform of Δv_C. To simplify the calculation we can treat the ripple waveform of current i_{LO} as a triangle waveform in Figure 2.35 because the ripple of the current i_{LO} is very small. Therefore, the peak-to-peak variation of voltage v_{CO} is calculated by the area (A) of the triangle with the width of $T/2$ and height $\Delta i_{LO}/2$:

$$\Delta v_{CO} = \frac{A}{C_O} = \frac{1}{2}\frac{T}{2}\frac{k}{16f^2 C C_O L_O}I_O = \frac{k}{64f^3 C C_O L_O}I_O \qquad (2.213)$$

The variation ratio of current v_{CO} is

$$\varepsilon = \frac{\Delta v_{CO}/2}{V_{CO}} = \frac{k}{128f^3 C C_O L_O}\frac{I_O}{V_O} = \frac{k}{128}\frac{1}{f^3 C C_O L_O R} \qquad (2.214)$$

Assuming that f = 50 kHz, $L = L_O$ = 100 μH, $C = C_O$ = 5 μF, R = 10 Ω and k = 0.6, we obtain

$$M_E = 1.5 \quad \zeta = 0.16 \quad \zeta = 0.03 \quad \rho = 0.12 \quad \text{and} \quad \varepsilon = 0.0015$$

The output voltage V_O is almost a real DC voltage with very small ripple. Since the load is resistive, the output current $i_o(t)$ is almost a real DC waveform with very small ripple as well, and it is equal to $I_O = V_O/R$.

2.4.1.4 Instantaneous Values of Currents and Voltages

Referring to Figure 2.35, the instantaneous current and voltage values are listed below:

$$v_S = \begin{cases} 0 & for \quad 0 < t \le kT \\ V_O & for \quad kT < t \le T \end{cases} \qquad (2.215)$$

$$v_D = \begin{cases} V_I + V_O & for \quad 0 < t \le kT \\ 0 & for \quad kT < t \le T \end{cases} \qquad (2.216)$$

$$v_L = \begin{cases} V_I & for \quad 0 < t \le kT \\ -V_O & for \quad kT < t \le T \end{cases} \qquad (2.217)$$

$$i_I = i_S = \begin{cases} i_L(0) + \dfrac{V_I}{L} t & for \quad 0 < t \le kT \\ 0 & for \quad kT < t \le T \end{cases} \qquad (2.218)$$

$$i_L = \begin{cases} i_L(0) + \dfrac{V_I}{L} t & for \quad 0 < t \le kT \\ i_L(kT) - \dfrac{V_O}{L}(t - kT) & for \quad kT < t \le T \end{cases} \qquad (2.219)$$

$$i_D = \begin{cases} 0 & for \quad 0 < t \le kT \\ i_L(kT) - \dfrac{V_O}{L}(t - kT) & for \quad kT < t \le T \end{cases} \qquad (2.220)$$

$$i_C \approx \begin{cases} -I_{C-on} & for \quad 0 < t \le kT \\ I_{C-off} & for \quad kT < t \le T \end{cases} \qquad (2.221)$$

where

$$i_L(0) = k\,I_I - k\,V_I/2f\,L$$

$$i_L(kT) = k\,I_I + k\,V_I/2f\,L$$

Since the instantaneous current i_{LO} and voltage v_{CO} are partial parabolas with very small ripples, they can be treated as a DC current and voltage.

2.4.1.5   *Discontinuous Mode*

Referring to Figure 2.33d, we can see that the diode current i_D becomes zero during switch off before next period switch on. The condition for discontinuous mode is

$$\zeta \ge 1$$

i.e.,

$$\frac{k^2}{M_E{}^2} \frac{R}{2fL} \ge 1$$

or

$$M_E \le k\sqrt{\frac{R}{2fL}} = k\sqrt{\frac{z_N}{2}} \qquad (2.222)$$

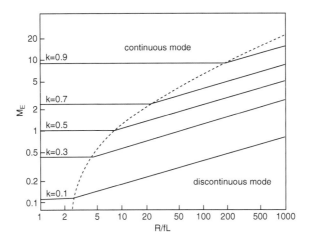

FIGURE 2.36
The boundary between continuous and discontinuous modes and output voltage vs. the normalized load $z_N = \sqrt{R/fL}$ (elementary circuit).

The graph of the boundary curve vs. the normalized load $z_N = R/fL$ is shown in Figure 2.36. It can be seen that the boundary curve is a monorising function of the parameter k.

In this case the current i_D exists in the period between kT and $t_1 = [k + (1 - k)m_E]T$, where m_E is the **filling efficiency** and it is defined as:

$$m_E = \frac{1}{\zeta} = \frac{M_E^2}{k^2 \dfrac{R}{2fL}} \tag{2.223}$$

Considering Equation (2.222), therefore $0 < m_E < 1$. Since the diode current i_D becomes zero at $t = kT + (1 - k)m_E T$, for the current i_L

$$kTV_I = (1 - k)m_E T V_C$$

or

$$V_C = \frac{k}{(1-k)m_E} V_I = k(1-k)\frac{R}{2fL}V_I$$

$$\text{with} \quad \sqrt{\frac{R}{2fL}} \geq \frac{1}{1-k}$$

and for the current i_{LO}

$$kT(V_I + V_C - V_O) = (1 - k)m_E T V_O$$

Therefore, output voltage in discontinuous mode is

$$V_O = \frac{k}{(1-k)m_E} V_I = k(1-k)\frac{R}{2fL}V_I \quad \text{with} \quad \sqrt{\frac{R}{2fL}} \geq \frac{1}{1-k} \qquad (2.224)$$

i.e., the output voltage will linearly increase during load resistance R increasing. The output voltage vs. the normalized load $z_N = R/fL$ is shown in Figure 2.36. Larger load resistance R may cause higher output voltage in discontinuous mode.

2.4.2 Self-Lift Circuit

Self-lift circuit, and its switch-on and -off equivalent circuits are shown in Figure 2.37, which is derived from the elementary circuit. It consists of eight passive components. They are one static switch S; two inductors L, L_O; three capacitors C, C_1, and C_O; and two diodes D, D_1. Comparing with Figure 2.33 and Figure 2.37, it can be seen that there are only one more capacitor C_1 and one more diode D_1 added into the self-lift circuit. Circuit C_1-D_1 is the lift circuit. Capacitor C_1 functions to lift the capacitor voltage V_C by a source voltage V_I. Current $i_{C1}(t)$ is an exponential function $\delta(t)$. It has a large value at the moment of power on, but it is small in the steady state because $V_{C1} = V_I$.

2.4.2.1   Circuit Description

When switch S is on, the equivalent circuit is shown in Figure 2.37b. In this case the source current $i_I = i_L + i_{C1}$. Inductor L absorbs energy from the source, and current i_I linearly increases with slope V_I/L. In the mean time the diode D_1 is conducted and capacitor C_1 is charged by the current i_{C1}. Inductor L_O keeps the output current I_O continuous and transfers energy from capacitor C to the load R, i.e., $i_{C-on} = i_{LO}$. When switch S is off, the equivalent circuit is shown in Figure 2.37c. In this case the source current $i_I = 0$. Current i_L flows through the free-wheeling diode D to charge capacitor C and enhances current i_{LO}. Inductor L transfers its stored energy via capacitor C_1 to capacitor C and load R (via inductor L_O), i.e., $i_L = i_{C1-off} = i_{C-off} + i_{LO}$. Thus, current i_L decreases.

2.4.2.2   Average Voltages and Currents

The output current $I_O = I_{LO}$ because the capacitor C_O does not consume any energy in the steady state. The average output current:

$$I_O = I_{LO} = I_{C-on} \qquad (2.225)$$

The charge of the capacitor C increases during switch-off:

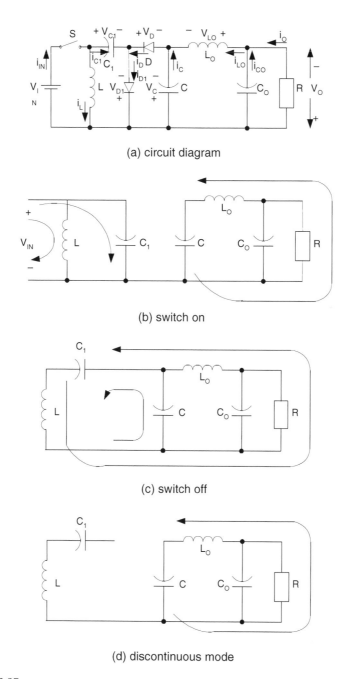

(a) circuit diagram

(b) switch on

(c) switch off

(d) discontinuous mode

FIGURE 2.37
Self-lift circuit: (a) circuit diagram; (b) switch on; (c) switch off; (d) discontinuous mode.

$$Q+ = (1-k)\, T\, I_{C\text{-}off}$$

And it decreases during switch-on:

$$Q- = k\, T\, I_{C\text{-}on} \tag{2.226}$$

In a whole repeating period T, $Q+ = Q-$.

Thus,

$$I_{C\text{-}off} = \frac{k}{1-k} I_{C\text{-}on} = \frac{k}{1-k} I_O$$

Therefore, the inductor current I_L is

$$I_L = I_{C\text{-}off} + I_O = \frac{I_O}{1-k} \tag{2.227}$$

From Figure 2.37,

$$I_{C1\text{-}off} = I_L = \frac{1}{1-k} I_O \tag{2.228}$$

and

$$I_{C1\text{-}on} = \frac{1-k}{k} I_{C1\text{-}off} = \frac{1}{k} I_O \tag{2.229}$$

In steady state we can use

$$V_{C1} = V_I$$

Investigate current i_L, it increases during switch–on with slope V_I/L and decreases during switch-off with slope $-(V_O - V_{C1})/L = -(V_O - V_I)/L$.
Therefore,

$$kV_I = (1-k)(V_O - V_I)$$

or

$$V_O = \frac{1}{1-k} V_I \tag{2.230}$$

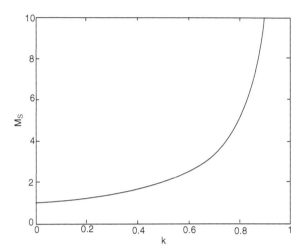

FIGURE 2.38
Voltage transfer gain M_S vs. k.

and

$$I_O = (1 - k)I_I \qquad (2.231)$$

The voltage transfer gain in continuous mode is

$$M_S = \frac{V_O}{V_I} = \frac{I_I}{I_O} = \frac{1}{1 - k} \qquad (2.232)$$

The curve of M_S vs. k is shown in Figure 2.38.

Circuit (C-L_O-C_O) is a "Π" type low-pass filter. Therefore,

$$V_C = V_O = \frac{k}{1 - k} V_I \qquad (2.233)$$

2.4.2.3 Variations of Currents and Voltages

To analyze the variations of currents and voltages, some voltage and current waveforms are shown in Figure 2.39.

Current i_L increases and is supplied by V_I during switch-on. Thus, its peak-to-peak variation is

$$\Delta i_L = \frac{kTV_I}{L}$$

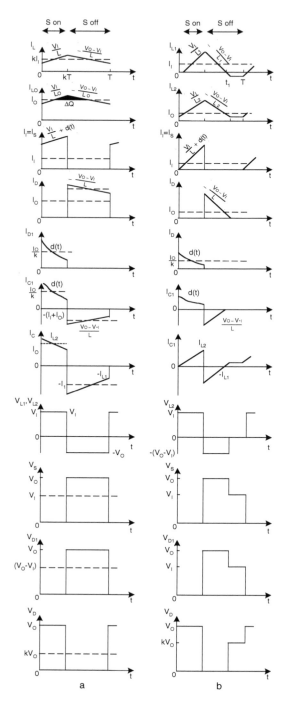

FIGURE 2.39
Some voltage and current waveforms of self-lift circuit.

Considering Equation (2.227) and $R = V_O/I_O$, the variation ratio of the current i_L is

$$\zeta = \frac{\Delta i_L / 2}{I_L} = \frac{k(1-k)V_I T}{2LI_O} = \frac{k(1-k)R}{2M_S fL} = \frac{k}{M_S^2}\frac{R}{2fL} \qquad (2.234)$$

Considering Equation (2.226), the peak-to-peak variation of voltage v_C is

$$\Delta v_C = \frac{Q-}{C} = \frac{k}{C}TI_O$$

The variation ratio of voltage v_C is

$$\rho = \frac{\Delta v_C / 2}{V_C} = \frac{kI_O T}{2CV_O} = \frac{k}{2}\frac{1}{fCR} \qquad (2.235)$$

The peak-to-peak variation of voltage v_{C1} is

$$\Delta v_{C1} = \frac{kT}{C_1}I_{C1-on} = \frac{1}{fC}I_O$$

The variation ratio of voltage v_{C1} is

$$\sigma_1 = \frac{\Delta v_{C1} / 2}{V_{C1}} = \frac{I_O}{2fC_1 V_I} = \frac{M_S}{2}\frac{1}{fC_1 R} \qquad (2.236)$$

Considering the Equation (2.211):

$$\Delta i_{LO} = \frac{1}{2}\frac{T}{2}\frac{k}{2CL_O}TI_O = \frac{k}{8f^2 CL_O}I_O$$

The variation ratio of current i_{LO} is

$$\xi = \frac{\Delta i_{LO} / 2}{I_{LO}} = \frac{k}{16}\frac{1}{f^2 CL_O} \qquad (2.237)$$

Considering Equation (2.213):

$$\Delta v_{CO} = \frac{B}{C_O} = \frac{1}{2}\frac{T}{2}\frac{k}{16f^2 CC_O L_O}I_O = \frac{k}{64f^3 CC_O L_O}I_O$$

The variation ratio of current v_{CO} is

$$\varepsilon = \frac{\Delta v_{CO}/2}{V_{CO}} = \frac{k}{128 f^3 C C_O L_O} \frac{I_O}{V_O} = \frac{k}{128} \frac{1}{f^3 C C_O L_O R} \tag{2.238}$$

Assuming that $f = 50$ kHz, $L = L_O = 100$ μH, $C = C_O = 5$ μF, $R = 10$ Ω and $k = 0.6$, we obtain

$$M_S = 2.5 \quad \zeta = 0.096 \quad \xi = 0.03 \quad \rho = 0.12 \quad \text{and} \quad \varepsilon = 0.0015$$

The output voltage V_O is almost a real DC voltage with very small ripple. Since the load is resistive, the output current $i_O(t)$ is almost a real DC waveform with very small ripple as well, and it is equal to $I_O = V_O/R$.

2.4.2.4   *Instantaneous Value of the Currents and Voltages*

Referring to Figure 2.39, the instantaneous values of the currents and voltages are listed below:

$$v_S = \begin{cases} 0 & for \quad 0 < t \le kT \\ V_O - V_I & for \quad kT < t \le T \end{cases} \tag{2.239}$$

$$v_D = \begin{cases} V_O & for \quad 0 < t \le kT \\ 0 & for \quad kT < t \le T \end{cases} \tag{2.240}$$

$$v_{D1} = \begin{cases} 0 & for \quad 0 < t \le kT \\ V_O & for \quad kT < t \le T \end{cases} \tag{2.241}$$

$$v_L = \begin{cases} V_I & for \quad 0 < t \le kT \\ -(V_O - V_I) & for \quad kT < t \le T \end{cases} \tag{2.242}$$

$$i_I = i_S = \begin{cases} i_{L1}(0) + \delta(t) + \dfrac{V_I}{L} t & for \quad 0 < t \le kT \\ 0 & for \quad kT < t \le T \end{cases} \tag{2.243}$$

$$i_L = \begin{cases} i_L(0) + \dfrac{V_I}{L} t & for \quad 0 < t \le kT \\ i_L(kT) - \dfrac{V_O - V_I}{L}(t - kT) & for \quad kT < t \le T \end{cases} \tag{2.244}$$

$$i_D = \begin{cases} 0 & \text{for} \quad 0 < t \le kT \\ i_{L1}(kT) - \dfrac{V_O - V_I}{L}(t - kT) & \text{for} \quad kT < t \le T \end{cases} \tag{2.245}$$

$$i_{D1} = \begin{cases} \delta(t) & \text{for} \quad 0 < t \le kT \\ 0 & \text{for} \quad kT < t \le T \end{cases} \tag{2.246}$$

$$i_{C1} = \begin{cases} \delta(t) & \text{for} \quad 0 < t \le kT \\ -i_L(kT) + \dfrac{V_O - V_I}{L}(t - kT) & \text{for} \quad kT < t \le T \end{cases} \tag{2.247}$$

$$i_C \approx \begin{cases} -I_{C-on} & \text{for} \quad 0 < t \le kT \\ I_{C-off} & \text{for} \quad kT < t \le T \end{cases} \tag{2.248}$$

where

$$i_L(0) = k\,I_I - k\,V_I/2fL$$

$$i_L(kT) = k\,I_I + k\,V_I/2fL$$

Since the instantaneous current i_{LO} and voltage v_{CO} are partial parabolas with very small ripples, they can be treated as a DC current and voltage.

2.4.2.5   Discontinuous Mode

Referring to Figure 2.37d, we can see that the diode current i_D becomes zero during switch off before next period switch on. The condition for discontinuous mode is

$$\zeta \ge 1$$

i.e.,

$$\frac{k}{M_S^{\,2}} \frac{R}{2fL} \ge 1$$

or

$$M_S \le \sqrt{k}\,\sqrt{\frac{R}{2fL}} = \sqrt{k}\,\sqrt{\frac{z_N}{2}} \tag{2.249}$$

The graph of the boundary curve vs. the normalized load $z_N = R/fL$ is shown in Figure 2.40. It can be seen that the boundary curve has a minimum value of 1.5 at $k = \frac{1}{3}$.

In this case the current i_D exists in the period between kT and $t_1 = [k + (1-k)m_S]T$, where m_S is the **filling efficiency** and it is defined as:

$$m_S = \frac{1}{\varsigma} = \frac{M_S^2}{k\dfrac{R}{2fL}} \tag{2.250}$$

Considering Equation (2.249), therefore $0 < m_S < 1$. Since the diode current i_D becomes 0 at $t = kT + (1-k)m_S T$, for the current i_L

$$kTV_I = (1-k)m_S T(V_C - V_I)$$

or

$$V_C = [1 + \frac{k}{(1-k)m_S}]V_I = [1 + k^2(1-k)\frac{R}{2fL}]V_I \quad \text{with} \quad \sqrt{k}\sqrt{\frac{R}{2fL}} \geq \frac{1}{1-k}$$

and for the current i_{LO}

$$kT(V_I + V_C - V_O) = (1-k)m_S T(V_O - V_I)$$

Therefore, output voltage in discontinuous mode is

$$V_O = [1 + \frac{k}{(1-k)m_S}]V_I = [1 + k^2(1-k)\frac{R}{2fL}]V_I \quad \text{with} \quad \sqrt{k}\sqrt{\frac{R}{2fL}} \geq \frac{1}{1-k} \tag{2.251}$$

i.e., the output voltage will linearly increase during load resistance R increasing. The output voltage V_O vs. the normalized load $z_N = R/fL$ is shown in Figure 2.40. Larger load resistance R causes higher output voltage in discontinuous mode.

2.4.3 Re-Lift Circuit

Re-lift circuit, and its switch-on and -off equivalent circuits are shown in Figure 2.41, which is derived from the self-lift circuit. It consists of one static switch S; three inductors L, L_1, and L_O; four capacitors C, C_1, C_2, and C_O; and diodes. From Figure 2.33, Figure 2.37, and Figure 2.41, it can be seen that there are one capacitor C_2, one inductor L_1 and two diodes D_2, D_{11} added into the re-lift circuit. Circuit C_1-D_1-D_{11}-L_1-C D_2 is the lift circuit. Capacitors C_1 and C_2 perform characteristics to lift the capacitor voltage V_C by twice

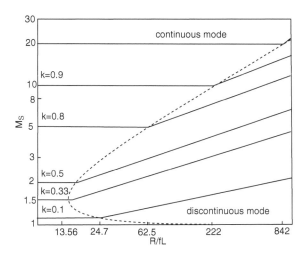

FIGURE 2.40
The boundary between continuous and discontinuous modes and output voltage vs. the normalized load $z_N = R/fL$ (self-lift circuit).

that of source voltage $2V_I$. Inductor L_1 performs the function as a ladder joint to link the two capacitors C_1 and C_2 and lift the capacitor voltage V_C up. Currents $i_{C1}(t)$ and $i_{C2}(t)$ are exponential functions $\delta_1(t)$ and $\delta_2(t)$. They have large values at the moment of power on, but they are small because $v_{C1} = v_{C2} \cong V_I$ is in steady state.

2.4.3.1 Circuit Description

When switch S is on, the equivalent circuit is shown in Figure 2.41b. In this case the source current $i_I = i_L + i_{C1} + i_{C2}$. Inductor L absorbs energy from the source, and current i_L linearly increases with slope V_I/L. In the mean time the diodes D_1, D_2 are conducted so that capacitors C_1 and C_2 are charged by the current i_{C1} and i_{C2}. Inductor L_O keeps the output current I_O continuous and transfers energy from capacitors C to the load R, i.e., $i_{C-on} = i_{LO}$. When switch S is off, the equivalent circuit is shown in Figure 2.41c. In this case the source current $i_I = 0$. Current i_L flows through the free-wheeling diode D, capacitors C_1 and C_2, inductor L_1 to charge capacitor C and enhances current i_{LO}. Inductor L transfers its stored energy to capacitor C and load R via inductor L_O, i.e., $i_L = i_{C1-off} = i_{C2-off} = i_{L1-off} = i_{C-off} + i_{LO}$. Thus, current i_L decreases.

2.4.3.2 Average Voltages and Currents

The output current $I_O = I_{LO}$ because the capacitor C_O does not consume any energy in the steady state. The average output current:

$$I_O = I_{LO} = I_{C-on} \tag{2.252}$$

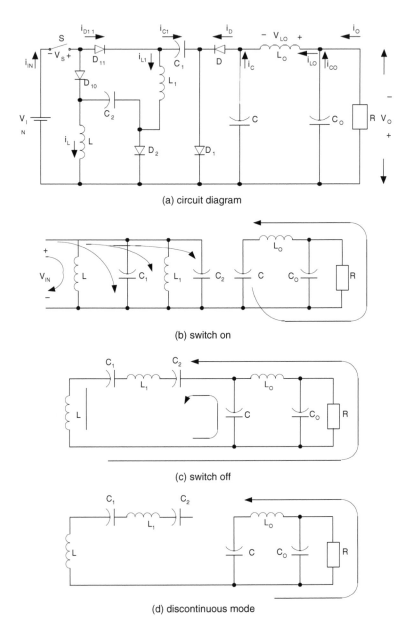

(a) circuit diagram

(b) switch on

(c) switch off

(d) discontinuous mode

FIGURE 2.41
Re-lift circuit: (a) circuit diagram; (b) switch on; (c) switch off; (d) discontinuous mode.

The charge of the capacitor C increases during switch-off:

$$Q+ = (1-k)\, T\, I_{C\text{-}off}$$

And it decreases during switch-on:

$$Q- = k\,T\,I_{C-on}$$

In a whole repeating period T, $Q+ = Q-$. Thus,

$$I_{C-off} = \frac{k}{1-k}I_{C-on} = \frac{k}{1-k}I_O$$

Therefore, the inductor current I_L is

$$I_L = I_{C-off} + I_O = \frac{I_O}{1-k} \qquad (2.253)$$

We know from Figure 2.48b that

$$I_{C1-off} = I_{C2-off} = I_{L1} = I_L = \frac{1}{1-k}I_O \qquad (2.254)$$

and

$$I_{C1-on} = \frac{1-k}{k}I_{C1-off} = \frac{1}{k}I_O \qquad (2.255)$$

and

$$I_{C2-on} = \frac{1-k}{k}I_{C2-off} = \frac{1}{k}I_O \qquad (2.256)$$

In steady state we can use

$$V_{C1} = V_{C2} = V_I$$

and

$$V_{L1-on} = V_I \qquad V_{L1-off} = \frac{k}{1-k}V_I$$

Investigate current i_L, it increases during switch-on with slope V_I/L and decreases during switch-off with slope $-(V_O - V_{C1} - V_{C2} - V_{L1-off})/L = -[V_O - 2V_I - k\,V_I/(1-k)]/L$. Therefore,

$$kTV_I = (1-k)T(V_O - 2V_I - \frac{k}{1-k}V_I)$$

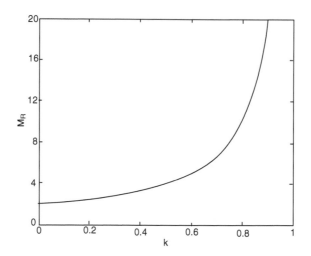

FIGURE 2.42
Voltage transfer gain M_R vs. k.

or

$$V_O = \frac{2}{1-k} V_I \qquad (2.257)$$

and

$$I_O = \frac{1-k}{2} I_I \qquad (2.258)$$

The voltage transfer gain in continuous mode is

$$M_R = \frac{V_O}{V_I} = \frac{I_I}{I_O} = \frac{2}{1-k} \qquad (2.259)$$

The curve of M_S vs. k is shown in Figure 2.42.
Circuit (C-L_O-C_O) is a "Π" type low-pass filter. Therefore,

$$V_C = V_O = \frac{2}{1-k} V_I \qquad (2.260)$$

2.4.3.3 *Variations of Currents and Voltages*

To analyze the variations of currents and voltages, some voltage and current waveforms are shown in Figure 2.43.

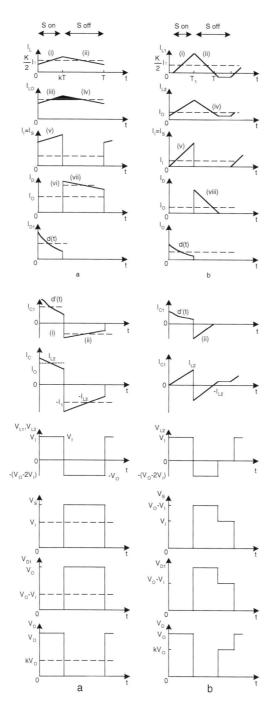

FIGURE 2.43
Some voltage and current waveforms of re-lift circuit.

Current i_L increases and is supplied by V_I during switch-on. Thus, its peak-to-peak variation is

$$\Delta i_L = \frac{kTV_I}{L}$$

Considering Equation (2.253) and $R = V_O/I_O$, the variation ratio of the current i_L is

$$\zeta = \frac{\Delta i_L / 2}{I_L} = \frac{k(1-k)V_I T}{2LI_O} = \frac{k(1-k)R}{2M_R fL} = \frac{k}{M_R^2} \frac{R}{fL} \tag{2.261}$$

The peak-to-peak variation of current i_{L1} is

$$\Delta i_{L1} = \frac{k}{L_1} TV_I$$

The variation ratio of current i_{L1} is

$$\chi_1 = \frac{\Delta i_{L1} / 2}{I_{L1}} = \frac{kTV_I}{2L_1 I_O}(1-k) = \frac{k(1-k)}{2M_R} \frac{R}{fL_1} \tag{2.262}$$

The peak-to-peak variation of voltage v_C is

$$\Delta v_C = \frac{Q-}{C} = \frac{k}{C} TI_O$$

The variation ratio of voltage v_C is

$$\rho = \frac{\Delta v_C / 2}{V_C} = \frac{kI_O T}{2CV_O} = \frac{k}{2} \frac{1}{fCR} \tag{2.263}$$

The peak-to-peak variation of voltage v_{C1} is

$$\Delta v_{C1} = \frac{kT}{C_1} I_{C1-on} = \frac{1}{fC} I_O$$

The variation ratio of voltage v_{C1} is

$$\sigma_1 = \frac{\Delta v_{C1} / 2}{V_{C1}} = \frac{I_O}{2fC_1 V_I} = \frac{M_R}{2} \frac{1}{fC_1 R} \tag{2.264}$$

Take the same operation, variation ratio of voltage v_{C2} is

$$\sigma_2 = \frac{\Delta v_{C2}/2}{V_{C2}} = \frac{I_O}{2fC_2V_I} = \frac{M_R}{2}\frac{1}{fC_2R} \tag{2.265}$$

Considering the Equation (2.211):

$$\Delta i_{LO} = \frac{1}{2}\frac{T}{2}\frac{k}{2CL_O}TI_O = \frac{k}{8f^2CL_O}I_O$$

The variation ratio of current i_{LO} is

$$\xi = \frac{\Delta i_{LO}/2}{I_{LO}} = \frac{k}{16}\frac{1}{f^2CL_O} \tag{2.266}$$

Considering the Equation (2.213):

$$\Delta v_{CO} = \frac{B}{C_O} = \frac{1}{2}\frac{T}{2}\frac{k}{16f^2CC_OL_O}I_O = \frac{k}{64f^3CC_OL_O}I_O$$

The variation ratio of current v_{CO} is

$$\varepsilon = \frac{\Delta v_{CO}/2}{V_{CO}} = \frac{k}{128f^3CC_OL_O}\frac{I_O}{V_O} = \frac{k}{128}\frac{1}{f^3CC_OL_OR} \tag{2.267}$$

Assuming that $f = 50$ kHz, $L = L_O = 100$ μH, $C = C_O = 5$ μF, $R = 10$ Ω and $k = 0.6$, we obtain

$$M_R = 5 \quad \zeta = 0.048 \quad \xi = 0.03 \quad \rho = 0.12 \quad \text{and} \quad \varepsilon = 0.0015$$

The output voltage V_O is almost a real DC voltage with very small ripple. Since the load is resistive, the output current $i_O(t)$ is almost a real DC waveform with very small ripple as well, and it is equal to $I_O = V_O/R$.

2.4.3.4 *Instantaneous Value of the Currents and Voltages*

Referring to Figure 2.43, the instantaneous current and voltage values are listed below:

$$v_S = \begin{cases} 0 & \text{for} \quad 0 < t \le kT \\ V_O - (2 - \dfrac{k}{1-k})V_I & \text{for} \quad kT < t \le T \end{cases} \tag{2.268}$$

$$v_D = \begin{cases} V_O & \text{for} \quad 0 < t \le kT \\ 0 & \text{for} \quad kT < t \le T \end{cases} \tag{2.269}$$

$$v_{D1} = v_{D2} = \begin{cases} 0 & \text{for} \quad 0 < t \le kT \\ V_O & \text{for} \quad kT < t \le T \end{cases} \tag{2.270}$$

$$v_{L1} = \begin{cases} V_I & \text{for} \quad 0 < t \le kT \\ -\dfrac{k}{1-k}V_I & \text{for} \quad kT < t \le T \end{cases} \tag{2.271}$$

$$v_L = \begin{cases} V_I & \text{for} \quad 0 < t \le kT \\ -[V_O - (2 - \dfrac{k}{1-k})V_I] & \text{for} \quad kT < t \le T \end{cases} \tag{2.272}$$

$$i_I = i_S = \begin{cases} i_L(0) + \dfrac{V_I}{L}t + \delta_1(t) + \delta_2(t) + i_{L1}(0) + \dfrac{V_I}{L_1}t & \text{for } 0 < t \le kT \\ 0 & \text{for } kT < t \le T \end{cases} \tag{2.273}$$

$$i_L = \begin{cases} i_L(0) + \dfrac{V_I}{L}t & \text{for} \quad 0 < t \le kT \\ i_L(kT) - \dfrac{V_O - (2 - \dfrac{k}{1-k})V_I}{L}(t - kT) & \text{for} \quad kT < t \le T \end{cases} \tag{2.274}$$

$$i_{L1} = \begin{cases} i_{L1}(0) + \dfrac{V_I}{L_1}t & \text{for} \quad 0 < t \le kT \\ i_{L1}(kT) - \dfrac{\dfrac{k}{1-k}V_I}{L_1}(t - kT) & \text{for} \quad kT < t \le T \end{cases} \tag{2.275}$$

$$i_D = \begin{cases} 0 & \text{for} \quad 0 < t \le kT \\ i_L(kT) - \dfrac{V_O - (2 - \dfrac{k}{1-k})V_I}{L}(t - kT) & \text{for} \quad kT < t \le T \end{cases} \tag{2.276}$$

$$i_{D1} = \begin{cases} \delta_1(t) & \text{for} \quad 0 < t \le kT \\ 0 & \text{for} \quad kT < t \le T \end{cases} \tag{2.277}$$

$$i_{D2} = \begin{cases} \delta_2(t) & \text{for} \quad 0 < t \le kT \\ 0 & \text{for} \quad kT < t \le T \end{cases} \tag{2.278}$$

$$i_{C1} = \begin{cases} \delta_1(t) & \text{for} \quad 0 < t \le kT \\ -i_{L1}(kT) + \dfrac{V_O - (2 - \dfrac{k}{1-k})V_I}{L}(t - kT) & \text{for} \quad kT < t \le T \end{cases} \qquad (2.279)$$

$$i_{C2} = \begin{cases} \delta_2(t) & \text{for} \quad 0 < t \le kT \\ -i_{L1}(kT) + \dfrac{V_O - (2 - \dfrac{k}{1-k})V_I}{L}(t - kT) & \text{for} \quad kT < t \le T \end{cases} \qquad (2.280)$$

$$i_C = \begin{cases} -I_{C-on} & \text{for} \quad 0 < t \le kT \\ I_{C-off} & \text{for} \quad kT < t \le T \end{cases} \qquad (2.281)$$

where

$$i_L(0) = k\,I_I - k\,V_I/2\,f\,L$$

$$i_L(kT) = k\,I_I + k\,V_I/2\,f\,L$$

and

$$i_{L1}(0) = k\,I_I - k\,V_I/2\,f\,L_1$$

$$i_{L1}(kT) = k\,I_I + k\,V_I/2\,f\,L_1$$

Since the instantaneous currents i_{LO} and i_{CO} are partial parabolas with very small ripples. They are very nearly DC current.

2.4.3.5 Discontinuous Mode

Referring to Figure 2.41d, we can see that the diode current i_D becomes zero during switch off before next period switch on. The condition for discontinuous mode is

$$\zeta \ge 1$$

i.e.,

$$\frac{k}{M_R^{\,2}}\frac{R}{fL} \ge 1$$

or

$$M_R \le \sqrt{k}\sqrt{\frac{R}{fL}} = \sqrt{k}\sqrt{z_N} \qquad (2.282)$$

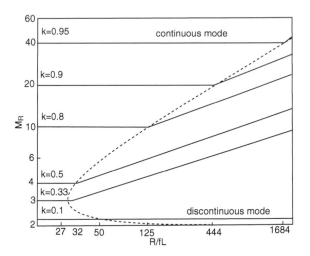

FIGURE 2.44
The boundary between continuous and discontinuous modes and output voltage vs. the normalized load $z_N = R/fL$ (re-lift circuit).

The graph of the boundary curve vs. the normalized load $z_N = R/fL$ is shown in Figure 2.44. It can be seen that the boundary curve has a minimum value of 3.0 at $k = 1/3$.

In this case the current i_D exists in the period between kT and $t_1 = [k + (1-k)m_R]T$, where m_R is the **filling efficiency** and it is defined as:

$$m_R = \frac{1}{\zeta} = \frac{M_R^2}{k\dfrac{R}{fL}} \qquad (2.283)$$

Considering Equation (2.282), therefore $0 < m_R < 1$. Because inductor current $i_{L1} = 0$ at $t = t_1$, so that

$$V_{L1-off} = \frac{k}{(1-k)m_R}V_I$$

Since the current i_D becomes zero at $t = t_1 = [k + (1-k)m_R]T$, for the current i_L

$$kTV_I = (1-k)m_RT(V_C - 2V_I - V_{L1-off})$$

or

$$V_C = [2 + \frac{2k}{(1-k)m_R}]V_I = [2 + k^2(1-k)\frac{R}{2fL}]V_I \quad \text{with} \quad \sqrt{k}\sqrt{\frac{R}{fL}} \geq \frac{2}{1-k}$$

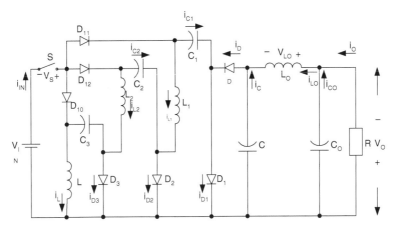

FIGURE 2.45
Triple-lift circuit.

and for the current i_{LO}

$$kT(V_I + V_C - V_O) = (1-k)m_RT(V_O - 2V_I - V_{L1-off})$$

Therefore, output voltage in discontinuous mode is

$$V_O = [2 + \frac{2k}{(1-k)m_R}]V_I = [2 + k^2(1-k)\frac{R}{2fL}]V_I \quad \text{with} \quad \sqrt{k}\sqrt{\frac{R}{fL}} \geq \frac{2}{1-k} \quad (2.284)$$

i.e., the output voltage will linearly increase during load resistance R increasing. The output voltage vs. the normalized load $z_N = R/fL$ is shown in Figure 2.44. Larger load resistance R may cause higher output voltage in discontinuous mode.

2.4.4 Multiple-Lift Circuits

Referring to Figure 2.45, it is possible to build a multiple-lift circuit just only using the parts $(L_1\text{-}C_2\text{-}D_2\text{-}D_{11})$ multiple times. For example, in Figure 2.16 the parts $(L_2\text{-}C_3\text{-}D_3\text{-}D_{12})$ were added in the triple-lift circuit. According to this principle, the triple-lift circuit and quadruple-lift circuit were built as shown in Figure 2.45 and Figure 2.48. In this book it is not necessary to introduce the particular analysis and calculations one by one to readers. However, their formulas are shown in this section.

2.4.4.1 Triple-Lift Circuit

Triple-lift circuit is shown in Figure 2.45. It consists of one static switch S; four inductors L, L_1, L_2, and L_O; and five capacitors C, C_1, C_2, C_3, and C_O; and

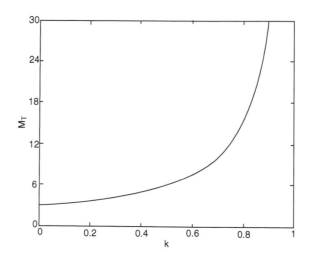

FIGURE 2.46
Voltage transfer gain of triple-lift circuit.

diodes. Circuit C_1-D_1-L_1-C_2-D_2-D_{11}-L_2-C_3-D_3-D_{12} is the lift circuit. Capacitors C_1, C_2, and C_3 perform characteristics to lift the capacitor voltage V_C by three times that of the source voltage V_I. L_1 and L_2 perform the function as ladder joints to link the three capacitors C_1, C_2, and C_3 and lift the capacitor voltage V_C up. Current $i_{C1}(t)$, $i_{C2}(t)$, and $i_{C3}(t)$ are exponential functions. They have large values at the moment of power on, but they are small because $v_{C1} = v_{C2} = v_{C3} \cong V_I$ in steady state.

The output voltage and current are

$$V_O = \frac{3}{1-k} V_I \tag{2.285}$$

and

$$I_O = \frac{1-k}{3} I_I \tag{2.286}$$

The voltage transfer gain in continuous mode is

$$M_T = V_O / V_I = \frac{3}{1-k} \tag{2.287}$$

The curve of M_T vs. k is shown in Figure 2.46.
Other average voltages:

$$V_C = V_O \quad V_{C1} = V_{C2} = V_{C3} = V_I$$

Other average currents:

$$I_{LO} = I_O \quad I_L = I_{L1} = I_{L2} = \frac{1}{1-k} I_O$$

Current variation ratios:

$$\zeta = \frac{k}{M_T{}^2} \frac{3R}{2fL} \quad \xi = \frac{k}{16} \frac{1}{f^2 C L_O}$$

$$\chi_1 = \frac{k(1-k)}{2M_T} \frac{R}{fL_1} \quad \chi_2 = \frac{k(1-k)}{2M_T} \frac{R}{fL_2}$$

Voltage variation ratios:

$$\rho = \frac{k}{2} \frac{1}{fCR} \quad \sigma_1 = \frac{M_T}{2} \frac{1}{fC_1 R}$$

$$\sigma_2 = \frac{M_T}{2} \frac{1}{fC_2 R} \quad \sigma_3 = \frac{M_T}{2} \frac{1}{fC_3 R}$$

The variation ratio of output voltage V_C is

$$\varepsilon = \frac{k}{128} \frac{1}{f^3 CC_O L_O R} \tag{2.288}$$

The output voltage ripple is very small. The boundary between continuous and discontinuous modes is

$$M_T \le \sqrt{k} \sqrt{\frac{3R}{2fL}} = \sqrt{\frac{3kz_N}{2}} \tag{2.289}$$

It can be seen that the boundary curve has a minimum value of M_T that is equal to 4.5, corresponding to $k = 1/3$. The boundary curve vs. the normalized load $z_N = R/fL$ is shown in Figure 2.47.

In discontinuous mode the current i_D exists in the period between kT and $[k + (1-k)m_T]T$, where m_T is the filling efficiency that is

$$m_T = \frac{1}{\zeta} = \frac{M_T{}^2}{k \dfrac{3R}{2fL}} \tag{2.290}$$

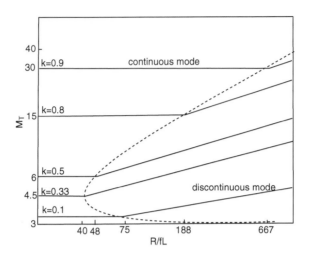

FIGURE 2.47
The boundary between continuous and discontinuous modes and output voltage vs. the normalized load $z_N = R/fL$ (triple-lift circuit).

Considering Equation (2.289), therefore $0 < m_T < 1$. Because inductor current $i_{L1} = i_{L2} = 0$ at $t = t_1$, so that

$$V_{L1\text{-}off} = V_{L2\text{-}off} = \frac{k}{(1-k)m_T} V_I$$

Since the current i_D becomes zero at $t = t_1 = [k + (1-k)m_T]T$, for the current i_L we have

$$kTV_I = (1-k)m_T T(V_C - 3V_I - V_{L1\text{-}off} - V_{L2\text{-}off})$$

or

$$V_C = [3 + \frac{3k}{(1-k)m_T}]V_I = [3 + k^2(1-k)\frac{R}{2fL}]V_I \quad \text{with} \quad \sqrt{k}\sqrt{\frac{3R}{2fL}} \geq \frac{3}{1-k}$$

and for the current i_{LO} we have

$$kT(V_I + V_C - V_O) = (1-k)m_T T(V_O - 2V_I - V_{L1\text{-}off} - V_{L2\text{-}off})$$

Therefore, output voltage in discontinuous mode is

$$V_O = [3 + \frac{3k}{(1-k)m_T}]V_I = [3 + k^2(1-k)\frac{R}{2fL}]V_I \quad \text{with} \quad \sqrt{k}\sqrt{\frac{3R}{2fL}} \geq \frac{3}{1-k} \quad (2.291)$$

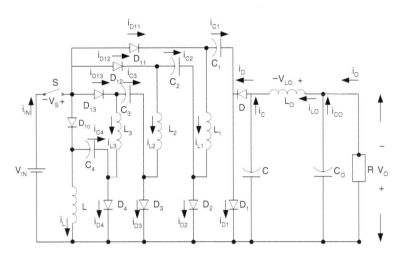

FIGURE 2.48
Quadruple-lift circuit.

i.e., the output voltage will linearly increase during load resistance R increasing. The output voltage vs. the normalized load $z_N = R/fL$ is shown in Figure 2.47. We can see that the output voltage will increase when the load resistance R increases.

2.4.4.2 Quadruple-Lift Circuit

Quadruple-lift circuit is shown in Figure 2.48. It consists of one static switch S; five inductors L, L_1, L_2, L_3, and L_O; and six capacitors C, C_1, C_2, C_3, C_4, and C_O. Capacitors C_1, C_2, C_3, and C_4 perform characteristics to lift the capacitor voltage V_C by four times of source voltage V_I. L_1, L_2, and L_3 perform the function as ladder joints to link the four capacitors C_1, C_2, C_3, and C_4 and lift the output capacitor voltage V_C up. Current $i_{C1}(t)$, $i_{C2}(t)$, $i_{C3}(t)$, and $i_{C4}(t)$ are exponential functions. They have large values at the moment of power on, but they are small because $v_{C1} = v_{C2} = v_{C3} = v_{C4} \cong V_I$ in steady state.

The output voltage and current are

$$V_O = \frac{4}{1-k} V_I \qquad (2.292)$$

and

$$I_O = \frac{1-k}{4} I_I \qquad (2.293)$$

The voltage transfer gain in continuous mode is

$$M_Q = V_O / V_I = \frac{4}{1-k} \qquad (2.294)$$

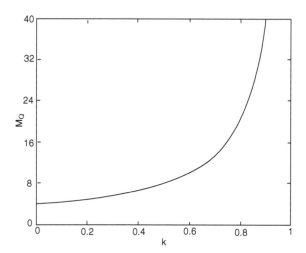

FIGURE 2.49
Voltage transfer gain of quadruple-lift circuit.

The curve of M_Q vs. k is shown in Figure 2.49. Other average voltages:

$$V_C = V_O \quad V_{C1} = V_{C2} = V_{C3} = V_{C4} = V_I$$

Other average currents:

$$I_{LO} = I_O \quad I_L = I_{L1} = I_{L2} = I_{L3} = \frac{1}{1-k} I_O$$

Current variation ratios:

$$\zeta = \frac{k}{M_Q^2} \frac{2R}{fL} \quad \xi = \frac{k}{16} \frac{1}{f^2 C L_O}$$

$$\chi_1 = \frac{k(1-k)}{2M_Q} \frac{R}{fL_1} \quad \chi_2 = \frac{k(1-k)}{2M_Q} \frac{R}{fL_2} \quad \chi_3 = \frac{k(1-k)}{2M_Q} \frac{R}{fL_3}$$

Voltage variation ratios:

$$\rho = \frac{k}{2} \frac{1}{fCR} \quad \sigma_1 = \frac{M_Q}{2} \frac{1}{fC_1 R}$$

$$\sigma_2 = \frac{M_Q}{2} \frac{1}{fC_2 R} \quad \sigma_3 = \frac{M_Q}{2} \frac{1}{fC_3 R} \quad \sigma_4 = \frac{M_Q}{2} \frac{1}{fC_4 R}$$

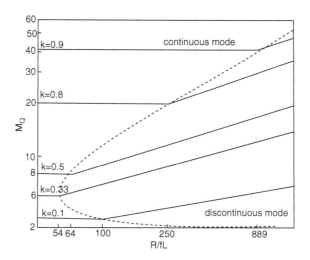

FIGURE 2.50
The boundary between continuous and discontinuous modes and output voltage vs. the normalized load $z_N = R/fL$ (quadruple-lift circuit).

The variation ratio of output voltage V_C is

$$\varepsilon = \frac{k}{128} \frac{1}{f^3 CC_O L_O R} \tag{2.295}$$

The output voltage ripple is very small. The boundary between continuous and discontinuous conduction modes is

$$M_Q \leq \sqrt{k} \sqrt{\frac{2R}{fL}} = \sqrt{2kz_N} \tag{2.296}$$

It can be seen that the boundary curve has a minimum value of M_Q that is equal to 6.0, corresponding to $k = 1/3$. The boundary curve is shown in Figure 2.50.

In discontinuous mode the current i_D exists in the period between kT and $[k + (1 - k)m_Q]T$, where m_Q is the filling efficiency that is

$$m_Q = \frac{1}{\zeta} = \frac{M_Q^2}{k\dfrac{2R}{fL}} \tag{2.297}$$

Considering Equation (2.296), therefore $0 < m_Q < 1$. Because inductor current $i_{L1} = i_{L2} = i_{L3} = 0$ at $t = t_1$, so that

$$V_{L1-off} = V_{L2-off} = V_{L3-off} = \frac{k}{(1-k)m_Q}V_I$$

Since the current i_D becomes zero at $t = t_1 = kT + (1-k)m_QT$, for the current i_L we have

$$kTV_I = (1-k)m_QT(V_C - 4V_I - V_{L1-off} - V_{L2-off} - V_{L3-off})$$

or

$$V_C = [4 + \frac{4k}{(1-k)m_Q}]V_I = [4 + k^2(1-k)\frac{R}{2fL}]V_I \quad \text{with} \quad \sqrt{k}\sqrt{\frac{2R}{fL}} \geq \frac{4}{1-k}$$

and for current i_{LO} we have

$$kT(V_I + V_C - V_O) = (1-k)m_QT(V_O - 2V_I - V_{L1-off} - V_{L2-off} - V_{L3-off})$$

Therefore, output voltage in discontinuous mode is

$$V_O = [4 + \frac{4k}{(1-k)m_Q}]V_I = [4 + k^2(1-k)\frac{R}{2fL}]V_I \quad \text{with} \quad \sqrt{k}\sqrt{\frac{2R}{fL}} \geq \frac{4}{1-k} \quad (2.298)$$

i.e., the output voltage will linearly increase while load resistance R increases. The output voltage vs. the normalized load $z_N = R/fL$ is shown in Figure 2.50. We can see that the output voltage will increase during load resistance while the load R increases.

2.4.5   Summary

From the analysis and calculation in previous sections, the common formulae can be obtained for all circuits:

$$M = \frac{V_O}{V_I} = \frac{I_I}{I_O} \qquad z_N = \frac{R}{fL} \qquad R = \frac{V_O}{I_O}$$

Current variation ratios:

$$\zeta = \frac{k(1-k)R}{2MfL} \qquad \xi = \frac{k}{16f^2CL_O}$$

$$\chi_j = \frac{k(1-k)R}{2MfL_j} \qquad (j = 1, 2, 3, \ldots)$$

Voltage variation ratios:

$$\rho = \frac{k}{2fCR} \qquad \varepsilon = \frac{k}{128f^3CC_OL_OR} \qquad \sigma_j = \frac{M}{2fC_jR} \qquad (j = 1, 2, 3, 4, \ldots)$$

In order to write common formulas for the boundaries between continuous and discontinuous modes and output voltage for all circuits, the circuits can be numbered. The definition is that subscript 0 means the elementary circuit, subscript 1 means the self-lift circuit, subscript 2 means the re-lift circuit, subscript 3 means the triple-lift circuit, subscript 4 means the quadruple-lift circuit, and so on. Therefore, the voltage transfer gain in continuous mode for all circuits is

$$M_j = \frac{k^{h(j)}[j + h(j)]}{1 - k} \qquad j = 0, 1, 2, 3, 4, \ldots \qquad (2.299)$$

The variation of the free-wheeling diode current i_D is

$$\zeta_j = \frac{k^{[1+h(j)]}}{M_j^2} \frac{j + h(j)}{2} z_N \qquad (2.300)$$

The boundaries are determined by the condition:

$$\zeta_j \geq 1$$

or

$$\frac{k^{[1+h(j)]}}{M_j^2} \frac{j + h(j)}{2} z_N \geq 1 \qquad j = 0, 1, 2, 3, 4, \ldots \qquad (2.301)$$

Therefore, the boundaries between continuous and discontinuous modes for all circuits are

$$M_j = k^{\frac{1+h(j)}{2}} \sqrt{\frac{j + h(j)}{2} z_N} \qquad j = 0, 1, 2, 3, 4, \ldots \qquad (2.302)$$

The filling efficiency is

$$m_j = \frac{1}{\zeta_j} = \frac{M_j^2}{k^{[1+h(j)]}} \frac{2}{j + h(j)} \frac{1}{z_N} \qquad (2.303)$$

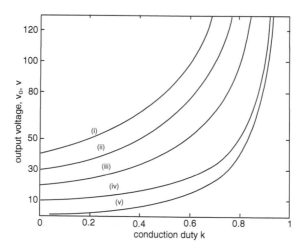

FIGURE 2.51
Output voltages of all negative output Luo-converters ($V_I = 10$ V).

The voltage across the capacitor C in discontinuous mode for all circuits

$$V_{C-j} = [j + k^{[2-h(j)]} \frac{1-k}{2} z_N] V_I \quad j = 0, 1, 2, 3, 4, \ldots \qquad (2.304)$$

The output voltage in discontinuous mode for all circuits

$$V_{O-j} = [j + k^{[2-h(j)]} \frac{1-k}{2} z_N] V_I \quad j = 0, 1, 2, 3, 4, \ldots \qquad (2.305)$$

where

$$h(j) = \begin{cases} 0 & if \quad j \geq 1 \\ 1 & if \quad j = 0 \end{cases} \quad \text{is the Hong Function}$$

The voltage transfer gains in continuous mode for all circuits are shown in Figure 2.51. The boundaries between continuous and discontinuous modes of all circuits are shown in Figure 2.52. The curves of all M vs. z_N state that the continuous mode area increases from M_E via M_S, M_R, M_T to M_Q. The boundary of elementary circuit is a monorising curve, but other curves are not monorising. There are minimum values of the boundaries of other curves, which of M_S, M_R, M_T, and M_Q correspond at $k = 1/3$.

Assuming that $f = 50$ kHz, $L = L_O = L_1 = L_2 = L_3 = L_4 = 100$ μH, $C = C_1 = C_2 = C_3 = C_4 = C_O = 5$ μF and the source voltage $V_I = 10$ V, the value of the output voltage V_O in various conduction duty k are shown in Figure 2.22. Typically, some values of the output voltage V_O in conduction duty $k = 0.33$,

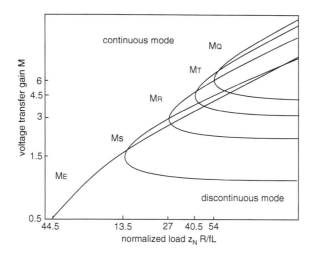

FIGURE 2.52
Boundaries between continuous and discontinuous modes of all negative output Luo-converters.

TABLE 2.2

Comparison among Five Negative Output Luo-Converters

Negative Output Luo-Converters	I_O	V_O	V_O (V_I = 10 V)			
			$k = 0.33$	$k = 0.5$	$k = 0.75$	$k = 0.9$
Elementary Circuit	$I_O = \dfrac{1-k}{k}I_I$	$V_O = \dfrac{k}{1-k}V_I$	5 V	10 V	30 V	90 V
Self-Lift Circuit	$I_O = (1-k)I_I$	$V_O = \dfrac{1}{1-k}V_I$	15 V	20 V	40 V	100 V
Re-Lift Circuit	$I_O = \dfrac{1-k}{2}I_I$	$V_O = \dfrac{2}{1-k}V_I$	30 V	40 V	80 V	200 V
Triple-Lift Circuit	$I_O = \dfrac{1-k}{3}I_I$	$V_O = \dfrac{3}{1-k}V_I$	45 V	60 V	120 V	300 V
Quadruple-Lift Circuit	$I_O = \dfrac{1-k}{4}I_I$	$V_O = \dfrac{4}{1-k}V_I$	60 V	80 V	160 V	400 V

0.5, 0.75, and 0.9 are listed in Table 2.2. The ripple of the output voltage is very small, say smaller than 1%. For example, using the above data and $R = 10 \ \Omega$, the variation ratio of the output voltage is $\varepsilon = 0.0025 \times k = 0.0008$, 0.0012, 0.0019, and 0.0023 respectively. From these data the fact we find is that the output voltage of all negative output Luo-converters is almost a real DC voltage with very small ripple.

2.5 Modified Positive Output Luo-Converters

Negative output Luo-converters perform the voltage conversion from positive to negative voltages using VL technique with only one switch S. This section introduces the technique to modify positive output Luo-converters that can employ only **one** switch for all circuits. Five circuits have been introduced in the literature. They are

- Elementary circuit
- Self-lift circuit
- Re-lift circuit
- Triple-lift circuit
- Quadruple-lift circuit

There are five circuits introduced in this section, namely the elementary circuit, self-lift circuit, re-lift circuit, and multiple-lift circuit (triple-lift and quadruple-list circuits). In all circuits the switch S is a PMOS. It is driven by a PWM switch signal with variable frequency f and conduction duty k. For all circuits, the load is usually resistive, $R = V_O/I_O$. We concentrate the absolute values rather than polarity in the following descriptions and calculations. The directions of all voltages and currents are defined and shown in the figures. We will assume that all the components are ideal and the capacitors are large enough. We also assume that the circuits operate in continuous conduction mode. The output voltage and current are V_O and I_O; the input voltage and current are V_I and I_I.

2.5.1 Elementary Circuit

Elementary circuit is shown in Figure 2.10. It is the elementary circuit of positive output Luo-converters. The output voltage and current and the voltage transfer gain are

$$V_O = \frac{k}{1-k}V_I$$

$$I_O = \frac{1-k}{k}I_I$$

$$M_E = \frac{k}{1-k}$$

Average voltage:

$$V_C = V_O$$

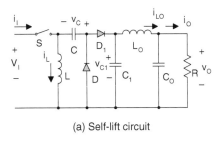

(a) Self-lift circuit

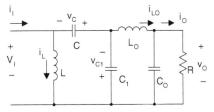

(b) Switch-on equivalent circuit

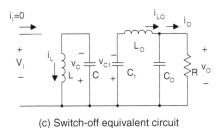

(c) Switch-off equivalent circuit

FIGURE 2.53
Modified self-lift circuit and its equivalent circuit. (a) Self-lift circuit. (b) Switch-on equivalent circuit. (c) Switch-off equivalent circuit.

Average currents:

$$I_{LO} = I_O \qquad I_L = \frac{k}{1-k} I_O$$

2.5.2 Self-Lift Circuit

Self-lift circuit is shown in Figure 2.53. It is derived from the elementary circuit. In steady state, the average inductor voltages over a period are zero. Thus

$$V_{C1} = V_{CO} = V_O \tag{2.306}$$

The inductor current i_L increases in the switch-on period and decreases in the switch-off period. The corresponding voltages across L are V_I and $-V_C$.

Therefore

$$kTV_I = (1-k)TV_C$$

Hence,

$$V_C = \frac{k}{1-k}V_I \tag{2.307}$$

During switch-on period, the voltage across capacitor C_1 are equal to the source voltage plus the voltage across C. Since we assume that C and C_1 are sufficiently large,

$$V_{C1} = V_I + V_C$$

Therefore,

$$V_{C1} = V_I + \frac{k}{1-k}V_I = \frac{1}{1-k}V_I$$

$$V_O = V_{CO} = V_{C1} = \frac{1}{1-k}V_I$$

The voltage transfer gain of continuous conduction mode (CCM) is

$$M = \frac{V_O}{V_I} = \frac{1}{1-k}$$

The output voltage and current and the voltage transfer gain are

$$V_O = \frac{1}{1-k}V_I$$

$$I_O = (1-k)I_I$$

$$M_S = \frac{1}{1-k} \tag{2.308}$$

Average voltages:

$$V_C = kV_O$$

$$V_{C1} = V_O$$

Average currents:

$$I_{LO} = I_O$$

$$I_L = \frac{1}{1-k}I_O$$

We also implement the breadboard prototype of the proposed self-lift circuit. NMOS IRFP460 is used as the semiconductor switch. The diode is MR824. The other parameters are

$$V_I = 0 \sim 30 \text{ V}, R = 30 \sim 340 \text{ } \Omega, k = 0.1 \sim 0.9$$

$$C = C_O = 100 \text{ } \mu\text{F and } L = 470 \text{ } \mu\text{H}$$

2.5.3 Re-Lift Circuit

Re-lift circuit and its equivalent circuits are shown in Figure 2.54. It is derived from the self-lift circuit. The function of capacitors C_2 is to lift the voltage v_C by source voltage V_I, the function of inductor L_1 acts like a hinge of the foldable ladder (capacitor C_2) to lift the voltage v_C during switch off.

In steady state, the average inductor voltages over a period are zero. Thus

$$V_{C1} = V_{CO} = V_O$$

Since we assume C_2 is large enough and C_2 is biased by the source voltage V_I during switch-on period, thus $V_{C2} = V_I$

From the switch-on equivalent circuit, another capacitor voltage equation can also be derived since we assume all the capacitors to be large enough,

$$V_O = V_{C1} = V_C + V_I$$

The inductor current i_L increases in the switch-on period and decreases in the switch-off period. The corresponding voltages across L are V_I and $-V_{L-OFF}$. Therefore

$$kTV_I = (1-k)TV_{L-OFF}$$

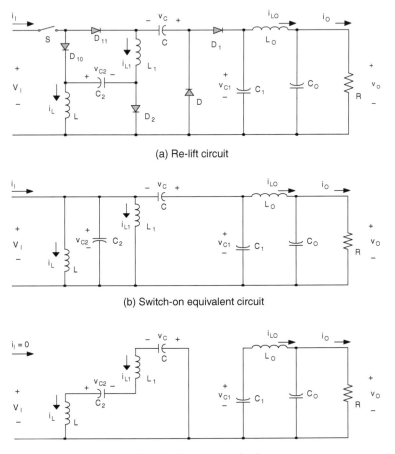

(a) Re-lift circuit

(b) Switch-on equivalent circuit

(c) Switch-off equivalent circuit

FIGURE 2.54
Modified re-lift circuit. (a) Re-lift circuit. (b) Switch-on equivalent circuit. (c) Switch-off equivalent circuit.

Hence,

$$V_{L-OFF} = \frac{k}{1-k}V_I$$

The inductor current i_{L1} increases in the switch-on period and decreases in the switch-off period. The corresponding voltages across L_1 are V_I and $-V_{L1-OFF}$.

Therefore

$$kTV_I = (1-k)TV_{L1-OFF}$$

Hence,

$$V_{L1-OFF} = \frac{k}{1-k}V_I$$

From the switch-off period equivalent circuit,

$$V_C = V_{C-OFF} = V_{L-OFF} + V_{L1-OFF} + V_{C2}$$

Therefore,

$$V_C = \frac{k}{1-k}V_I + \frac{k}{1-k}V_I + V_I = \frac{1+k}{1-k}V_I \qquad (2.309)$$

$$V_O = \frac{1+k}{1-k}V_I + V_I = \frac{2}{1-k}V_I$$

Then we get the voltage transfer ratio in CCM,

$$M = M_R = \frac{2}{1-k} \qquad (2.310)$$

The following is a brief summary of the main equations for the re-lift circuit. The output voltage and current and gain are

$$V_O = \frac{2}{1-k}V_I$$

$$I_O = \frac{1-k}{2}I_I$$

$$M_R = \frac{2}{1-k}$$

Average voltages:

$$V_C = \frac{1+k}{1-k}V_I$$

$$V_{C1} = V_{CO} = V_O$$

$$V_{C2} = V_I$$

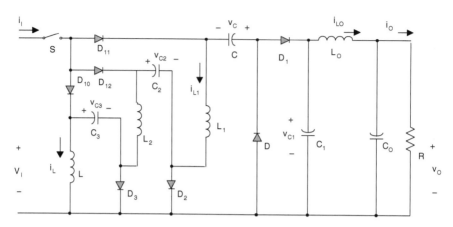

FIGURE 2.55
Modified triple-lift circuit.

Average currents:

$$I_{LO} = I_O$$

$$I_L = I_{L1} = \frac{1}{1-k} I_O$$

2.5.4 Multi-Lift Circuit

Multiple-lift circuits are derived from re-lift circuits by repeating the section of L_1-C_1-D_1 multiple times. For example, triple-list circuit is shown in Figure 2.55. The function of capacitors C_2 and C_3 is to lift the voltage v_C across capacitor C by twice the source voltage $2V_I$, the function of inductors L_1 and L_2 acts like hinges of the foldable ladder (capacitors C_2 and C_3) to lift the voltage v_C during switch off.

The output voltage and current and voltage transfer gain are

$$V_O = \frac{3}{1-k} V_I$$

and

$$I_O = \frac{1-k}{3} I_I$$

$$M_T = \frac{3}{1-k} \tag{2.311}$$

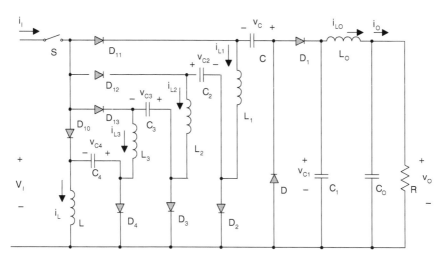

FIGURE 2.56
Modified quadruple-lift circuit.

Other average voltages:

$$V_C = \frac{2+k}{1-k}V_I$$

and

$$V_{C1} = V_O \quad V_{C2} = V_{C3} = V_I$$

Other average currents:

$$I_{LO} = I_O$$

$$I_{L1} = I_{L2} = I_L = \frac{1}{1-k}I_O$$

The quadruple-lift circuit is shown in Figure 2.56. The function of capacitors C_2, C_3, and C_4 is to lift the voltage v_C across capacitor C by three times the source voltage $3V_I$. The function of inductors L_1, L_2, and L_3 acts like hinges of the foldable ladder (capacitors C_2, C_3, and C_4) to lift the voltage v_C during switch off. The output voltage and current and voltage transfer gain are

$$V_O = \frac{4}{1-k}V_I$$

and

$$I_O = \frac{1-k}{4} I_I$$

$$M_Q = \frac{4}{1-k} \tag{2.312}$$

Average voltages:

$$V_C = \frac{3+k}{1-k} V_I$$

and

$$V_{C1} = V_O$$

$$V_{C2} = V_{C3} = V_{C4} = V_I$$

Average currents:

$$I_{LO} = I_O$$

and

$$I_L = \frac{k}{1-k} I_O$$

$$I_{L1} = I_{L2} = I_{L3} + I_L + I_{LO} = \frac{1}{1-k} I_O$$

2.5.5 Application

A high-efficiency, widely adjustable high voltage regulated power supply (HVRPS) is designed to use these Luo-converters in a high voltage test rig. The proposed HVRPS is shown in Figure 2.57. The HVRPS was constructed by using a PWM IC TL494 to implement closed-loop control together with the modified positive output Luo-converters. Its output voltage is basically a DC value with small ripple and can be widely adjustable. The source voltage is 24 V DC and the output voltage can vary from 36 V to 1000 V DC. The measured experimental results show that the efficiency can be as high as 95% and the source effect ratio is about 0.001 and load effect ratio is about 0.005.

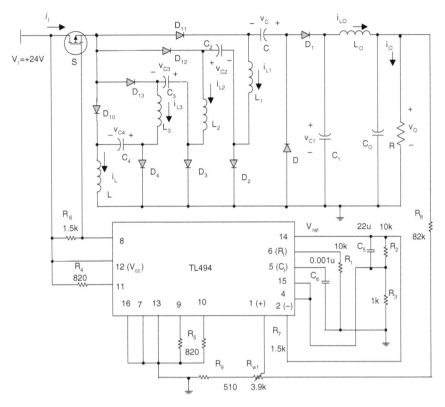

FIGURE 2.57
A high voltage testing power supply.

2.6 Double Output Luo-Converters

Mirror-symmetrical double output voltages are specially required in industrial applications and computer periphery circuits. Double output DC-DC Luo-converters can convert the positive input source voltage to positive and negative output voltages. It consists of two conversion paths. Double output Luo-converters perform from positive to positive and negative DC-DC voltage increasing conversion with high power density, high efficiency, and cheap topology in simple structure.

Double output DC-DC Luo-converters consist of two conversion paths. Usually, mirror-symmetrical double output voltages are required in industrial applications and computer periphery circuits such as operational amplifiers, computer periphery power supplies, differential servo-motor drives, and some symmetrical voltage medical equipment. In recent years the DC-DC conversion technique has been greatly developed. The main objective is to reach a high efficiency, high power density and cheap topology in simple structure.

The elementary circuit can perform step-down and step-up DC-DC conversion. The other double output Luo-converters are derived from this elementary circuit, they are the self-lift circuit, re-lift circuit, and multiple-lift circuits (e.g., triple-lift and quadruple-lift circuits). Switch S in these circuits is a PMOS. It is driven by a PWM switch signal with repeating frequency f and conduction duty k. In this paper the switch repeating period is $T = 1/f$, so that the switch-on period is kT and switch-off period is $(1 - k)T$. For all circuits, the loads are usually resistive, i.e., $R = V_{O+}/I_{O+}$ and $R_1 = V_{O-}/I_{O-}$; the normalized loads are $z_{N+} = R/fL$ (where $L = L_1$ and $L = L_1L_2/L_1 + L_2$ for elementary circuits) and $z_{N-} = R_1/fL_{11}$. In order to keep the positive and negative output voltages to be symmetrically equal to each other, usually, we purposely select that $L = L_{11}$ and $z_{N+} = z_{N-}$.

Each converter has two conversion paths. The positive path consists of a positive pump circuit $S\text{-}L_1\text{-}D_0\text{-}C_1$ and a "Π"-type filter $(C_2)\text{-}L_2\text{-}C_O$, and a lift circuit (except elementary circuits). The pump inductor L_1 absorbs energy from source during switch-on and transfers the stored energy to capacitor C_1 during switch-off. The energy on capacitor C_1 is then delivered to load R during switch-on. Therefore, a high voltage V_{C1} will correspondingly cause a high output voltage V_{O+}.

The negative path consists of a negative pump circuit $S\text{-}L_{11}\text{-}D_{10}\text{-}(C_{11})$ and a "Π"-type filter $C_{11}\text{-}L_{12}\text{-}C_{10}$, and a lift circuit (except elementary circuits). The pump inductor L_{11} absorbs the energy from source during switch-on and transfers the stored energy to capacitor C_{11} during switch-off. The energy on capacitor C_{11} is then delivered to load R_1 during switch-on. Hence, a high voltage V_{C11} will correspondingly cause a high output voltage V_{O-}.

When switch S is turned off, the currents flowing though the freewheeling diodes D_0 and D_{10} are existing. If the currents i_{D0} and i_{D10} do not fall to zero before switch S is turned on again, we define this working state to be continuous conduction mode. If the currents i_{D0} and i_{D10} become zero before switch S is turned on again, we define that working state to be discontinuous conduction mode.

The output voltages and currents are V_{O+}, V_{O-} and I_{O+}, I_{O-}; the input voltage and current are V_I and $I_I = I_{I+} + I_{I-}$. Assuming that the power loss can be ignored, $P_I = P_O$, or $V_II_I = V_{O+} I_{O+} + V_{O-} I_{O-}$. For general description, we have the following definitions in continuous mode: The voltage transfer gain in the continuous mode:

$$M_+ = \frac{V_{O+}}{V_I} \quad \text{and} \quad M_- = \frac{V_{O-}}{V_I}$$

Variation ratio of the diode's currents:

$$\zeta_+ = \frac{\Delta i_{D0}/2}{I_{D0}} \quad \text{and} \quad \zeta_- = \frac{\Delta i_{D10}/2}{I_{L11}}$$

Variation ratio of pump inductor's currents:

$$\xi_{1+} = \frac{\Delta i_{L1}/2}{I_{L1}} \quad \text{and} \quad \zeta_- = \frac{\Delta i_{L11}/2}{I_{L11}}$$

Variation ratio of filter inductor's currents:

$$\xi_{2+} = \frac{\Delta i_{L2}/2}{I_{L2}} \quad \text{and} \quad \xi_- = \frac{\Delta i_{L12}/2}{I_{L12}}$$

Variation ratio of lift inductor's currents:

$$\chi_{j+} = \frac{\Delta i_{L2+j}/2}{I_{L2+j}} \quad \text{and} \quad \chi_{j-} = \frac{\Delta i_{L12+j}/2}{I_{L12+j}} \quad j = 1, 2, 3, \ldots$$

Variation ratio of pump capacitor's voltages:

$$\rho_+ = \frac{\Delta v_{C1}/2}{V_{C1}} \quad \text{and} \quad \rho_- = \frac{\Delta v_{C11}/2}{V_{C11}}$$

Variation ratio of lift capacitor's voltages:

$$\sigma_{j+} = \frac{\Delta v_{C1+j}/2}{V_{C1+j}} \quad \text{and} \quad \sigma_{j-} = \frac{\Delta v_{C11+j}/2}{V_{C11+j}} \quad j = 1, 2, 3, 4, \ldots$$

Variation ratio of output voltages:

$$\varepsilon_+ = \frac{\Delta v_{O+}/2}{V_{O+}} \quad \text{and} \quad \varepsilon_- = \frac{\Delta v_{O-}/2}{V_{O-}}$$

2.6.1 Elementary Circuit

The elementary circuit is shown in Figure 2.58. Since the positive Luo-converters and negative Luo-converters have been published, this section can be simplified.

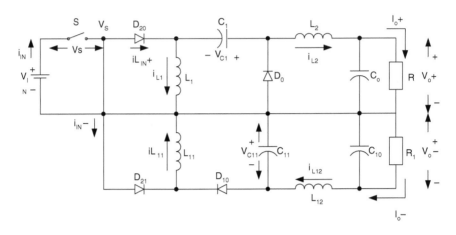

FIGURE 2.58
Elementary circuit.

2.6.1.1 Positive Conversion Path

The equivalent circuit during switch-on is shown in Figure 2.59a, and the equivalent circuit during switch-off in Figure 2.59b. The relations of the average currents and voltages are

$$I_{L2} = \frac{1-k}{k} I_{L1} \quad \text{and} \quad I_{L2} = I_{O+}$$

Positive path input current is

$$I_{I+} = k \times i_{I+} = k(i_{L1} + i_{L2}) = k(1 + \frac{1-k}{k})I_{L1} = I_{L1} \tag{2.313}$$

The output current and voltage are

$$I_{O+} = \frac{1-k}{k} I_{I+} \quad \text{and} \quad V_{O+} = \frac{k}{1-k} V_I$$

The voltage transfer gain in continuous mode is

$$M_{E+} = \frac{V_{O+}}{V_I} = \frac{k}{1-k} \tag{2.314}$$

The average voltage across capacitor C_1 is

$$V_{C1} = \frac{k}{1-k} V_I = V_{O+}$$

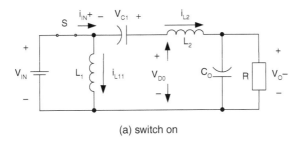

(a) switch on

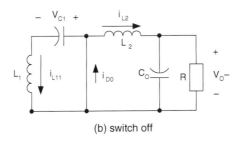

(b) switch off

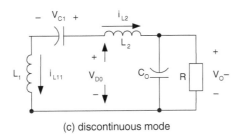

(c) discontinuous mode

FIGURE 2.59
Equivalent circuits of elementary circuit positive path: (a) switch on; (b) switch off; (c) discontinuous conduction mode.

The variation ratios of the parameters are

$$\xi_{1+} = \frac{\Delta i_{L1}/2}{I_{L1}} = \frac{kTV_I}{2L_1 I_{1+}} = \frac{1-k}{2M_E} \frac{R}{fL_1} \quad \text{and} \quad \xi_{2+} = \frac{\Delta i_{L2}/2}{I_{L2}} = \frac{kTV_I}{2L_2 I_{O+}} = \frac{k}{2M_E} \frac{R}{fL_2}$$

The variation ratio of current i_{D0} is

$$\zeta_+ = \frac{\Delta i_{D0}/2}{I_{D0}} = \frac{(1-k)^2 TV_{O+}}{2LI_{O+}} = \frac{k(1-k)R}{2M_E fL} = \frac{k^2}{M_E^2} \frac{R}{2fL} \tag{2.315}$$

The variation ratio of v_{C1} is

$$\rho_+ = \frac{\Delta v_{C1}/2}{V_{C1}} = \frac{(1-k)TI_{I+}}{2C_1 V_{O+}} = \frac{k}{2}\frac{1}{fC_1 R}$$

The variation ratio of output voltage v_{O+} is

$$\varepsilon_+ = \frac{\Delta v_{O+}/2}{V_{O+}} = \frac{kT^2}{8C_0 L_2}\frac{V_I}{V_{O+}} = \frac{k}{8M_E}\frac{1}{f^2 C_0 L_2} \qquad (2.316)$$

If $L_1 = L_2 = 1$ mH, $C_1 = C_0 = 20$ μF, $R = 10$ Ω, $f = 50$ kHz and $k = 0.5$, we get $\xi_{1+} = 0.05$, $\xi_{2+} = 0.05$, $\zeta_+ = 0.05$, $\rho_+ = 0.025$, and $\varepsilon = 0.00125$. Therefore, the variations of i_{L1}, i_{L2} and v_{C1} are small. The output voltage V_{O+} is almost a real DC voltage with very small ripple. Because of the resistive load, the output current $i_{O+}(t)$ is almost a real DC waveform with very small ripple as well, and is equal to $I_{O+} = V_{O+}/R$.

2.6.1.2 Negative Conversion Path

The equivalent circuit during switch-on is shown in Figure 2.60(a) the equivalent circuit during switch-off is shown in Figure 2.60(b). The relations of the average currents and voltages are

$$I_{O-} = I_{L12} \quad \text{and} \quad I_{O-} = I_{L12} = I_{C11-on}$$

Since

$$I_{C11-off} = \frac{k}{1-k}I_{C11-on} = \frac{k}{1-k}I_{O-}$$

the inductor current I_{L11} is

$$I_{L11} = I_{C11-off} + I_{O-} = \frac{I_{O-}}{1-k} \qquad (2.317)$$

So that

$$I_{I-} = k \times i_{I-} = ki_{L11} = kI_{L11} = \frac{k}{1-k}I_{O-}$$

The output current and voltage are

$$I_{O-} = \frac{1-k}{k}I_{I-} \quad \text{and} \quad V_{O-} = \frac{k}{1-k}V_I$$

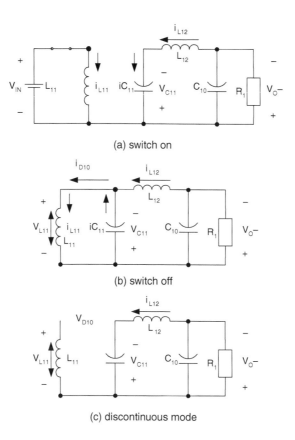

FIGURE 2.60
Equivalent circuits of elementary circuit negative path: (a) switch on; (b) switch off; (c) discontinuous conduction mode.

The voltage transfer gain in continuous mode is

$$M_{E-} = \frac{V_{O-}}{V_I} = \frac{k}{1-k}$$

(2.318)

and

$$V_{C11} = V_{O-} = \frac{k}{1-k} V_I$$

From Equations (2.314) and (2.318), we can define that $M_E = M_{E+} = M_{E-}$. The curve of M_E vs. k is shown in Figure 2.61.

The variation ratios of the parameters are

$$\xi_- = \frac{\Delta i_{L12}/2}{I_{L12}} = \frac{k}{16} \frac{1}{f^2 C_{10} L_{12}} \quad \text{and} \quad \rho_- = \frac{\Delta v_{C11}/2}{V_{C11}} = \frac{k I_{O-} T}{2 C_{11} V_{O-}} = \frac{k}{2} \frac{1}{f C_{11} R_1}$$

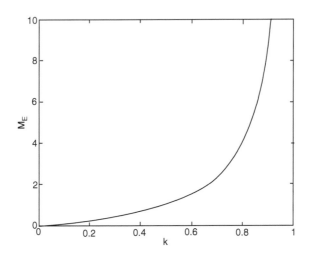

FIGURE 2.61
Voltage transfer gain M_E vs. k.

The variation ratio of current i_{L11} and i_{D10} is

$$\zeta_- = \frac{\Delta i_{L11}/2}{I_{L11}} = \frac{k(1-k)V_I T}{2L_{11}I_{O-}} = \frac{k(1-k)R_1}{2M_E fL_{11}} = \frac{k^2}{M_E^2} \frac{R_1}{2fL_{11}} \qquad (2.319)$$

The variation ratio of current v_{C10} is

$$\varepsilon_- = \frac{\Delta v_{C10}/2}{V_{C10}} = \frac{k}{128f^3 C_{11}C_{10}L_{12}} \frac{I_{O-}}{V_{O-}} = \frac{k}{128} \frac{1}{f^3 C_{11}C_{10}L_{12}R_1} \qquad (2.320)$$

Assuming that $f = 50$ kHz, $L_{11} = L_{12} = 0.5$ mH, $C = C_O = 20$ µF, $R_1 = 10$ Ω and $k = 0.5$, obtained $M_E = 1$, $\zeta = 0.05$, $\rho = 0.025$, $\xi = 0.00125$ and $\varepsilon = 0.0000156$. The output voltage V_{O-} is almost a real DC voltage with very small ripple. Since the load is resistive, the output current $i_{O-}(t)$ is almost a real DC waveform with very small ripple as well, and it is equal to $I_{O-} = V_{O-}/R_1$.

2.6.1.3 Discontinuous Mode

The equivalent circuits of the discontinuous mode's operation are shown in Figure 2.59c and Figure 2.60c. In order to obtain the mirror-symmetrical double output voltages, select:

$$L = \frac{L_1 L_2}{L_1 + L_2} = L_{11}$$

and $R = R_1$. Thus, we define

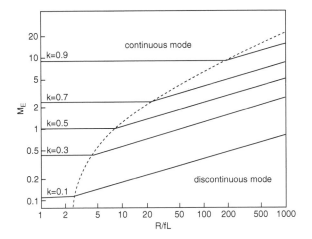

FIGURE 2.62
The boundary between continuous and discontinuous modes and the output voltage vs. the normalized load $z_N = R/fL$ (elementary circuit).

$$V_O = V_{O+} = |V_{O-}| \quad M_E = M_{E+} = M_{E-} = \frac{V_O}{V_I} = \frac{k}{1-k}$$

$$z_N = z_{N+} = z_{N-} \quad \text{and} \quad \zeta = \zeta_+ = \zeta_-$$

The free-wheeling diode currents i_{D0} and i_{D10} become zero during switch off before next period switch on. The boundary between continuous and discontinuous modes is

$$\zeta \geq 1$$

i.e.,

$$\frac{k^2}{M_E^2} \frac{z_N}{2} \geq 1$$

or

$$M_E \leq k\sqrt{\frac{z_N}{2}} \tag{2.321}$$

The boundary curve is shown in Figure 2.62.

In this case the free-wheeling diode's diode current exists in the period between kT and $[k + (1-k)m_E]T$, where m_E is the **filling efficiency** and it is defined as:

$$m_E = \frac{1}{\zeta} = \frac{2M_E^2}{k^2 z_N} \tag{2.322}$$

Considering the Equation (2.321), therefore, $0 < m_E < 1$. Since the diode current i_{D0} becomes zero at $t = kT + (1 - k)m_E T$, for the current i_{L1} $kTV_I = (1 - k)m_E TV_C$ or

$$V_{C1} = \frac{k}{(1-k)m_E} V_I = k(1-k)\frac{z_N}{2} V_I \quad \text{with} \quad \sqrt{\frac{z_N}{2}} \geq \frac{1}{1-k} \tag{2.323}$$

and for the current i_{L2}

$$kT(V_I + V_{C1} - V_{O+}) = (1 - k)m_E TV_{O+}$$

Therefore, the positive output voltage in discontinuous mode is

$$V_{O+} = \frac{k}{(1-k)m_E} V_I = k(1-k)\frac{z_N}{2} V_I \quad \text{with} \quad \sqrt{\frac{z_N}{2}} \geq \frac{1}{1-k} \tag{2.324}$$

For the current i_{L11} we have

$$kTV_I = (1 - k)m_E TV_{C11}$$

or

$$V_{C11} = \frac{k}{(1-k)m_E} V_I = k(1-k)\frac{z_N}{2} V_I \quad \text{with} \quad \sqrt{\frac{z_N}{2}} \geq \frac{1}{1-k} \tag{2.325}$$

and for the current i_{L12} we have

$$kT(V_I + V_{C11} - V_{O-}) = (1 - k)m_E TV_{O-}$$

Therefore, the negative output voltage in discontinuous mode is

$$V_{O-} = \frac{k}{(1-k)m_E} V_I = k(1-k)\frac{z_N}{2} V_I \quad \text{with} \quad \sqrt{\frac{z_N}{2}} \geq \frac{1}{1-k} \tag{2.326}$$

We then have

$$V_O = V_{O+} = V_{O-} = k(1-k)\frac{z_N}{2} V_I$$

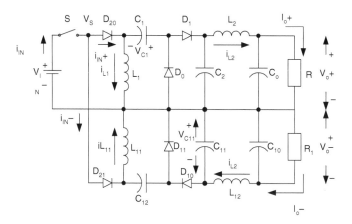

FIGURE 2.63
Self-lift circuit.

i.e., the output voltage will linearly increase while load resistance increases. It can be seen that larger load resistance may cause higher output voltage in discontinuous mode as shown in Figure 2.62.

2.6.2 Self-Lift Circuit

Self-lift circuit shown in Figure 2.63 is derived from the elementary circuit. The positive conversion path consists of a pump circuit S-L_1-D_0-C_1, a filter (C_2)-L_2-C_O, and a lift circuit D_1-C_2. The negative conversion path consists of a pump circuit S-L_{11}-D_{10}-(C_{11}), a "Π"-type filter C_{11}-L_{12}-C_{10}, and a lift circuit D_{11}-C_{12}.

2.6.2.1 Positive Conversion Path

The equivalent circuit during switch-on is shown in Figure 2.64a, and the equivalent circuit during switch-off in Figure 2.64b. The voltage across inductor L_1 is equal to V_I during switch-on, and $-V_{C1}$ during switch-off. We have the relations:

$$V_{C1} = \frac{k}{1-k} V_I$$

Hence,

$$V_O = V_{CO} = V_{C2} = V_I + V_{C1} = \frac{1}{1-k} V_I$$

and

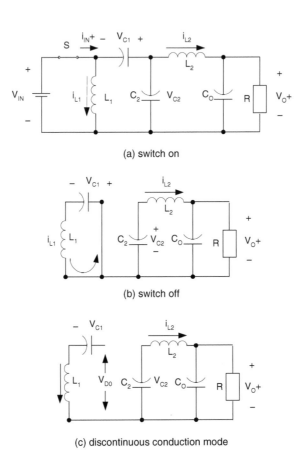

(a) switch on

(b) switch off

(c) discontinuous conduction mode

FIGURE 2.64
Equivalent circuits of self-lift circuit positive path: (a) switch on; (b) switch off; (c) discontinuous conduction mode.

$$V_{O+} = \frac{1}{1-k} V_I$$

The output current is

$$I_{O+} = (1-k)I_{I+}$$

Other relations are

$$I_{I+} = k\, i_{I+} \quad i_{I+} = I_{L1} + i_{C1-on} \quad i_{C1-off} = \frac{k}{1-k} i_{C1-on}$$

and

$$I_{L1} = i_{C1-off} = k i_{I+} = I_{I+} \tag{2.327}$$

Therefore, the voltage transfer gain in continuous mode is

$$M_{S+} = \frac{V_{O+}}{V_I} = \frac{1}{1-k} \tag{2.328}$$

The variation ratios of the parameters are

$$\xi_{2+} = \frac{\Delta i_{L2}/2}{I_{L2}} = \frac{k}{16} \frac{1}{f^2 C_2 L_2} \qquad \rho_+ = \frac{\Delta v_{C1}/2}{V_{C1}} = \frac{(1-k)I_{I+}}{2fC_1 \dfrac{k}{1-k} V_I} = \frac{1}{2kfC_1 R}$$

and

$$\sigma_{1+} = \frac{\Delta v_{C2}/2}{V_{C2}} = \frac{k}{2fC_2 R}$$

The variation ratio of the currents i_{D0} and i_{L1} is

$$\zeta_+ = \xi_{1+} = \frac{\Delta i_{L1}/2}{I_{L1}} = \frac{kV_I T}{2L_1 I_{I+}} = \frac{k}{M_S^2} \frac{R}{2fL_1} \tag{2.329}$$

The variation ratio of output voltage v_{O+} is

$$\varepsilon_+ = \frac{\Delta v_{O+}/2}{V_{O+}} = \frac{k}{128} \frac{1}{f^3 C_2 C_O L_2 R} \tag{2.330}$$

If $L_1 = L_2 = 0.5$ mH, $C_1 = C_2 = C_O = 20$ µF, $R = 40$ Ω, $f = 50$ kHz, and $k = 0.5$, obtained that $\xi_{1+} = \zeta = 0.1$, $\rho+ = 0.00625$; and $\sigma_{1+} = 0.00625$, $\xi_{2+} = 0.00125$ and $\varepsilon = 0.000004$. Therefore, the variations of i_{L1}, v_{C1}, i_{L2} and v_{C2} are small. The output voltage V_{O+} is almost a real DC voltage with very small ripple. Because of the resistive load, the output current $i_{O+}(t)$ is almost a real DC waveform with very small ripple as well, and $I_{O+} = V_{O+}/R$.

2.6.2.2 Negative Conversion Path

The equivalent circuit during switch-on is shown in Figure 2.65a, and the equivalent circuit during switch-off in Figure 2.65b. The relations of the average currents and voltages are

$$I_{O-} = I_{L12} = I_{C11-on} \qquad I_{C11-off} = \frac{k}{1-k} \qquad I_{C11-on} = \frac{k}{1-k} I_{O-}$$

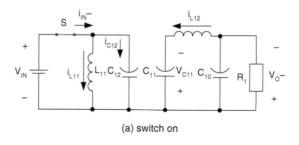

(a) switch on

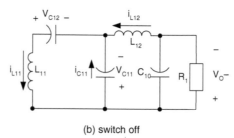

(b) switch off

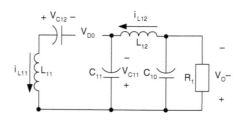

(c) discontinuous conduction mode

FIGURE 2.65
Equivalent circuits of self-lift circuit negative path: (a) switch on; (b) switch off; (c) discontinuous
conduction mode.

and

$$I_{L11} = I_{C11-off} + I_{O-} = \frac{I_{O-}}{1-k} \tag{2.331}$$

We know that

$$I_{C12-off} = I_{L11} = \frac{1}{1-k}I_{O-} \quad \text{and} \quad I_{C12-on} = \frac{1-k}{k}I_{C12-off} = \frac{1}{k}I_{O-}$$

So that

$$V_{O-} = \frac{1}{1-k}V_I \quad \text{and} \quad I_{O-} = (1-k)I_I$$

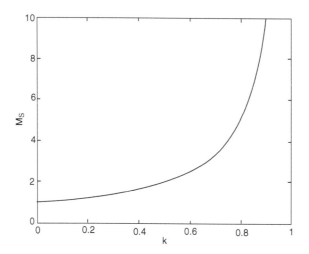

FIGURE 2.66
Voltage transfer gain M_S vs. k.

The voltage transfer gain in continuous mode:

$$M_{S-} = \frac{V_{O-}}{V_I} = \frac{1}{1-k} \tag{2.332}$$

Circuit (C_{11}-L_{12}-C_{10}) is a "Π" type low-pass filter. Therefore,

$$V_{C11} = V_{O-} = \frac{k}{1-k} V_I$$

From Equation (2.328) and Equation (2.332), we define $M_S = M_{S+} = M_{S-}$. The curve of M_S vs. k is shown in Figure 2.66.

The variation ratios of the parameters are

$$\xi_- = \frac{\Delta i_{L12}/2}{I_{L12}} = \frac{k}{16} \frac{1}{f^2 C_{10} L_{12}}$$

$$\rho_- = \frac{\Delta v_{C11}/2}{V_{C11}} = \frac{k I_{O-} T}{2 C_{11} V_{O-}} = \frac{k}{2} \frac{1}{f C_{11} R_1}$$

$$\sigma_{1-} = \frac{\Delta v_{C12}/2}{V_{C12}} = \frac{I_{O-}}{2 f C_{12} V_I} = \frac{M_S}{2} \frac{1}{f C_{12} R_1}$$

The variation ratio of currents i_{D10} and i_{L11} is

$$\zeta_{-} = \frac{\Delta i_{L11}/2}{I_{L11}} = \frac{k(1-k)V_1 T}{2L_{11}I_{O-}} = \frac{k(1-k)R_1}{2M_S fL_{11}} = \frac{k}{M_S^2} \frac{R_1}{2fL_{11}} \tag{2.333}$$

the variation ratio of current v_{C10} is

$$\varepsilon_{-} = \frac{\Delta v_{C10}/2}{V_{C10}} = \frac{k}{128f^3 C_{11}C_{10}L_{12}} \frac{I_{O-}}{V_{O-}} = \frac{k}{128} \frac{1}{f^3 C_{11}C_{10}L_{12}R_1} \tag{2.334}$$

Assuming that $f = 50$ kHz, $L_{11} = L_{12} = 0.5~\mu$H, $C_{11} = C_{10} = 20~\mu$F, $R_1 = 40~\Omega$ and $k = 0.5$, we obtain $M_S = 2$, $\zeta_{-} = 0.2$, $\rho_{-} = 0.006$, $\sigma_{1-} = 0.025$, $\xi_{-} = 0.0006$ and $\varepsilon_{-} = 0.000004$. The output voltage V_{O-} is almost a real DC voltage with very small ripple. Since the load is resistive, the output current $i_{O-}(t)$ is almost a real DC waveform with very small ripple as well, and it is equal to $I_{O-} = V_{O-}/R_1$.

2.6.2.3   *Discontinuous Conduction Mode*

The equivalent circuits of the discontinuous conduction mode's operation are shown in Figure 2.64 and Figure 2.65 Since we select $z_N = z_{N+} = z_{N-}$, $M_S = M_{S+} = M_{S-}$ and $\zeta = \zeta_{+} = \zeta_{-}$ The boundary between continuous and discontinuous conduction modes is

$$\zeta \geq 1$$

or

$$\frac{k}{M_S^2} \frac{z_N}{2} \geq 1$$

Hence,

$$M_S \leq \sqrt{k}\sqrt{\frac{z}{2}} = \sqrt{\frac{kz_N}{2}} \tag{2.335}$$

This boundary curve is shown in Figure 2.67. Compared with Equation (2.321) and Equation (2.335), this boundary curve has a minimum value of M_S that is equal to 1.5 at $k = 1/3$.

The filling efficiency is defined as:

$$m_S = \frac{1}{\zeta} = \frac{2M_S^2}{kz_N} \tag{2.336}$$

For the current i_{L1} we have

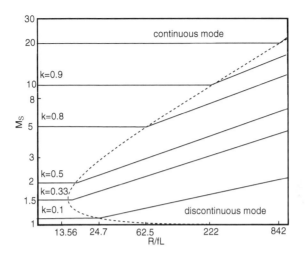

FIGURE 2.67
The boundary between continuous and discontinuous conduction modes and the output voltage vs. the normalized load $z_N = R/fL$ (self-lift circuit).

$$kTV_I = (1 - k)m_{S+}TV_{C1}$$

or

$$V_{C1} = \frac{k}{(1-k)m_S}V_I = k^2(1-k)\frac{z_N}{2}V_I \quad \text{with} \quad \sqrt{\frac{kz_N}{2}} \geq \frac{1}{1-k} \qquad (2.337)$$

Therefore, the positive output voltage in discontinuous mode is

$$V_{O+} = V_{C1} + V_I = [1 + \frac{k}{(1-k)m_S}]V_I = [1 + k^2(1-k)\frac{z_N}{2}]V_I$$

with

$$\sqrt{\frac{kz_N}{2}} \geq \frac{1}{1-k} \qquad (2.338)$$

For the current i_{L11} we have

$$kTV_I = (1 - k)m_S T(V_{C11} - V_I)$$

or

$$V_{C11} = [1 + \frac{k}{(1-k)m_S}]V_I = [1 + k^2(1-k)\frac{z_N}{2}]V_I$$

with
$$\sqrt{\frac{kz_N}{2}} \geq \frac{1}{1-k} \qquad (2.339)$$

and for the current i_{L12} we have

$$kT(V_I + V_{C11} - V_{O-}) = (1-k)m_S T(V_{O-} - V_I)$$

Therefore, the negative output voltage in discontinuous conduction mode is

$$V_{O-} = [1 + \frac{k}{(1-k)m_S}]V_I = [1 + k^2(1-k)\frac{z_N}{2}]V_I$$

with
$$\sqrt{\frac{kz_N}{2}} \geq \frac{1}{1-k} \qquad (2.340)$$

We then have

$$V_O = V_{O+} = V_{O-} = [1 + k^2(1-k)\frac{z_N}{2}]V_I$$

i.e., the output voltage will linearly increase during load resistance increasing. Larger load resistance causes higher output voltage in discontinuous conduction mode as shown in Figure 2.67.

2.6.3 Re-Lift Circuit

Re-lift circuit shown in Figure 2.68 is derived from self-lift circuit. The positive conversion path consists of a pump circuit S-L_1-D_0-C_1 and a filter (C_2)-L_2-C_O, and a lift circuit D_1-C_2-D_3-L_3-D_2-C_3. The negative conversion path consists of a pump circuit S-L_{11}-D_{10}-(C_{11}) and a "Π"-type filter C_{11}-L_{12}-C_{10}, and a lift circuit D_{11}-C_{12}-L_{13}-D_{22}-C_{13}-D_{12}.

2.6.3.1 Positive Conversion Path

The equivalent circuit during switch-on is shown in Figure 2.69a, and the equivalent circuit during switch-off in Figure 2.69b. The voltage across inductors L_1 and L_3 is equal to V_I during switch-on, and $-(V_{C1} - V_I)$ during switch-off. We have the relations:

$$V_{C1} = \frac{1+k}{1-k}V_I \quad \text{and} \quad V_O = V_{CO} = V_{C2} = V_I + V_{C1} = \frac{2}{1-k}V_I$$

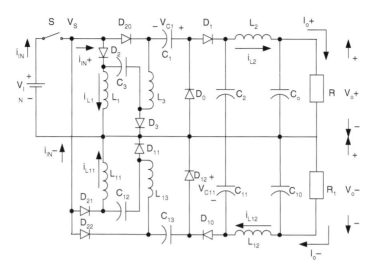

FIGURE 2.68
Re-lift circuit.

Thus,

$$V_{O+} = \frac{2}{1-k} V_I$$

and

$$I_{O+} = \frac{1-k}{2} I_{I+}$$

The other relations are

$$I_{I+} = k\, i_{I+} \quad i_{I+} = I_{L1} + I_{L3} + i_{C3-on} + i_{C1-on} \quad i_{C1-off} = \frac{k}{1-k} i_{C1-on}$$

and

$$I_{L1} = I_{L3} = i_{C1-off} = i_{C3-off} = \frac{k}{2} i_{I+} = \frac{1}{2} I_{I+} \tag{2.341}$$

The voltage transfer gain in continuous mode is

$$M_{R+} = \frac{V_{O+}}{V_I} = \frac{2}{1-k} \tag{2.342}$$

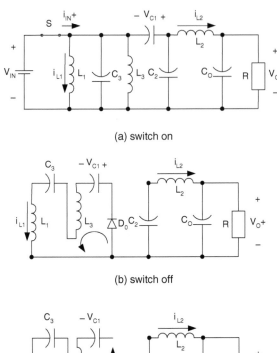

(a) switch on

(b) switch off

(c) discontinuous mode

FIGURE 2.69
Equivalent circuits of re-lift circuit positive path: (a) switch on; (b) switch off; (c) discontinuous mode.

The variation ratios of the parameters are

$$\xi_{2+} = \frac{\Delta i_{L2}/2}{I_{L2}} = \frac{k}{16}\frac{1}{f^2 C_2 L_2} \quad \text{and} \quad \chi_{1+} = \frac{\Delta i_{L3}/2}{I_{L3}} = \frac{kV_I T}{2L_3 \frac{1}{2} I_{I+}} = \frac{k}{M_R^2}\frac{R}{fL_3}$$

$$\rho_+ = \frac{\Delta v_{C1}/2}{V_{C1}} = \frac{(1-k)TI_I}{4C_1 \frac{1+k}{1-k} V_I} = \frac{1}{(1+k)fC_1 R}$$

$$\sigma_{1+} = \frac{\Delta v_{C2}/2}{V_{C2}} = \frac{k}{2fC_2 R} \qquad \sigma_{2+} = \frac{\Delta v_{C3}/2}{V_{C3}} = \frac{1-k}{4fC_3}\frac{I_{I+}}{V_I} = \frac{M_R}{2fC_3 R}$$

The variation ratio of currents i_{D0} and i_{L1} is

$$\zeta_+ = \xi_{1+} = \frac{\Delta i_{D0}/2}{I_{D0}} = \frac{kV_I T}{L_1 I_{I+}} = \frac{k}{M_R^2} \frac{R}{fL_1} \tag{2.343}$$

and the variation ratio of output voltage v_{O+} is

$$\varepsilon_+ = \frac{\Delta v_{O+}/2}{V_{O+}} = \frac{k}{128} \frac{1}{f^3 C_2 C_0 L_2 R} \tag{2.344}$$

If $L_1 = L_2 = L_3 = 0.5$ mH, $C_1 = C_2 = C_3 = C_0 = 20$ µF, $R = 160$ Ω, $f = 50$ kHz, and $k = 0.5$, we obtained that $\xi_1 = \zeta_+ = 0.2$, $\chi_1 = 0.2$, $\sigma_\square = 0.0125$, $\rho = 0.004$; and $\sigma_1 = 0.00156$, $\xi_2 = 0.0125$ and $\varepsilon = 0.000001$. Therefore, the variations of i_{L1}, i_{L2} and i_{L3} are small, and the ripples of v_{C1}, v_{C3} and v_{C2} are small. The output voltage v_{O+} (and v_{CO}) is almost a real DC voltage with very small ripple. Because of the resistive load, the output current i_{O+} is almost a real DC waveform with very small ripple as well, and $I_{O+} = V_{O+}/R$.

2.6.3.2 Negative Conversion Path

The equivalent circuit during switch-on is shown in Figure 2.70a, and the equivalent circuit during switch-off is shown in Figure 2.70b. The relations of the average currents and voltages are

$$I_{O-} = I_{L12} = I_{C11-on} \quad I_{C11-off} = \frac{k}{1-k} I_{C11-on} = \frac{k}{1-k} I_{O-}$$

and

$$I_{L11} = I_{C11-off} + I_{O-} = \frac{I_{O-}}{1-k} \tag{2.345}$$

$$I_{C12-off} = I_{C13-off} = I_{L11} = \frac{1}{1-k} I_{O-} \quad I_{C12-on} = \frac{1-k}{k} I_{C12-off} = \frac{1}{k} I_{O-}$$

$$I_{C13-on} = \frac{1-k}{k} I_{C13-off} = \frac{1}{k} I_{O-}$$

In steady state we have:

$$V_{C12} = V_{C13} = V_I \quad V_{L13-on} = V_I \quad \text{and} \quad V_{L13-off} = \frac{k}{1-k} V_I$$

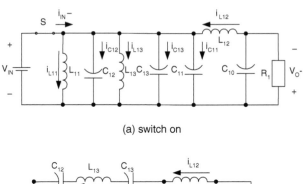

(a) switch on

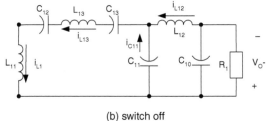

(b) switch off

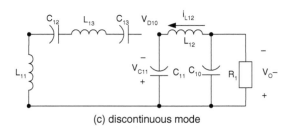

(c) discontinuous mode

FIGURE 2.70
Equivalent circuits of re-lift circuit negative path: (a) switch on; (b) switch off; (c) discontinuous mode.

$$V_{O-} = \frac{2}{1-k}V_I \quad \text{and} \quad I_{O-} = \frac{1-k}{2}I_{I-}$$

The voltage transfer gain in continuous mode is

$$M_{R-} = \frac{V_{O-}}{V_I} = \frac{I_{I-}}{I_{O-}} = \frac{2}{1-k} \tag{2.346}$$

Circuit $(C_{11}\text{-}L_{12}\text{-}C_{10})$ is a "Π" type low-pass filter.

Therefore,

$$V_{C11} = V_{O-} = \frac{2}{1-k}V_I$$

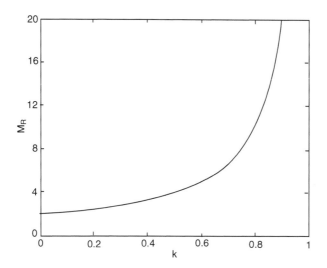

FIGURE 2.71
Voltage transfer gain M_R vs. k.

From Equation (2.342) and Equation (2.346) we define $M_R = M_{R+} = M_{R-}$. The curve of M_R vs. k is shown in Figure 2.71.

The variation ratios of the parameters are

$$\xi_- = \frac{\Delta i_{L12}/2}{I_{L12}} = \frac{k}{16} \frac{1}{f^2 C_{10} L_{12}}$$

and

$$\chi_{1-} = \frac{\Delta i_{L13}/2}{I_{L13}} = \frac{kTV_I}{2L_{13}I_{O-}}(1-k) = \frac{k(1-k)}{2M_R} \frac{R_1}{fL_{13}}$$

$$\rho_- = \frac{\Delta v_{C11}/2}{V_{C11}} = \frac{kI_{O-}T}{2C_{11}V_{O-}} = \frac{k}{2} \frac{1}{fC_{11}R_1}$$

$$\sigma_{1-} = \frac{\Delta v_{C12}/2}{V_{C12}} = \frac{I_{O-}}{2fC_{12}V_I} = \frac{M_R}{2} \frac{1}{fC_{12}R_1}$$

$$\sigma_{2-} = \frac{\Delta v_{C13}/2}{V_{C13}} = \frac{I_{O-}}{2fC_{13}V_I} = \frac{M_R}{2} \frac{1}{fC_{13}R_1}$$

The variation ratio of the current i_{D10} and i_{L11} is

$$\zeta_- = \frac{\Delta i_{L11}/2}{I_{L11}} = \frac{k(1-k)V_I T}{2L_{11}I_{O-}} = \frac{k(1-k)R_1}{2M_R fL_{11}} = \frac{k}{M_R^2} \frac{R_1}{fL_{11}} \qquad (2.347)$$

The variation ratio of current v_{C10} is

$$\varepsilon_- = \frac{\Delta v_{C10}/2}{V_{C10}} = \frac{k}{128f^3 C_{11} C_{10} L_{12}} \frac{I_{O-}}{V_{O-}} = \frac{k}{128} \frac{1}{f^3 C_{11} C_{10} L_{12} R_1} \quad (2.348)$$

Assuming that $f = 50$ kHz, $L_{11} = L_{12} = 0.5$ mH, $C = C_O = 20$ μF, $R_1 = 160$ Ω and $k = 0.5$, we obtain $M_R = 4$, $\zeta_- = 0.2$, $\rho_- = 0.0016$, $\sigma_{1-} = \sigma_{2-} = 0.0125$, $\xi_- = 0.00125$ and $\varepsilon_- = 10^{-6}$. The output voltage V_{O-} is almost a real DC voltage with very small ripple. Since the load is resistive, the output current $i_{O-}(t)$ is almost a real DC waveform with very small ripple as well, and it is equal to $I_{O-} = V_{O-}/R_1$.

2.6.3.3   Discontinuous Conduction Mode

The equivalent circuits of the discontinuous conduction mode are shown in Figure 2.69 and Figure 2.70. In order to obtain the mirror-symmetrical double output voltages, we purposely select $z_N = z_{N+} = z_{N-}$ and $\zeta = \zeta_+ = \zeta_-$. The freewheeling diode currents i_{D0} and i_{D10} become zero during switch off before next period switch on. The boundary between continuous and discontinuous conduction modes is

$$\zeta \geq 1$$

or

$$\frac{k}{M_R^{\,2}} z_N \geq 1$$

Hence,

$$M_R \leq \sqrt{kz_N} \quad (2.349)$$

This boundary curve is shown in Figure 2.71. Comparing with Equations (2.321), (2.335), and (2.349), it can be seen that the boundary curve has a minimum value of M_R that is equal to 3.0, corresponding to $k = 1/3$. The filling efficiency m_R is

$$m_R = \frac{1}{\zeta} = \frac{M_R^{\,2}}{kz_N} \quad (2.350)$$

So

$$V_{C1} = [1 + \frac{2k}{(1-k)m_R}]V_I = [1 + k^2(1-k)\frac{z_N}{2}]V_I$$

with

$$\sqrt{kz_N} \geq \frac{2}{1-k} \tag{2.351}$$

Therefore, the positive output voltage in discontinuous conduction mode is

$$V_{O+} = V_{C1} + V_I = [2 + \frac{2k}{(1-k)m_R}]V_I = [2 + k^2(1-k)\frac{z_N}{2}]V_I$$

with

$$\sqrt{kz_N} \geq \frac{2}{1-k} \tag{2.352}$$

For the current i_{L11} because inductor current $i_{L13} = 0$ at $t = t_1$, so that

$$V_{L13-off} = \frac{k}{(1-k)m_R}V_I$$

for the current i_{L11} we have

$$kTV_I = (1-k)m_RT(V_{C11} - 2V_I - V_{L13-off})$$

or

$$V_{C11} = [2 + \frac{2k}{(1-k)m_R}]V_I = [2 + k^2(1-k)\frac{z_N}{2}]V_I \quad \text{with} \quad \sqrt{kz_N} \geq \frac{2}{1-k} \tag{2.353}$$

and for the current i_{L12}

$$kT(V_I + V_{C11} - V_{O-}) = (1-k)m_RT(V_{O-} - 2V_I - V_{L13-off})$$

Therefore, the negative output voltage in discontinuous conduction mode is

$$V_{O-} = [2 + \frac{2k}{(1-k)m_R}]V_I = [2 + k^2(1-k)\frac{z_N}{2}]V_I \quad \text{with} \quad \sqrt{kz_N} \geq \frac{2}{1-k} \tag{2.354}$$

So

$$V_O = V_{O+} = V_{O-} = [2 + k^2(1-k)\frac{z_N}{2}]V_I$$

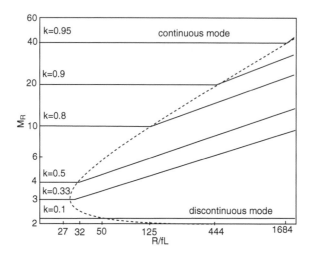

FIGURE 2.72

The boundary between continuous and discontinuous modes and the output voltage vs. the normalized load $z_N = R/fL$ (re-lift circuit).

i.e., the output voltage will linearly increase during load resistance increasing. Larger load resistance may cause higher output voltage in discontinuous mode as shown in Figure 2.72.

2.6.4 Multiple-Lift Circuit

Referring to Figure 2.68, it is possible to build a multiple-lifts circuit only using the parts $(L_3\text{-}D_{20}\text{-}C_3\text{-}D_3)$ multiple times in the positive conversion path, and using the parts $(D_{22}\text{-}L_{13}\text{-}C_{13}\text{-}D_{12})$ multiple times in the negative conversion path. For example, in Figure 2.73 the parts $(L_4\text{-}D_4\text{-}C_4\text{-}D_5)$ and parts $(D_{23}\text{-}L_{14}\text{-}C_{14}\text{-}D_{13})$ were added in the triple-lift circuit. According to this principle, triple-lift circuits and quadruple-lift circuits have been built as shown in Figure 2.73 and Figure 2.76. In this book it is not necessary to introduce the particular analysis and calculations one by one to readers. However, their calculation formulas are shown in this section.

2.6.4.1 Triple-Lift Circuit

Triple-lift circuit is shown in Figure 2.73. The positive conversion path consists of a pump circuit $S\text{-}L_1\text{-}D_0\text{-}C_1$ and a filter $(C_2)\text{-}L_2\text{-}C_O$, and a lift circuit $D_1\text{-}C_2\text{-}D_2\text{-}C_3\text{-}D_3\text{-}L_3\text{-}D_4\text{-}C_4\text{-}D_5\text{-}L_4$. The negative conversion path consists of a pump circuit $S\text{-}L_{11}\text{-}D_{10}\text{-}(C_{11})$ and a "Π"-type filter $C_{11}\text{-}L_{12}\text{-}C_{10}$, and a lift circuit $D_{11}\text{-}C_{12}\text{-}D_{22}\text{-}C_{13}\text{-}L_{13}\text{-}D_{12}\text{-}D_{23}\text{-}L_{14}\text{-}C_{14}\text{-}D_{13}$.

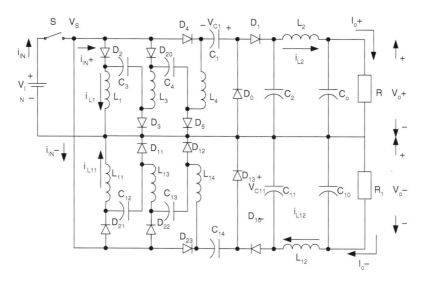

FIGURE 2.73
Triple-lift circuit.

2.6.4.1.1 *Positive Conversion Path*

The lift circuit is D_1-C_2-D_2-C_3-D_3-L_3-D_4-C_4-D_5-L_4. Capacitors C_2, C_3, and C_4 perform characteristics to lift the capacitor voltage V_{C1} by three times of source voltage V_I. L_3 and L_4 perform the function as ladder joints to link the three capacitors C_3 and C_4 and lift the capacitor voltage V_{C1} up. Current $i_{C2}(t)$, $i_{C3}(t)$, and $i_{C4}(t)$ are exponential functions. They have large values at the moment of power on, but they are small because $v_{C3} = v_{C4} = V_I$ and $v_{C2} = V_{O+}$ in steady state.

The output voltage and current are

$$V_{O+} = \frac{3}{1-k}V_I \quad \text{and} \quad I_{O+} = \frac{1-k}{3}I_{I+}$$

The voltage transfer gain in continuous mode is

$$M_{T+} = \frac{V_{O+}}{V_I} = \frac{3}{1-k} \tag{2.355}$$

Other average voltages:

$$V_{C1} = \frac{2+k}{1-k}V_I \quad V_{C3} = V_{C4} = V_I \quad V_{CO} = V_{C2} = V_{O+}$$

Other average currents:

$$I_{L2} = I_{O+} \quad I_{L1} = I_{L3} = I_{L4} = \frac{1}{3}I_{I+} = \frac{1}{1-k}I_{O+}$$

Current variations:

$$\xi_{1+} = \zeta_+ = \frac{k(1-k)R}{2M_T fL} = \frac{k}{M_T{}^2}\frac{3R}{2fL} \quad \xi_{2+} = \frac{k}{16}\frac{1}{f^2C_2L_2}$$

$$\chi_{1+} = \frac{k}{M_T{}^2}\frac{3R}{2fL_3} \quad \chi_{2+} = \frac{k}{M_T{}^2}\frac{3R}{2fL_4}$$

Voltage variations:

$$\rho_+ = \frac{3}{2(2+k)fC_1R} \quad \sigma_{1+} = \frac{k}{2fC_2R}$$

$$\sigma_{2+} = \frac{M_T}{2fC_3R} \quad \sigma_{3+} = \frac{M_T}{2fC_4R}$$

The variation ratio of output voltage V_{C0} is

$$\varepsilon_+ = \frac{k}{128}\frac{1}{f^3C_2C_0L_2R} \tag{2.356}$$

2.6.4.1.2 Negative Conversion Path

Circuit C_{12}-D_{11}-L_{13}-D_{22}-C_{13}-D_{12}-L_{14}-D_{23}-C_{14}-D_{13} is the lift circuit. Capacitors C_{12}, C_{13}, and C_{14} perform characteristics to lift the capacitor voltage V_{C11} by three times the source voltage V_I. L_{13} and L_{14} perform the function as ladder joints to link the three capacitors C_{12}, C_{13}, and C_{14} and lift the capacitor voltage V_{C11} up. Current $i_{C12}(t)$, $i_{C13}(t)$ and $i_{C14}(t)$ are exponential functions. They have large values at the moment of power on, but they are small because $v_{C12} = v_{C13} = v_{C14} \cong V_I$ in steady state.

The output voltage and current are

$$V_{O-} = \frac{3}{1-k}V_I \quad \text{and} \quad I_{O-} = \frac{1-k}{3}I_{I-}$$

The voltage transfer gain in continuous mode is

$$M_{T-} = V_{O-}/V_I = \frac{3}{1-k} \tag{2.357}$$

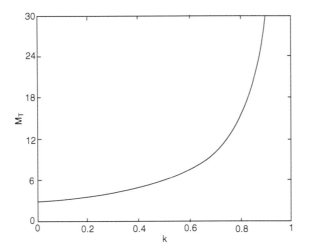

FIGURE 2.74
Voltage transfer gain M_T vs. k.

From Equation (2.355) and Equation (2.357) we define $M_T = M_{T+} = M_{T-}$. The curve of M_T vs. k is shown in Figure 2.74.
Other average voltages:

$$V_{C11} = V_{O-} \quad V_{C12} = V_{C13} = V_{C14} = V_I$$

Other average currents:

$$I_{L12} = I_{O-} \quad I_{L11} = I_{L13} = I_{L14} = \frac{1}{1-k} I_{O-}$$

Current variation ratios:

$$\zeta_- = \frac{k}{M_T^2} \frac{3R_1}{2fL_{11}} \quad \xi_{2-} = \frac{k}{16} \frac{1}{f^2 C_{10} L_{12}}$$

$$\chi_{1-} = \frac{k(1-k)}{2M_T} \frac{R_1}{fL_{13}} \quad \chi_{2-} = \frac{k(1-k)}{2M_T} \frac{R_1}{fL_{14}}$$

Voltage variation ratios:

$$\rho_- = \frac{k}{2} \frac{1}{fC_{11}R_1} \quad \sigma_{1-} = \frac{M_T}{2} \frac{1}{fC_{12}R_1}$$

$$\sigma_{2-} = \frac{M_T}{2} \frac{1}{fC_{13}R_1} \quad \sigma_{3-} = \frac{M_T}{2} \frac{1}{fC_{14}R_1}$$

The variation ratio of output voltage V_{C10} is

$$\varepsilon_- = \frac{k}{128} \frac{1}{f^3 C_{11} C_{10} L_{12} R_1} \tag{2.358}$$

2.6.4.1.3 Discontinuous Mode

To obtain the mirror-symmetrical double output voltages, we purposely select: $L_1 = L_{11}$ and $R = R_1$.
Define:

$$V_O = V_{O+} = V_{O-} \quad M_T = M_{T+} = M_{T-} = \frac{V_O}{V_I} = \frac{3}{1-k} \quad z_N = z_{N+} = z_{N-}$$

and

$$\zeta = \zeta_+ = \zeta_-$$

The free-wheeling diode currents i_{D0} and i_{D10} become zero during switch off before next period switch on. The boundary between continuous and discontinuous conduction modes is

$$\zeta \geq 1$$

Then

$$M_T \leq \sqrt{\frac{3kz_N}{2}} \tag{2.359}$$

This boundary curve is shown in Figure 2.75. Comparing Equation (2.321), Equation (2.335), Equation (2.349), and Equation (2.359), it can be seen that the boundary curve has a minimum value of M_T that is equal to 4.5, corresponding to $k = 1/3$.

In discontinuous mode the currents i_{D0} and i_{D10} exist in the period between kT and $[k + (1-k)m_T]T$, where m_T is the filling efficiency that is

$$m_T = \frac{1}{\zeta} = \frac{2M_T^2}{3kz_N} \tag{2.360}$$

Considering Equation (2.359), therefore $0 < m_T < 1$. Since the current i_{D0} becomes zero at $t = t_1 = [k + (1-k)m_T]T$, for the current i_{L1}, i_{L3} and i_{L4}

$$3kTV_I = (1-k)m_T T(V_{C1} - 2V_I)$$

or

$$V_{C1} = [2 + \frac{3k}{(1-k)m_T}]V_I = [2 + k^2(1-k)\frac{z_N}{2}]V_I$$

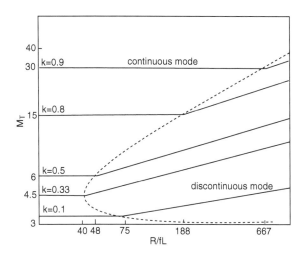

FIGURE 2.75
The boundary between continuous and discontinuous modes and the output voltage vs. the normalized load $z_N = R/fL$ (triple-lift circuit).

with
$$\sqrt{\frac{3kz_N}{2}} \geq \frac{3}{1-k} \qquad (2.361)$$

Therefore, the positive output voltage in discontinuous mode is

$$V_{O+} = V_{C1} + V_I = [3 + \frac{3k}{(1-k)m_T}]V_I = [3 + k^2(1-k)\frac{z_N}{2}]V_I$$

with
$$\sqrt{\frac{3kz_N}{2}} \geq \frac{3}{1-k} \qquad (2.362)$$

Because inductor current $i_{L11} = 0$ at $t = t_1$, so that

$$V_{L13-off} = V_{L14-off} = \frac{k}{(1-k)m_T}V_I$$

Since i_{D10} becomes 0 at $t_1 = [k + (1-k)m_T]T$, for the current i_{L11},

$$kTV_I = (1-k)m_{T-}T(V_{C11} - 3V_I - V_{L13-off} - V_{L14-off})$$

$$V_{C11} = [3 + \frac{3k}{(1-k)m_T}]V_I = [3 + k^2(1-k)\frac{z_N}{2}]V_I$$

with
$$\sqrt{\frac{3kz_N}{2}} \geq \frac{3}{1-k} \qquad (2.363)$$

for the current i_{L12}

$$kT(V_I + V_{C14} - V_{O-}) = (1-k)m_{T-}T(V_{O-} - 2V_I - V_{L13-off} - V_{L14-off})$$

Therefore, the negative output voltage in discontinuous mode is

$$V_{O-} = [3 + \frac{3k}{(1-k)m_T}]V_I = [3 + k^2(1-k)\frac{z_N}{2}]V_I$$

with
$$\sqrt{\frac{3kz_N}{2}} \geq \frac{3}{1-k} \qquad (2.364)$$

So

$$V_O = V_{O+} = V_{O-} = [3 + k^2(1-k)\frac{z_N}{2}]V_I$$

i.e., the output voltage will linearly increase during load resistance increasing, as shown in Figure 2.75.

2.6.4.2   *Quadruple-Lift Circuit*

Quadruple-lift circuit is shown in Figure 2.76. The positive conversion path consists of a pump circuit S-L_1-D_0-C_1 and a filter (C_2)-L_2-C_O, and a lift circuit D_1-C_2-L_3-D_2-C_3-D_3-L_4-D_4-C_4-D_5-L_5-D_6-C_5-S_1. The negative conversion path consists of a pump circuit S-L_{11}-D_{10}-(C_{11}) and a "Π"-type filter C_{11}-L_{12}-C_{10}, and a lift circuit D_{11}-C_{12}-D_{22}-L_{13}-C_{13}-D_{12}-D_{23}-L_{14}-C_{14}-D_{13}-D_{24}-L_{15}-C_{15}-D_{14}.

2.6.4.2.1 *Positive Conversion Path*

Capacitors C_2, C_3, C_4, and C_5 perform characteristics to lift the capacitor voltage V_{C1} by four times the source voltage V_I. L_3, L_4, and L_5 perform the function as ladder joints to link the four capacitors C_2, C_3, C_4, and C_5, and lift the output capacitor voltage V_{C1} up. Current $i_{C2}(t)$, $i_{C3}(t)$, $i_{C4}(t)$ and $i_{C5}(t)$ are exponential functions. They have large values at the moment of power on, but they are small because $v_{C3} = v_{C4} = v_{C5} = V_I$ and $v_{C2} = V_{O+}$ in steady state.

The output voltage and current are

$$V_{O+} = \frac{4}{1-k}V_I \quad \text{and} \quad I_{O+} = \frac{1-k}{4}I_{I+}$$

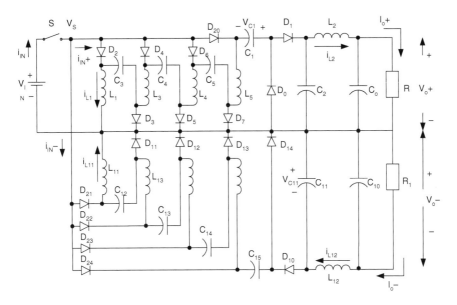

FIGURE 2.76
Quadruple-lift circuit.

The voltage transfer gain in continuous mode is

$$M_{Q+} = \frac{V_{O+}}{V_I} = \frac{4}{1-k} \tag{2.365}$$

Other average voltages:

$$V_{C1} = \frac{3+k}{1-k} V_I \quad V_{C3} = V_{C4} = V_{C5} = V_I \quad V_{CO} = V_{C2} = V_O$$

Other average currents:

$$I_{L2} = I_{O+} \quad I_{L1} = I_{L3} = I_{L4} = I_{L5} = \frac{1}{4} I_{I+} = \frac{1}{1-k} I_{O+}$$

Current variations:

$$\xi_{1+} = \zeta_+ = \frac{k(1-k)R}{2M_Q fL} = \frac{k}{M_Q{}^2} \frac{2R}{fL} \quad \xi_{2+} = \frac{k}{16} \frac{1}{f^2 C_2 L_2}$$

$$\chi_{1+} = \frac{k}{M_Q{}^2} \frac{2R}{fL_3} \quad \chi_{2+} = \frac{k}{M_Q{}^2} \frac{2R}{fL_4} \quad \chi_{3+} = \frac{k}{M_Q{}^2} \frac{2R}{fL_5}$$

Voltage variations:

$$\rho_+ = \frac{2}{(3+2k)fC_1R} \qquad \sigma_{1+} = \frac{M_Q}{2fC_2R}$$

$$\sigma_{2+} = \frac{M_Q}{2fC_3R} \qquad \sigma_{3+} = \frac{M_Q}{2fC_4R} \qquad \sigma_{4+} = \frac{M_Q}{2fC_5R}$$

The variation ratio of output voltage V_{C0} is

$$\varepsilon_+ = \frac{k}{128} \frac{1}{f^3 C_2 C_0 L_2 R} \qquad (2.366)$$

2.6.4.2.2 *Negative Conversion Path*

Capacitors C_{12}, C_{13}, C_{14}, and C_{15} perform characteristics to lift the capacitor voltage V_{C11} by four times the source voltage V_I. L_{13}, L_{14}, and L_{15} perform the function as ladder joints to link the four capacitors C_{12}, C_{13}, C_{14}, and C_{15} and lift the output capacitor voltage V_{C11} up. Current $i_{C12}(t)$, $i_{C13}(t)$, $i_{C14}(t)$, and $i_{C15}(t)$ are exponential functions. They have large values at the moment of power on, but they are small because $v_{C12} = v_{C13} = v_{C14} = v_{C15} \cong V_I$ in steady state.

The output voltage and current are

$$V_{O-} = \frac{4}{1-k}V_I \quad \text{and} \quad I_{O-} = \frac{1-k}{4}I_{I-}$$

The voltage transfer gain in continuous mode is

$$M_{Q-} = V_{O-}/V_I = \frac{4}{1-k} \qquad (2.367)$$

From Equation (2.365) and Equation (2.367) we define $M_Q = M_{Q+} = M_{Q-}$. The curve of M_Q vs. k is shown in Figure 2.77.
Other average voltages:

$$V_{C10} = V_{O-} \quad V_{C12} = V_{C13} = V_{C14} = V_{C15} = V_I$$

Other average currents:

$$I_{L12} = I_{O-} \quad I_{L11} = I_{L13} = I_{L14} = I_{L15} = \frac{1}{1-k}I_{O-}$$

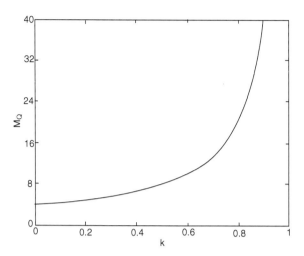

FIGURE 2.77
Voltage transfer gain M_Q vs. k.

Current variation ratios:

$$\zeta_- = \frac{k}{M_Q{}^2}\frac{2R_1}{fL_{11}} \qquad \xi_- = \frac{k}{16}\frac{1}{f^2 CL_{12}}$$

$$\chi_{1-} = \frac{k(1-k)}{2M_Q}\frac{R_1}{fL_{13}} \qquad \chi_{2-} = \frac{k(1-k)}{2M_Q}\frac{R_1}{fL_{14}} \qquad \chi_{3-} = \frac{k(1-k)}{2M_Q}\frac{R_1}{fL_{15}}$$

Voltage variation ratios:

$$\rho_- = \frac{k}{2}\frac{1}{fC_{11}R_1} \qquad \sigma_{1-} = \frac{M_Q}{2}\frac{1}{fC_{12}R_1}$$

$$\sigma_{2-} = \frac{M_Q}{2}\frac{1}{fC_{13}R_1} \qquad \sigma_{3-} = \frac{M_Q}{2}\frac{1}{fC_{14}R_1} \qquad \sigma_{4-} = \frac{M_Q}{2}\frac{1}{fC_{15}R_1}$$

The variation ratio of output voltage V_{C10} is

$$\varepsilon_- = \frac{k}{128}\frac{1}{f^3 C_{11}C_{10}L_{12}R_1} \qquad (2.368)$$

The output voltage ripple is very small.

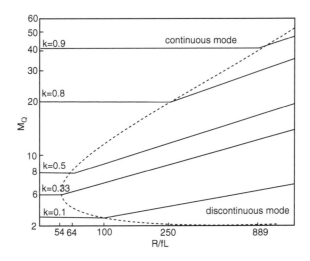

FIGURE 2.78
The boundary between continuous and discontinuous modes and the output voltage vs. the normalized load $z_N = R/fL$ (quadruple-lift circuit).

2.6.4.2.3 Discontinuous Conduction Mode

In order to obtain the mirror-symmetrical double output voltages, we purposely select: $L_1 = L_{11}$ and $R = R_1$. Therefore, we may define

$$V_O = V_{O+} = V_{O-} \quad M_Q = M_{Q+} = M_{Q-} = \frac{V_O}{V_I} = \frac{4}{1-k} \quad z_N = z_{N+} = z_{N-}$$

and
$$\zeta = \zeta_+ = \zeta_-$$

The free-wheeling diode currents i_{D0} and i_{D10} become zero during switch off before next period switch on. The boundary between continuous and discontinuous conduction modes is

$$\zeta \geq 1$$

or

$$M_Q \leq \sqrt{2kz_N} \qquad (2.369)$$

This boundary curve is shown in Figure 2.78. Comparing Equations (2.321), (2.335), (2.349), (2.359), and (2.369), it can be seen that this boundary curve has a minimum value of M_Q that is equal to 6.0, corresponding to $k = 1/3$.

In discontinuous mode the currents i_{D0} and i_{D10} exist in the period between kT and $[k + (1 - k)m_Q]T$, where m_Q is the filling efficiency that is

$$m_Q = \frac{1}{\zeta} = \frac{M_Q^2}{2kz_N} \tag{2.370}$$

Considering Equation (2.369), therefore $0 < m_Q < 1$. Since the current i_{D0} becomes zero at $t = t_1 = kT + (1 - k)m_Q T$, for the current i_{L1}, i_{L3}, i_{L4} and i_{L5}

$$4kTV_I = (1 - k)m_Q T(V_{C1} - 3V_I)$$

$$V_{C1} = [3 + \frac{4k}{(1 - k)m_Q}]V_I = [3 + k^2(1 - k)\frac{z_N}{2}]V_I$$

with $$\sqrt{2kz_N} \geq \frac{4}{1 - k} \tag{2.371}$$

Therefore, the positive output voltage in discontinuous conduction mode is

$$V_{O+} = V_{C1} + V_I = [4 + \frac{4k}{(1 - k)m_Q}]V_I = [4 + k^2(1 - k)\frac{z_N}{2}]V_I$$

with $$\sqrt{2kz_N} \geq \frac{4}{1 - k} \tag{2.372}$$

Because inductor current $i_{L11} = 0$ at $t = t_1$, so that

$$V_{L13-off} = V_{L14-off} = V_{L15-off} = \frac{k}{(1 - k)m_Q}V_I$$

Since the current i_{D10} becomes zero at $t = t_1 = kT + (1 - k)m_Q T$, for the current i_{L11} we have

$$kTV_I = (1 - k)m_Q T(V_{C11} - 4V_I - V_{L13-off} - V_{L14-off} - V_{L15-off})$$

So

$$V_{C11} = [4 + \frac{4k}{(1 - k)m_Q}]V_I = [4 + k^2(1 - k)\frac{z_N}{2}]V_I$$

with $$\sqrt{2kz_N} \geq \frac{4}{1 - k} \tag{2.373}$$

TABLE 2.3

Comparison among Five Circuits of Double Output Luo-Converters

Double Output Luo-Converters	I_O	V_O	$V_O(V_s = 10\ V)$			
			$k = 0.33$	$k = 0.5$	$k = 0.75$	$k = 0.9$
Elementary Circuit	$I_O = \dfrac{1-k}{k}I_S$	$V_O = \dfrac{k}{1-k}V_S$	5 V	10 V	30 V	90 V
Self-Lift Circuit	$I_O = (1-k)I_S$	$V_O = \dfrac{1}{1-k}V_S$	15 V	20 V	40 V	100 V
Re-Lift Circuit	$I_O = \dfrac{1-k}{2}I_S$	$V_O = \dfrac{2}{1-k}V_S$	30 V	40 V	80 V	200 V
Triple-Lift Circuit	$I_O = \dfrac{1-k}{3}I_S$	$V_O = \dfrac{3}{1-k}V_S$	45 V	60 V	120 V	300 V
Quadruple-Lift Circuit	$I_O = \dfrac{1-k}{4}I_S$	$V_O = \dfrac{4}{1-k}V_S$	60 V	80 V	160 V	400 V

For the current i_{L12}

$$kT(V_I + V_{C15} - V_{O-}) = (1-k)m_Q T(V_{O-} - 2V_I - V_{L13\text{-}off} - V_{L14\text{-}off} - V_{L15\text{-}off})$$

Therefore, the negative output voltage in discontinuous conduction mode is

$$V_{O-} = [4 + \frac{4k}{(1-k)m_Q}]V_I = [4 + k^2(1-k)\frac{z_N}{2}]V_I$$

with

$$\sqrt{2kz_N} \geq \frac{4}{1-k} \qquad (2.374)$$

So

$$V_O = V_{O+} = V_{O-} = [4 + k^2(1-k)\frac{z_N}{2}]V_I$$

i.e., the output voltage will linearly increase during load resistance increasing, as shown in Figure 2.78.

2.6.5 Summary

2.6.5.1 *Positive Conversion Path*

From the analysis and calculation in previous sections, the common formulae can be obtained for all circuits:

$$M = \frac{V_{O+}}{V_I} = \frac{I_{I+}}{I_{O+}} \qquad z_N = \frac{R}{fL} \qquad R = \frac{V_{O+}}{I_{O+}}$$

$$L = \frac{L_1 L_2}{L_1 + L_2}$$

for elementary circuits only;

$$L = L_1$$

for other lift circuit's current variations:

$$\xi_{1+} = \frac{1-k}{2M_E} \frac{R}{fL_1} \quad \text{and} \quad \xi_{2+} = \frac{k}{2M_E} \frac{R}{fL_2}$$

for elementary circuit only;

$$\xi_{1+} = \zeta_+ = \frac{k(1-k)R}{2MfL} \quad \text{and} \quad \xi_{2+} = \frac{k}{16} \frac{1}{f^2 C_2 L_2}$$

for other lift circuits

$$\zeta_+ = \frac{k(1-k)R}{2MfL} \qquad \chi_{j+} = \frac{k}{M^2} \frac{R}{fL_{j+2}} \qquad (j = 1, 2, 3, \dots)$$

Voltage variations are

$$\rho_+ = \frac{k}{2fC_1 R} \qquad \varepsilon_+ = \frac{k}{8M_E} \frac{1}{f^2 C_0 L_2}$$

for elementary circuit only;

$$\rho_+ = \frac{M}{M-1} \frac{1}{2fC_1 R} \qquad \varepsilon_+ = \frac{k}{128} \frac{1}{f^3 C_2 C_0 L_2 R}$$

for other lift circuits

$$\sigma_{1+} = \frac{k}{2fC_2 R} \qquad \sigma_{j+} = \frac{M}{2fC_{j+1} R} \qquad (j = 2, 3, 4, \dots)$$

2.6.5.2 *Negative Conversion Path*

From the analysis and calculation in previous sections, the common formulae can be obtained for all circuits:

$$M = \frac{V_{O-}}{V_I} = \frac{I_{I-}}{I_{O-}} \qquad z_{N-} = \frac{R_1}{fL_{11}} \qquad R_1 = \frac{V_{O-}}{I_{O-}}$$

Current variation ratios:

$$\zeta_- = \frac{k(1-k)R_1}{2MfL_{11}} \qquad \xi_- = \frac{k}{16f^2C_{11}L_{12}} \qquad \chi_{j-} = \frac{k(1-k)R_1}{2MfL_{j+2}} \quad (j = 1, 2, 3, \dots)$$

Voltage variation ratios:

$$\rho_- = \frac{k}{2fC_{11}R_1} \qquad \varepsilon_- = \frac{k}{128f^3C_{11}C_{10}L_{12}R_1} \qquad \sigma_{j-} = \frac{M}{2fC_{j+11}R_1} \quad (j = 1, 2, 3, 4, \dots)$$

2.6.5.3 *Common Parameters*

Usually, we select the loads $R = R_1$, $L = L_{11}$, so that we have got $z_N = z_{N+} = z_{N-}$. In order to write common formulas for the boundaries between continuous and discontinuous modes and output voltage for all circuits, the circuits can be numbered. The definition is that subscript 0 means the elementary circuit, subscript 1 means the self-lift circuit, subscript 2 means the re-lift circuit, subscript 3 means the triple-lift circuit, subscript 4 means the quadruple-lift circuit, and so on. The voltage transfer gain is

$$M_j = \frac{k^{h(j)}[j + h(j)]}{1 - k} \qquad j = 0, 1, 2, 3, 4, \dots$$

The characteristics of output voltage of all circuits are shown in Figure 2.79. The free-wheeling diode current's variation is

$$\zeta_j = \frac{k^{[1+h(j)]}}{M_j^2} \frac{j + h(j)}{2} z_N$$

The boundaries are determined by the condition:

$$\zeta_j \geq 1$$

or

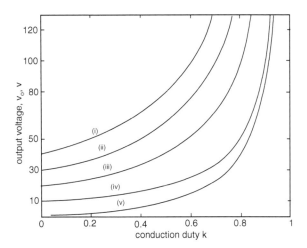

FIGURE 2.79
Output voltages of all double output Luo-converters ($V_i = 10$ V).

$$\frac{k^{[1+h(j)]}}{M_j^{\,2}} \frac{j+h(j)}{2} z_N \geq 1 \quad j = 0, 1, 2, 3, 4, \ldots$$

Therefore, the boundaries between continuous and discontinuous modes for all circuits are

$$M_j = k^{\frac{1+h(j)}{2}} \sqrt{\frac{j+h(j)}{2} z_N} \quad j = 0, 1, 2, 3, 4, \ldots$$

The filling efficiency is

$$m_j = \frac{1}{\zeta_j} = \frac{M_j^{\,2}}{k^{[1+h(j)]}} \frac{2}{j+h(j)} \frac{1}{z_N} \quad j = 0, 1, 2, 3, 4, \ldots$$

The output voltage in discontinuous mode for all circuits

$$V_{O-j} = [j + k^{[2-h(j)]} \frac{1-k}{2} z_N]V_I$$

where

$$h(j) = \begin{cases} 0 & if \quad j \geq 1 \\ 1 & if \quad j = 0 \end{cases} \quad j = 0, 1, 2, 3, 4, \ldots \quad h(j) \text{ is the } \textbf{Hong Function}$$

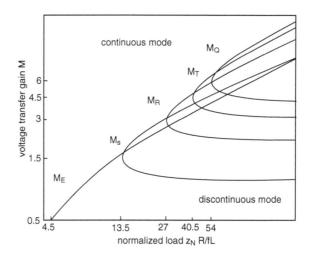

FIGURE 2.80
Boundaries between continuous and discontinuous modes of all double output Luo-Converters.

The boundaries between continuous and discontinuous modes of all circuits are shown in Figure 2.80. The curves of all M vs. z_N state that the continuous mode area increases from M_E via M_S, M_R, M_T to M_Q. The boundary of elementary circuit is a monorising curve, but other curves are not monorising. There are minimum values of the boundaries of other curves, which of M_S, M_R, M_T, and M_Q correspond at $k = 1/3$.

Bibliography

Chen, X.F., Luo, F.L., and Ye, H., Modified positive output Luo converters, in *Proceedings of the IEEE International Conference PEDS'99*, Hong Kong, 1999, p. 450.

Farkas, T. and Schecht, M.F., Viability of active EMI filters for utility applications, *IEEE Transactions on Power Electronics*, 9, 328, 1994.

Hart, D., *Introduction to Power Electronics*, Prentice Hall, New York, 1997.

Jozwik, J.J. and Kazimerczuk, M.K., Dual sepic PWM switching-mode DC/DC power converter, *IEEE Transactions on Industrial Electronics*, 36, 64, 1989.

Krein, P., *Elements of Power Electronics*, Oxford University Press, 1998.

LaWhite, I.E. and Schlecht, M.F., Active filters for 1 MHz power circuits with strict input/output ripple requirements, *IEEE Transactions on Power Electronics*, 2, 282, 1987.

Luo, F.L., Positive output Luo-converters: voltage lift technique, *IEE Proceedings on Electric Power Applications*, 146, 415, 1999.

Luo, F.L., Negative output Luo-Converters: voltage lift technique, *IEE Proceedings on Electric Power Applications*, 146, 208, 1999.

Luo, F.L., Double output Luo-converters: advanced voltage lift technique, *IEE-Proceedings on Electric Power Applications*, 147, 469, 2000.

Luo, F.L., Six self-lift DC/DC converters: voltage-lift technique, *IEEE- Transactions on Industrial Electronics*, 48, 1268, 2001.

Luo, F.L., Seven self-lift DC/DC converters: voltage-lift technique, *IEE Proceedings on Electric Power Applications*, 148, 329, 2001.

Martins, D.C., Application of the Zeta converter in switch-mode power supplies, in *Proceedings of IEEE APEC'93*, U.S., 1993, p. 214.

Maksimovic, D. and Cúk, S., A general approach to synthesis and analysis of quasi-resonant converters, *IEEE Transactions on Power Electronics*, 6, 127, 1991.

Maksimovic, D. and Cúk, S., Constant-frequency control of quasi-resonant converters, *IEEE Transactions on Power Electronics*, 6, 141, 1991.

Massey, R.P. and Snyder, E.C., High voltage single-ended DC-DC converter, in *Proceedings of IEEE PESC'77*, Record. 1977, p. 156.

Newell, W.E., Power electronics-emerging from limbo, in *Proceedings of IEEE Power Electronics Specialists Conference*, U.S., 1993, p. 6.

Walker, J., Design of practical and effective active EMI filters, in *Proceedings of Poweron 11*, U.S., 1984, p. 1.

Zhu, W., Perreault, D.J., Caliskan, V., Neugebauer, T.C., Guttowski, S., and Kassakian, J.G., Design and evaluaion of an active ripple filter with Rogowski-coil current sensing, in *Proceedings of the 30th Annual IEEE-PESC'99*, U.S., 1999, p. 874.

3

Positive Output Super-Lift Luo-Converters

Voltage lift (VL) technique has been successfully employed in design of DC/DC converters, e.g., three series Luo-converters. However, the output voltage increases in arithmetic progression. Super-lift (SL) technique is more powerful than VL technique, its voltage transfer gain can be a very large number. SL technique implements the output voltage increasing in geometric progression. It effectively enhances the voltage transfer gain in power series.

3.1 Introduction

This chapter introduces positive output super lift Luo-converters. In order to differentiate these converters from existing VL converters, these converters are called *positive output super-lift Luo-converters*. There are several sub-series:

- Main series — Each circuit of the main series has only one switch S, n inductors for n^{th} stage circuit, $2n$ capacitors, and $(3n - 1)$ diodes.

- Additional series — Each circuit of the additional series has one switch S, n inductors for n^{th} stage circuit, $2(n + 1)$ capacitors, and $(3n + 1)$ diodes.

- Enhanced series — Each circuit of the enhanced series has one switch S, n inductors for n^{th} stage circuit, $4n$ capacitors, and $(5n - 1)$ diodes.

- Re-enhanced series — Each circuit of the re-enhanced series has one switch S, n inductors for n^{th} stage circuit, $6n$ capacitors, and $(7n - 1)$ diodes.

- Multiple (j)-enhanced series — Each circuit of the multiple (j times)-enhanced series has one switch S, n inductors for n^{th} stage circuit, $2(1 + j)n$ capacitors and $[(3 + 2j)n - 1]$ diodes.

In order to concentrate the voltage enlargement, assume the converters are working in steady state with continuous conduction mode (CCM). The conduction duty ratio is k, switch frequency is f, switch period is $T = 1/f$, the load is resistive load R. The input voltage and current are V_{in} and I_{in}, out voltage and current are V_O and I_O. Assume no power losses during the conversion process, $V_{in} \times I_{in} = V_O \times I_O$. The voltage transfer gain is G:

$$G = \frac{V_O}{V_{in}}$$

3.2   Main Series

The first three stages of positive output super-lift Luo-converters — main series — are shown in Figure 3.1 through Figure 3.3. For convenience, they are called elementary circuits, re-lift circuit, and triple-lift circuit respectively, and are numbered as $n = 1$, 2, and 3.

3.2.1   Elementary Circuit

The elementary circuit and its equivalent circuits during switch-on and -off are shown in Figure 3.1. The voltage across capacitor C_1 is charged to V_{in}. The current i_{L1} flowing through inductor L_1 increases with voltage V_{in} during switch-on period kT and decreases with voltage $-(V_O - 2V_{in})$ during switch-off period $(1 - k)T$. Therefore, the ripple of the inductor current i_{L1} is

$$\Delta i_{L1} = \frac{V_{in}}{L_1} kT = \frac{V_O - 2V_{in}}{L_1}(1-k)T \tag{3.1}$$

$$V_O = \frac{2-k}{1-k} V_{in} \tag{3.2}$$

The voltage transfer gain is

$$G = \frac{V_O}{V_{in}} = \frac{2-k}{1-k} \tag{3.3}$$

The input current i_{in} is equal to $(i_{L1} + i_{C1})$ during switch-on, and only equal to i_{L1} during switch-off. Capacitor current i_{C1} is equal to i_{L1} during switch-off. In steady–state, the average charge across capacitor C_1 should not change. The following relations are obtained:

$$i_{in-off} = i_{L1-off} = i_{C1-off} \quad i_{in-on} = i_{L1-on} + i_{C1-on} \quad kTi_{C1-on} = (1-k)Ti_{C1-off}$$

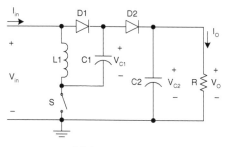

(a) Circuit diagram

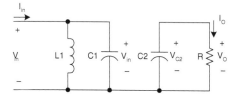

(b) Equivalent circuit during switching-on

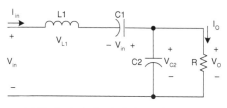

(c) Equivalent circuit during switching-off

FIGURE 3.1
Elementary circuit.

If inductance L_1 is large enough, i_{L1} is nearly equal to its average current I_{L1}. Therefore,

$$i_{in-off} = i_{C1-off} = I_{L1} \qquad i_{in-on} = I_{L1} + \frac{1-k}{k}I_{L1} = \frac{I_{L1}}{k} \qquad i_{C1-on} = \frac{1-k}{k}I_{L1}$$

and average input current

$$I_{in} = ki_{in-on} + (1-k)i_{in-off} = I_{L1} + (1-k)I_{L1} = (2-k)I_{L1} \tag{3.4}$$

Considering

$$\frac{V_{in}}{I_{in}} = (\frac{1-k}{2-k})^2 \frac{V_O}{I_O} = (\frac{1-k}{2-k})^2 R$$

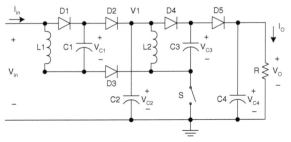

(a) Circuit diagram

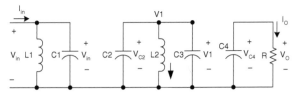

(b) Equivalent circuit during switching-on

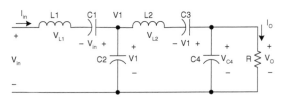

(c) Equivalent circuit during switching-off

FIGURE 3.2
Re-lift circuit.

the variation ratio of current i_{L1} through inductor L_1 is

$$\xi_1 = \frac{\Delta i_{L1}/2}{I_{L1}} = \frac{k(2-k)TV_{in}}{2L_1 I_{in}} = \frac{k(1-k)^2}{2(2-k)} \frac{R}{fL_1} \tag{3.5}$$

Usually ξ_1 is small (much lower than unity), it means this converter normally works in the continuous mode.

The ripple voltage of output voltage v_O is

$$\Delta v_O = \frac{\Delta Q}{C_2} = \frac{I_O kT}{C_2} = \frac{k}{fC_2} \frac{V_O}{R}$$

Therefore, the variation ratio of output voltage v_O is

$$\varepsilon = \frac{\Delta v_O/2}{V_O} = \frac{k}{2RfC_2} \tag{3.6}$$

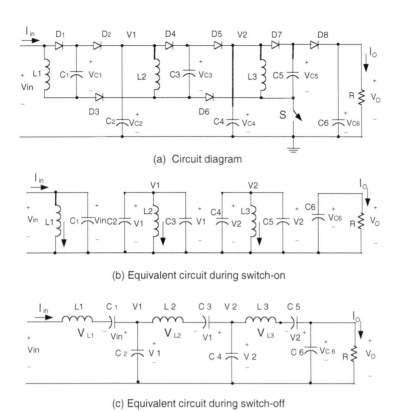

(a) Circuit diagram

(b) Equivalent circuit during switch-on

(c) Equivalent circuit during switch-off

FIGURE 3.3
Triple-lift circuit.

Usually R is in $k\Omega$, f in 10 kHz, and C_2 in μF, this ripple is smaller than 1%.

3.2.2 Re-Lift Circuit

The re-lift circuit is derived from elementary circuit by adding the parts (L_2-D_3-D_4-D_5-C_3-C_4). Its circuit diagram and equivalent circuits during switch-on and -off are shown in Figure 3.2. The voltage across capacitor C_1 is charged to V_{in}. As described in previous section the voltage V_1 across capacitor C_2 is

$$V_1 = \frac{2-k}{1-k} V_{in}$$

The voltage across capacitor C_3 is charged to V_1. The current flowing through inductor L_2 increases with voltage V_1 during switch-on period kT and decreases with voltage $-(V_O - 2V_1)$ during switch-off period $(1-k)T$. Therefore, the ripple of the inductor current i_{L2} is

$$\Delta i_{L2} = \frac{V_1}{L_2}kT = \frac{V_O - 2V_1}{L_2}(1-k)T \tag{3.7}$$

$$V_O = \frac{2-k}{1-k}V_1 = (\frac{2-k}{1-k})^2 V_{in} \tag{3.8}$$

The voltage transfer gain is

$$G = \frac{V_O}{V_{in}} = (\frac{2-k}{1-k})^2 \tag{3.9}$$

Similarly, the following relations are obtained:

$$\Delta i_{L1} = \frac{V_{in}}{L_1}kT \qquad\qquad I_{L1} = \frac{I_{in}}{2-k}$$

$$\Delta i_{L2} = \frac{V_1}{L_2}kT \qquad\qquad I_{L2} = (\frac{2-k}{1-k}-1)I_O = \frac{I_O}{1-k}$$

Therefore, the variation ratio of current i_{L1} through inductor L_1 is

$$\xi_1 = \frac{\Delta i_{L1}/2}{I_{L1}} = \frac{k(2-k)TV_{in}}{2L_1 I_{in}} = \frac{k(1-k)^4}{2(2-k)^3}\frac{R}{fL_1} \tag{3.10}$$

The variation ratio of current i_{L2} through inductor L_2 is

$$\xi_2 = \frac{\Delta i_{L2}/2}{I_{L2}} = \frac{k(1-k)TV_1}{2L_2 I_O} = \frac{k(1-k)^2 TV_O}{2(2-k)L_2 I_O} = \frac{k(1-k)^2}{2(2-k)}\frac{R}{fL_2} \tag{3.11}$$

and the variation ratio of output voltage v_O is

$$\varepsilon = \frac{\Delta v_O/2}{V_O} = \frac{k}{2RfC_4} \tag{3.12}$$

3.2.3 Triple-Lift Circuit

Triple-lift circuit is derived from re-lift circuit by double adding the parts (L_2-D_3-D_4-D_5-C_3-C_4). Its circuit diagram and equivalent circuits during switch-on and -off are shown in Figure 3.3. The voltage across capacitor C_1 is charged to V_{in}. As described before the voltage V_1 across capacitor C_2 is $V_1 = (2-k/1-k)V_{in}$, and voltage V_2 across capacitor C_4 is $V_2 = (2-k/1-k)^2 V_{in}$.

The voltage across capacitor C_5 is charged to V_2. The current flowing through inductor L_3 increases with voltage V_2 during switch-on period kT and decreases with voltage $-(V_O - 2V_2)$ during switch-off $(1 - k)T$. Therefore, the ripple of the inductor current i_{L2} is

$$\Delta i_{L3} = \frac{V_2}{L_3} kT = \frac{V_O - 2V_2}{L_3} (1-k)T \tag{3.13}$$

$$V_O = \frac{2-k}{1-k} V_2 = (\frac{2-k}{1-k})^2 V_1 = (\frac{2-k}{1-k})^3 V_{in} \tag{3.14}$$

The voltage transfer gain is

$$G = \frac{V_O}{V_{in}} = (\frac{2-k}{1-k})^3 \tag{3.15}$$

Analogously,

$$\Delta i_{L1} = \frac{V_{in}}{L_1} kT \qquad\qquad I_{L1} = \frac{I_{in}}{2-k}$$

$$\Delta i_{L2} = \frac{V_1}{L_2} kT \qquad\qquad I_{L2} = \frac{2-k}{(1-k)^2} I_O$$

$$\Delta i_{L3} = \frac{V_2}{L_3} kT \qquad\qquad I_{L3} = \frac{I_O}{1-k}$$

Therefore, the variation ratio of current i_{L1} through inductor L_1 is

$$\xi_1 = \frac{\Delta i_{L1}/2}{I_{L1}} = \frac{k(2-k)TV_{in}}{2L_1 I_{in}} = \frac{k(1-k)^6}{2(2-k)^5} \frac{R}{fL_1} \tag{3.16}$$

The variation ratio of current i_{L2} through inductor L_2 is

$$\xi_2 = \frac{\Delta i_{L2}/2}{I_{L2}} = \frac{k(1-k)^2 TV_1}{2(2-k)L_2 I_O} = \frac{kT(2-k)^4 V_O}{2(1-k)^3 L_2 I_O} = \frac{k(2-k)^4}{2(1-k)^3} \frac{R}{fL_2} \tag{3.17}$$

The variation ratio of current i_{L3} through inductor L_3 is

$$\xi_3 = \frac{\Delta i_{L3}/2}{I_{L3}} = \frac{k(1-k)TV_2}{2L_3 I_O} = \frac{k(1-k)^2 TV_O}{2(2-k)L_2 I_O} = \frac{k(1-k)^2}{2(2-k)} \frac{R}{fL_3} \tag{3.18}$$

and the variation ratio of output voltage v_O is

$$\varepsilon = \frac{\Delta v_O / 2}{V_O} = \frac{k}{2RfC_6} \tag{3.19}$$

3.2.4 Higher Order Lift Circuit

Higher order lift circuit can be designed by just multiple repeating the parts $(L_2\text{-}D_3\text{-}D_4\text{-}D_5\text{-}C_3\text{-}C_4)$. For nth order lift circuit, the final output voltage across capacitor C_{2n} is

$$V_O = (\frac{2-k}{1-k})^n V_{in}$$

The voltage transfer gain is

$$G = \frac{V_O}{V_{in}} = (\frac{2-k}{1-k})^n \tag{3.20}$$

The variation ratio of current i_{Li} through inductor L_i ($i = 1, 2, 3, \ldots n$) is

$$\xi_i = \frac{\Delta i_{Li} / 2}{I_{Li}} = \frac{k(1-k)^{2(n-i+1)}}{2(2-k)^{2(n-i)+1}} \frac{R}{fL_i} \tag{3.21}$$

and the variation ratio of output voltage v_O is

$$\varepsilon = \frac{\Delta v_O / 2}{V_O} = \frac{k}{2RfC_{2n}} \tag{3.22}$$

3.3   Additional Series

Using two diodes and two capacitors $(D_{11}\text{-}D_{12}\text{-}C_{11}\text{-}C_{12})$, a circuit called *double/enhance circuit* (DEC) can be constructed, which is shown in Figure 3.4, which is same as the Figure 1.22 but with components renumbered. If the input voltage is V_{in}, the output voltage V_O can be $2V_{in}$, or other value that is higher than V_{in}. The DEC is very versatile to enhance DC/DC converter's voltage transfer gain.

All circuits of positive output super-lift Luo-converters-additional series are derived from the corresponding circuits of the main series by adding a DEC. The first three stages of this series are shown in Figure 3.5 to Figure 3.7.

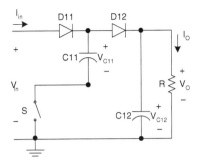

FIGURE 3.4
Double/enhanced circuit (DEC).

For convenience they are called elementary additional circuit, re-lift additional circuit, and triple-lift additional circuit respectively, and numbered as $n = 1, 2$, and 3.

3.3.1 Elementary Additional Circuit

This circuit is derived from elementary circuit by adding a DEC. Its circuit and switch-on and -off equivalent circuits are shown in Figure 3.5. The voltage across capacitor C_1 is charged to V_{in} and voltage across capacitor C_2 and C_{11} is charged to V_1. The current i_{L1} flowing through inductor L_1 increases with voltage V_{in} during switch-on period kT and decreases with voltage $-(V_O - 2V_{in})$ during switch-off $(1 - k)T$. Therefore,

$$V_1 = \frac{2-k}{1-k} V_{in} \tag{3.23}$$

and

$$V_{L1} = \frac{k}{1-k} V_{in} \tag{3.24}$$

The output voltage is

$$V_O = V_{in} + V_{L1} + V_1 = \frac{3-k}{1-k} V_{in} \tag{3.25}$$

The voltage transfer gain is

$$G = \frac{V_O}{V_{in}} = \frac{3-k}{1-k} \tag{3.26}$$

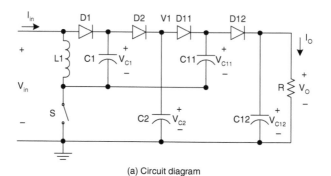

(a) Circuit diagram

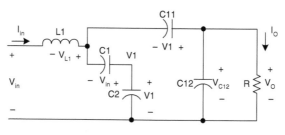

(b) Equivalent circuit during switching–on

(c) Equivalent circuit during switching–off

FIGURE 3.5
Elementary additional (enhanced) circuit.

The following relations are derived:

$$i_{in-off} = I_{L1} = i_{C11-off} + i_{C1-off} = \frac{2I_O}{1-k} \qquad\qquad i_{in-on} = i_{L1-on} + i_{C1-on} = I_{L1} + \frac{I_O}{k}$$

$$i_{C1-on} = \frac{1-k}{k} i_{C1-off} = \frac{I_O}{k} \qquad\qquad\qquad i_{C1-off} = i_{C2-off} = \frac{I_O}{1-k}$$

$$i_{C2-off} = \frac{k}{1-k} i_{C2-on} = \frac{k}{1-k} i_{C11-on} = \frac{I_O}{1-k} \qquad i_{C11-on} = \frac{1-k}{k} i_{C11-off} = \frac{I_O}{k}$$

$$i_{C11-off} = I_O + i_{C12-off} = I_O + \frac{k}{1-k} i_{C12-on} = \frac{I_O}{1-k} \qquad i_{C12-off} = \frac{k}{1-k} i_{C12-on} = \frac{kI_O}{1-k}$$

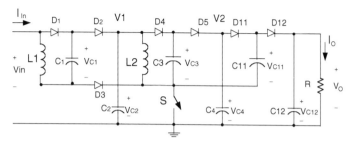

(a) Circuit diagram

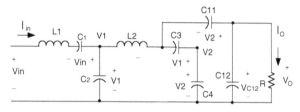

(b) Equivalent circuit during switching–on

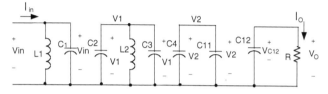

(c) Equivalent circuit during switching–off

FIGURE 3.6
Re-lift additional circuit.

If inductance L_1 is large enough, i_{L1} is nearly equal to its average current I_{L1}. Therefore,

$$i_{in-off} = I_{L1} = \frac{2I_O}{1-k} \qquad i_{in-on} = I_{L1} + \frac{I_O}{k} = (\frac{2}{1-k} + \frac{1}{k})I_O = \frac{1+k}{k(1-k)}I_O$$

Verification:

$$I_{in} = ki_{in-on} + (1-k)i_{in-off} = (\frac{1+k}{1-k} + 2)I_O = \frac{3-k}{1-k}I_O$$

Considering

$$\frac{V_{in}}{I_{in}} = (\frac{1-k}{2-k})^2 \frac{V_O}{I_O} = (\frac{1-k}{2-k})^2 R$$

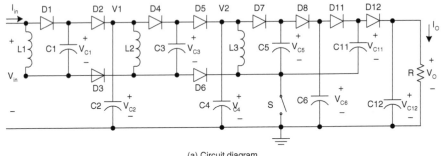

(a) Circuit diagram

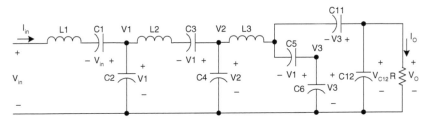

(b) Equivalent circuit during switching-on

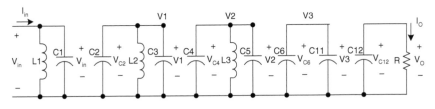

(c) Equivalent circuit during switching-off

FIGURE 3.7
Triple-lift additional circuit.

The variation of current i_{L1} is

$$\Delta i_{L1} = \frac{kTV_{in}}{L_1}$$

Therefore, the variation ratio of current i_{L1} through inductor L_1 is

$$\xi_1 = \frac{\Delta i_{L1}/2}{I_{L1}} = \frac{k(1-k)TV_{in}}{4L_1 I_O} = \frac{k(1-k)^2}{4(3-k)} \frac{R}{fL_1} \tag{3.27}$$

The ripple voltage of output voltage v_O is

$$\Delta v_O = \frac{\Delta Q}{C_{12}} = \frac{I_O kT}{C_{12}} = \frac{k}{fC_{12}} \frac{V_O}{R}$$

Therefore, the variation ratio of output voltage v_O is

$$\varepsilon = \frac{\Delta v_O / 2}{V_O} = \frac{k}{2RfC_{12}} \tag{3.28}$$

3.3.2 Re-Lift Additional Circuit

This circuit is derived from the re-lift circuit by adding a DEC. Its circuit diagram and switch-on and switch-off equivalent circuits are shown in Figure 3.6. The voltage across capacitor C_1 is charged to V_{in}. As described in previous section the voltage across C_2 is

$$V_1 = \frac{2-k}{1-k} V_{in}$$

The voltage across capacitor C_3 is charged to V_1 and voltage across capacitor C_4 and C_{11} is charged to V_2. The current flowing through inductor L_2 increases with voltage V_1 during switch-on period kT and decreases with voltage $-(V_O - 2V_1)$ during switch-off $(1-k)T$. Therefore,

$$V_2 = \frac{2-k}{1-k} V_1 = \left(\frac{2-k}{1-k}\right)^2 V_{in} \tag{3.29}$$

and

$$V_{L2} = \frac{k}{1-k} V_1 \tag{3.30}$$

The output voltage is

$$V_O = V_1 + V_{L2} + V_2 = \frac{2-k}{1-k} \frac{3-k}{1-k} V_{in} \tag{3.31}$$

The voltage transfer gain is

$$G = \frac{V_O}{V_{in}} = \frac{2-k}{1-k} \frac{3-k}{1-k} \tag{3.32}$$

Analogously,

$$\Delta i_{L1} = \frac{V_{in}}{L_1} kT \qquad\qquad I_{L1} = \frac{3-k}{(1-k)^2} I_O$$

$$\Delta i_{L2} = \frac{V_1}{L_2} kT \qquad\qquad I_{L2} = \frac{2I_O}{1-k}$$

Therefore, the variation ratio of current i_{L1} through inductor L_1 is

$$\xi_1 = \frac{\Delta i_{L1}/2}{I_{L1}} = \frac{k(1-k)^2 T V_{in}}{2(3-k)L_1 I_O} = \frac{k(1-k)^4}{2(2-k)(3-k)^2} \frac{R}{fL_1} \tag{3.33}$$

and the variation ratio of current i_{L2} through inductor L_2 is

$$\xi_2 = \frac{\Delta i_{L2}/2}{I_{L2}} = \frac{k(1-k)T V_1}{4 L_2 I_O} = \frac{k(1-k)^2}{4(3-k)} \frac{R}{fL_2} \tag{3.34}$$

The ripple voltage of output voltage v_O is

$$\Delta v_O = \frac{\Delta Q}{C_{12}} = \frac{I_O kT}{C_{12}} = \frac{k}{fC_{12}} \frac{V_O}{R}$$

Therefore, the variation ratio of output voltage v_O is

$$\varepsilon = \frac{\Delta v_O/2}{V_O} = \frac{k}{2RfC_{12}} \tag{3.35}$$

3.3.3 Triple-Lift Additional Circuit

This circuit is derived from the triple-lift circuit by adding a DEC. Its circuit diagram and equivalent circuits during switch-on and -off are shown in Figure 3.7. The voltage across capacitor C_1 is charged to V_{in}. As described in previous section the voltage across C_2 is

$$V_1 = \frac{2-k}{1-k} V_{in}$$

and voltage across C_4 is

$$V_2 = \frac{2-k}{1-k} V_1 = (\frac{2-k}{1-k})^2 V_{in}$$

The voltage across capacitor C_5 is charged to V_2 and voltage across capacitor C_6 and C_{11} is charged to V_3. The current flowing through inductor L_3 increases with voltage V_2 during switch-on period kT and decreases with voltage $-(V_O - 2V_2)$ during switch-off $(1-k)T$. Therefore,

$$V_3 = \frac{2-k}{1-k} V_2 = (\frac{2-k}{1-k})^2 V_1 = (\frac{2-k}{1-k})^3 V_{in} \tag{3.36}$$

and

$$V_{L3} = \frac{k}{1-k} V_2 \qquad (3.37)$$

The output voltage is

$$V_O = V_2 + V_{L3} + V_3 = (\frac{2-k}{1-k})^2 \frac{3-k}{1-k} V_{in} \qquad (3.38)$$

The voltage transfer gain is

$$G = \frac{V_O}{V_{in}} = (\frac{2-k}{1-k})^2 \frac{3-k}{1-k} \qquad (3.39)$$

Analogously,

$$\Delta i_{L1} = \frac{V_{in}}{L_1} kT \qquad\qquad I_{L1} = \frac{(2-k)(3-k)}{(1-k)^3} I_O$$

$$\Delta i_{L2} = \frac{V_1}{L_2} kT \qquad\qquad I_{L2} = \frac{3-k}{(1-k)^2} I_O$$

$$\Delta i_{L3} = \frac{V_2}{L_3} kT \qquad\qquad I_{L3} = \frac{2I_O}{1-k}$$

Considering

$$\frac{V_{in}}{I_{in}} = (\frac{1-k}{2-k})^2 \frac{V_O}{I_O} = (\frac{1-k}{2-k})^2 R$$

the variation ratio of current i_{L1} through inductor L_1 is

$$\xi_1 = \frac{\Delta i_{L1}/2}{I_{L1}} = \frac{k(1-k)^3 TV_{in}}{2(2-k)(3-k)L_1 I_O}$$

$$= \frac{k(1-k)^3 T}{2(2-k)(3-k)L_1 I_O} \frac{(1-k)^3}{(2-k)^2(3-k)} V_O = \frac{k(1-k)^6}{2(2-k)^3(3-k)^2} \frac{R}{fL_1} \qquad (3.40)$$

and the variation ratio of current i_{L2} through inductor L_2 is

$$\xi_2 = \frac{\Delta i_{L2}/2}{I_{L2}} = \frac{k(1-k)^2 TV_1}{2(3-k)L_2 I_O}$$

$$= \frac{k(1-k)^2 T}{2(3-k)L_2 I_O} \frac{(1-k)^2}{(2-k)(3-k)} V_O = \frac{k(1-k)^4}{2(2-k)(3-k)^2} \frac{R}{fL_2}$$

(3.41)

and the variation ratio of current i_{L3} through inductor L_3 is

$$\xi_3 = \frac{\Delta i_{L3}/2}{I_{L3}} = \frac{k(1-k)TV_2}{4L_3 I_O} = \frac{k(1-k)T}{4L_3 I_O}\frac{1-k}{3-k}V_O = \frac{k(1-k)^2}{4(3-k)}\frac{R}{fL_3}$$ (3.42)

The ripple voltage of output voltage v_O is

$$\Delta v_O = \frac{\Delta Q}{C_{12}} = \frac{I_O kT}{C_{12}} = \frac{k}{fC_{12}}\frac{V_O}{R}$$

Therefore, the variation ratio of output voltage v_O is

$$\varepsilon = \frac{\Delta v_O/2}{V_O} = \frac{k}{2RfC_{12}}$$

(3.43)

3.3.4 Higher Order Lift Additional Circuit

The higher order lift additional circuit is derived from the corresponding circuit of the main series by adding a DEC. For n^{th} order lift additional circuit, the final output voltage is

$$V_O = (\frac{2-k}{1-k})^{n-1}\frac{3-k}{1-k}V_{in}$$

The voltage transfer gain is

$$G = \frac{V_O}{V_{in}} = (\frac{2-k}{1-k})^{n-1}\frac{3-k}{1-k}$$

(3.44)

Analogously, the variation ratio of current i_{Li} through inductor L_i ($i = 1, 2, 3, \ldots n$) is

$$\xi_i = \frac{\Delta i_{Li}/2}{I_{Li}} = \frac{k(1-k)^{2(n-i+1)}}{2[2(2-k)]^{h(n-i)}(2-k)^{2(n-i)+1}(3-k)^{2u(n-i-1)}}\frac{R}{fL_i}$$ (3.45)

where

$$h(x) = \begin{cases} 0 & x > 0 \\ 1 & x \leq 0 \end{cases} \text{ is the \textbf{Hong function}}$$

and

$$u(x) = \begin{cases} 1 & x \geq 0 \\ 0 & x < 0 \end{cases} \text{ is the \textbf{unit-step function}}$$

and the variation ratio of output voltage v_O is

$$\varepsilon = \frac{\Delta v_O / 2}{V_O} = \frac{k}{2RfC_{12}} \tag{3.46}$$

3.4 Enhanced Series

All circuits of positive output super-lift Luo-converters-enhanced series — are derived from the corresponding circuits of the main series by adding a DEC in each stage circuit. The first three stages of this series are shown in Figures 3.5, 3.8, and 3.9. For convenience they are called elementary enhanced circuit, re-lift enhanced circuit, and triple-lift enhanced circuit respectively, and numbered as $n = 1$, 2 and 3.

3.4.1 Elementary Enhanced Circuit

This circuit is same as the elementary additional circuit shown in Figure 3.5. The output voltage is

$$V_O = V_{in} + V_{L1} + V_1 = \frac{3-k}{1-k} V_{in} \tag{3.25}$$

The voltage transfer gain is

$$G = \frac{V_O}{V_{in}} = \frac{3-k}{1-k} \tag{3.26}$$

The variation of current i_{L1} is

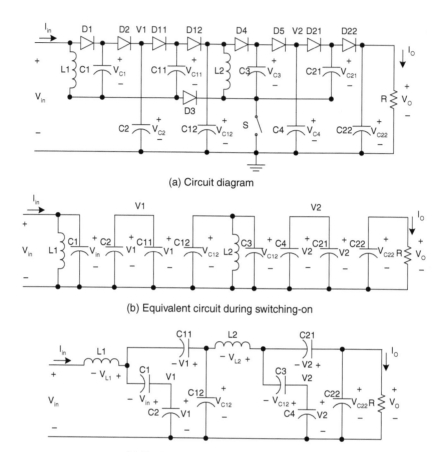

(a) Circuit diagram

(b) Equivalent circuit during switching-on

(c) Equivalent circuit during switching-off

FIGURE 3.8
Re-lift enhanced circuit.

$$\Delta i_{L1} = \frac{kTV_{in}}{L_1}$$

Therefore, the variation ratio of current i_{L1} through inductor L_1 is

$$\xi_1 = \frac{\Delta i_{L1}/2}{I_{L1}} = \frac{k(1-k)TV_{in}}{4L_1 I_O} = \frac{k(1-k)^2}{4(3-k)} \frac{R}{fL_1} \tag{3.27}$$

The ripple voltage of output voltage v_O is

$$\Delta v_O = \frac{\Delta Q}{C_{12}} = \frac{I_O kT}{C_{12}} = \frac{k}{fC_{12}} \frac{V_O}{R}$$

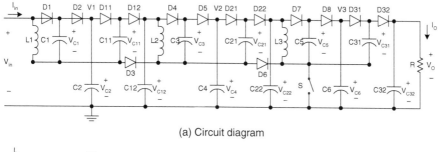

(a) Circuit diagram

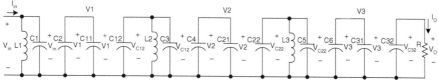

(b) Equivalent circuit during switching–on

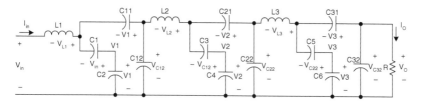

(c) Equivalent circuit during switching–off

FIGURE 3.9
Triple-lift enhanced circuit.

Therefore, the variation ratio of output voltage v_O is

$$\varepsilon = \frac{\Delta v_O / 2}{V_O} = \frac{k}{2RfC_{12}} \tag{3.28}$$

3.4.2 Re-Lift Enhanced Circuit

This circuit is derived from the re-lift circuit of the main series by adding the DEC in each stage circuit. Its circuit diagram and switch-on and switch-off equivalent circuits are shown in Figure 3.8. As described in the previous section the voltage across capacitor C_{12} is charged to

$$V_{C12} = \frac{3-k}{1-k} V_{in}$$

The voltage across capacitor C_3 is charged to V_{C12} and voltage across capacitor C_4 and C_{21} is charged to V_{C4},

$$V_{C4} = \frac{2-k}{1-k}V_{C12} = \frac{2-k}{1-k}\frac{3-k}{1-k}V_{in} \tag{3.47}$$

The current flowing through inductor L_2 increases with voltage V_{C12} during switch-on period kT and decreases with voltage $-(V_O - V_{C4} - V_{C12})$ during switch-off $(1-k)T$. Therefore,

$$\Delta i_{L2} = \frac{k}{L_2}V_{C12} = \frac{1-k}{L_2}(V_O - V_{C4} - V_{C12}) \tag{3.48}$$

$$V_O = \frac{3-k}{1-k}V_{C12} = (\frac{3-k}{1-k})^2 V_{in} \tag{3.49}$$

The voltage transfer gain is

$$G = \frac{V_O}{V_{in}} = (\frac{3-k}{1-k})^2 \tag{3.50}$$

Analogously,

$$\Delta i_{L1} = \frac{V_{in}}{L_1}kT \qquad\qquad I_{L1} = \frac{3-k}{(1-k)^2}I_O$$

$$\Delta i_{L2} = \frac{V_1}{L_2}kT \qquad\qquad I_{L2} = \frac{2I_O}{1-k}$$

Therefore, the variation ratio of current i_{L1} through inductor L_1 is

$$\xi_1 = \frac{\Delta i_{L1}/2}{I_{L1}} = \frac{k(1-k)^2 T V_{in}}{2(3-k)L_1 I_O} = \frac{k(1-k)^4}{2(2-k)(3-k)^2}\frac{R}{fL_1} \tag{3.51}$$

and the variation ratio of current i_{L2} through inductor L_2 is

$$\xi_2 = \frac{\Delta i_{L2}/2}{I_{L2}} = \frac{k(1-k)TV_1}{4L_2 I_O} = \frac{k(1-k)^2}{4(3-k)}\frac{R}{fL_2} \tag{3.52}$$

The ripple voltage of output voltage v_O is

$$\Delta v_O = \frac{\Delta Q}{C_{22}} = \frac{I_O kT}{C_{22}} = \frac{k}{fC_{22}}\frac{V_O}{R}$$

Therefore, the variation ratio of output voltage v_O is

$$\varepsilon = \frac{\Delta v_O/2}{V_O} = \frac{k}{2RfC_{22}} \tag{3.53}$$

3.4.3 Triple-Lift Enhanced Circuit

This circuit is derived from triple-lift circuit of the main series by adding the DEC in each stage circuit. Its circuit diagram and equivalent circuits during switch-on and -off are shown in Figure 3.9. As described in the previous section the voltage across capacitor C_{12} is charged to $V_{C12} = (3 - k/1 - k)V_{in}$, and the voltage across capacitor C_{22} is charged to $V_{C22} = (3 - k/1 - k)^2 V_{in}$.

The voltage across capacitor C_5 is charged to V_{C22} and voltage across capacitor C_6 and C_{31} is charged to V_{C6}.

$$V_{C6} = \frac{2-k}{1-k} V_{C22} = \frac{2-k}{1-k}(\frac{3-k}{1-k})^2 V_{in} \tag{3.54}$$

The current flowing through inductor L_3 increases with voltage V_{C22} during switch-on period kT and decreases with voltage $-(V_O - V_{C6} - V_{C22})$ during switch-off $(1 - k)T$.

Therefore,

$$\Delta i_{L3} = \frac{k}{L_3} V_{C22} = \frac{1-k}{L_3}(V_O - V_{C6} - V_{C22}) \tag{3.55}$$

$$V_O = \frac{3-k}{1-k} V_{C22} = (\frac{3-k}{1-k})^3 V_{in} \tag{3.56}$$

The voltage transfer gain is

$$G = \frac{V_O}{V_{in}} = (\frac{3-k}{1-k})^3 \tag{3.57}$$

Analogously,

$$\Delta i_{L1} = \frac{V_{in}}{L_1} kT \qquad\qquad I_{L1} = \frac{(2-k)(3-k)}{(1-k)^3} I_O$$

$$\Delta i_{L2} = \frac{V_1}{L_2} kT \qquad\qquad I_{L2} = \frac{3-k}{(1-k)^2} I_O$$

$$\Delta i_{L3} = \frac{V_2}{L_3} kT \qquad\qquad I_{L3} = \frac{2I_O}{1-k}$$

Considering

$$\frac{V_{in}}{I_{in}} = (\frac{1-k}{2-k})^2 \frac{V_O}{I_O} = (\frac{1-k}{2-k})^2 R$$

the variation ratio of current i_{L1} through inductor L_1 is

$$\xi_1 = \frac{\Delta i_{L1}/2}{I_{L1}} = \frac{k(1-k)^3 T V_{in}}{2(2-k)(3-k)L_1 I_O}$$

$$= \frac{k(1-k)^3 T}{2(2-k)(3-k)L_1 I_O} \frac{(1-k)^3}{(2-k)^2(3-k)} V_O = \frac{k(1-k)^6}{2(2-k)^3(3-k)^2} \frac{R}{fL_1}$$

(3.58)

The variation ratio of current i_{L2} through inductor L_2 is

$$\xi_2 = \frac{\Delta i_{L2}/2}{I_{L2}} = \frac{k(1-k)^2 T V_1}{2(3-k)L_2 I_O}$$

$$= \frac{k(1-k)^2 T}{2(3-k)L_2 I_O} \frac{(1-k)^2}{(2-k)(3-k)} V_O = \frac{k(1-k)^4}{2(2-k)(3-k)^2} \frac{R}{fL_2}$$

(3.59)

and the variation ratio of current i_{L3} through inductor L_3 is

$$\xi_3 = \frac{\Delta i_{L3}/2}{I_{L3}} = \frac{k(1-k)T V_2}{4L_3 I_O}$$

$$= \frac{k(1-k)T}{4L_3 I_O} \frac{1-k}{3-k} V_O = \frac{k(1-k)^2}{4(3-k)} \frac{R}{fL_3}$$

(3.60)

The ripple voltage of output voltage v_O is

$$\Delta v_O = \frac{\Delta Q}{C_{32}} = \frac{I_O kT}{C_{32}} = \frac{k}{fC_{32}} \frac{V_O}{R}$$

Therefore, the variation ratio of output voltage v_O is

$$\varepsilon = \frac{\Delta v_O/2}{V_O} = \frac{k}{2RfC_{32}}$$

(3.61)

3.4.4 Higher Order Lift Enhanced Circuit

The higher order lift enhanced circuit is derived from the corresponding circuit of the main series by adding the DEC in each stage circuit. For the nth order lift enhanced circuit, the final output voltage is $V_O = (3 - k/1 - k)^n V_{in}$. The voltage transfer gain is

$$G = \frac{V_O}{V_{in}} = (\frac{3-k}{1-k})^n \tag{3.62}$$

Analogously, the variation ratio of current i_{Li} through inductor L_i (i = 1, 2, 3, ...n) is

$$\xi_i = \frac{\Delta i_{Li}/2}{I_{Li}} = \frac{k(1-k)^{2(n-i+1)}}{2[2(2-k)]^{h(n-i)}(2-k)^{2(n-i)+1}(3-k)^{2u(n-i-1)}} \frac{R}{fL_i} \tag{3.63}$$

where

$$h(x) = \begin{cases} 0 & x > 0 \\ 1 & x \le 0 \end{cases} \quad \text{is the \textbf{Hong function}}$$

and

$$u(x) = \begin{cases} 1 & x \ge 0 \\ 0 & x < 0 \end{cases} \quad \text{is the \textbf{unit-step function}}$$

and the variation ratio of output voltage v_O is

$$\varepsilon = \frac{\Delta v_O/2}{V_O} = \frac{k}{2RfC_{n2}} \tag{3.64}$$

3.5 Re-Enhanced Series

All circuits of positive output super-lift Luo-converters-re-enhanced series — are derived from the corresponding circuits of the main series by adding the DEC twice in each stage circuit.

The first three stages of this series are shown in Figure 3.10 to Figure 3.12. For convenience they are named elementary re-enhanced circuits, re-lift re-enhanced circuits, and triple-lift re-enhanced circuits respectively, and numbered as n = 1, 2 and 3.

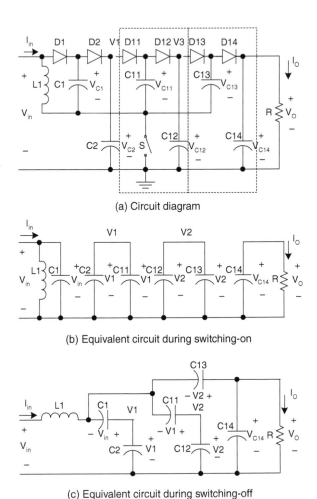

(a) Circuit diagram

(b) Equivalent circuit during switching-on

(c) Equivalent circuit during switching-off

FIGURE 3.10
Elementary re-enhanced circuit.

3.5.1 Elementary Re-Enhanced Circuit

This circuit is derived from the elementary circuit by adding the DEC twice. Its circuit and switch-on and -off equivalent circuits are shown in Figure 3.10.

The output voltage is

$$V_O = V_{in} + V_{L1} + V_{C12} = \frac{4-k}{1-k} V_{in} \qquad (3.65)$$

The voltage transfer gain is

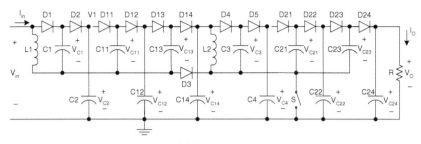

(a) Circuit diagram

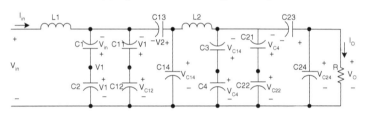

(b) Equivalent circuit during switching–on

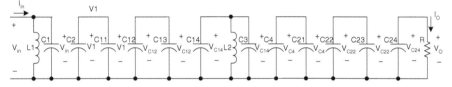

(c) Equivalent circuit during switching–off

FIGURE 3.11
Re-lift re-enhanced circuit.

$$G = \frac{V_O}{V_{in}} = \frac{4-k}{1-k} \tag{3.66}$$

where

$$V_{C2} = \frac{2-k}{1-k} V_{in} \tag{3.67}$$

$$V_{C12} = \frac{3-k}{1-k} V_{in} \tag{3.68}$$

and

$$V_{L1} = \frac{k}{1-k} V_{in} \tag{3.69}$$

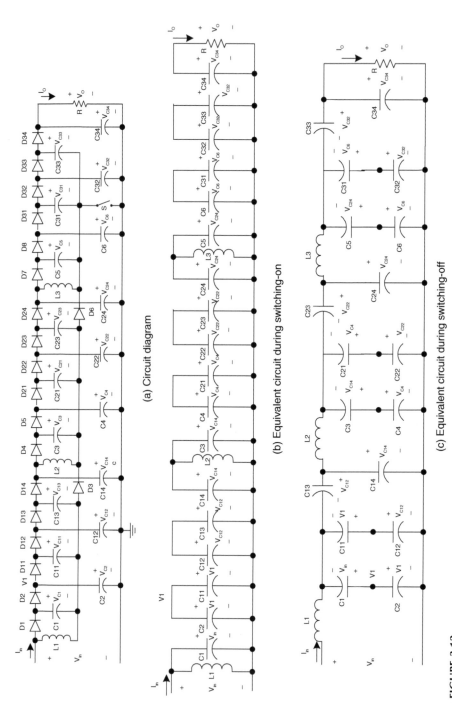

(a) Circuit diagram

(b) Equivalent circuit during switching-on

(c) Equivalent circuit during switching-off

FIGURE 3.12
Triple-lift re-enhanced circuit.

The following relations are obtained:

$$i_{in-off} = I_{L1} = i_{C11-off} + i_{C1-off} = \frac{2I_O}{1-k} \qquad i_{in-on} = i_{L1-on} + i_{C1-on} = I_{L1} + \frac{I_O}{k}$$

$$i_{C1-on} = \frac{1-k}{k} i_{C1-off} = \frac{I_O}{k} \qquad i_{C1-off} = i_{C2-off} = \frac{I_O}{1-k}$$

$$i_{C2-off} = \frac{k}{1-k} i_{C2-on} = \frac{k}{1-k} i_{C11-on} = \frac{I_O}{1-k} \qquad i_{C11-on} = \frac{1-k}{k} i_{C11-off} = \frac{I_O}{k}$$

$$i_{C11-off} = I_O + i_{C12-off} = I_O + \frac{k}{1-k} i_{C12-on} = \frac{I_O}{1-k} \qquad i_{C12-off} = \frac{k}{1-k} i_{C12-on} = \frac{kI_O}{1-k}$$

If inductance L_1 is large enough, i_{L1} is nearly equal to its average current I_{L1}. Therefore,

$$i_{in-off} = I_{L1} = \frac{2I_O}{1-k} \qquad i_{in-on} = I_{L1} + \frac{I_O}{k} = (\frac{2}{1-k} + \frac{1}{k})I_O = \frac{1+k}{k(1-k)} I_O$$

Verification:

$$I_{in} = ki_{in-on} + (1-k)i_{in-off} = (\frac{1+k}{1-k} + 2)I_O = \frac{3-k}{1-k} I_O$$

Considering

$$\frac{V_{in}}{I_{in}} = (\frac{1-k}{2-k})^2 \frac{V_O}{I_O} = (\frac{1-k}{2-k})^2 R$$

the variation of current i_{L1} is

$$\Delta i_{L1} = \frac{kTV_{in}}{L_1}$$

Therefore, the variation ratio of current i_{L1} through inductor L_1 is

$$\xi_1 = \frac{\Delta i_{L1}/2}{I_{L1}} = \frac{k(1-k)TV_{in}}{4L_1 I_O} = \frac{k(1-k)^2}{4(3-k)} \frac{R}{fL_1} \qquad (3.70)$$

The ripple voltage of output voltage v_O is

$$\Delta v_O = \frac{\Delta Q}{C_{14}} = \frac{I_O kT}{C_{14}} = \frac{k}{fC_{14}} \frac{V_O}{R}$$

Therefore, the variation ratio of output voltage v_O is

$$\varepsilon = \frac{\Delta v_O / 2}{V_O} = \frac{k}{2RfC_{14}} \tag{3.71}$$

3.5.2 Re-Lift Re-Enhanced Circuit

This circuit is derived from the re-lift circuit of the main series by adding the DEC twice in each stage circuit. Its circuit and switch-on and -off equivalent circuits are shown in Figure 3.11. The voltage across capacitor C_{14} is

$$V_{C14} = \frac{4-k}{1-k} V_{in} \tag{3.72}$$

By the same analysis

$$V_O = \frac{4-k}{1-k} V_{C14} = \left(\frac{4-k}{1-k}\right)^2 V_{in} \tag{3.73}$$

The voltage transfer gain is

$$G = \frac{V_O}{V_{in}} = \left(\frac{4-k}{1-k}\right)^2 \tag{3.74}$$

Analogously,

$$\Delta i_{L1} = \frac{V_{in}}{L_1} kT \qquad\qquad I_{L1} = \frac{3-k}{(1-k)^2} I_O$$

$$\Delta i_{L2} = \frac{V_1}{L_2} kT \qquad\qquad I_{L2} = \frac{2I_O}{1-k}$$

Therefore, the variation ratio of current i_{L1} through inductor L_1 is

$$\xi_1 = \frac{\Delta i_{L1} / 2}{I_{L1}} = \frac{k(1-k)^2 TV_{in}}{2(3-k)L_1 I_O} = \frac{k(1-k)^4}{2(2-k)(3-k)^2} \frac{R}{fL_1} \tag{3.75}$$

The variation ratio of current i_{L2} through inductor L_2 is

$$\xi_2 = \frac{\Delta i_{L2}/2}{I_{L2}} = \frac{k(1-k)TV_1}{4L_2 I_O} = \frac{k(1-k)^2}{4(3-k)} \frac{R}{fL_2} \tag{3.76}$$

The ripple voltage of output voltage v_O is

$$\Delta v_O = \frac{\Delta Q}{C_{24}} = \frac{I_O kT}{C_{24}} = \frac{k}{fC_{24}} \frac{V_O}{R}$$

Therefore, the variation ratio of output voltage v_O is

$$\varepsilon = \frac{\Delta v_O / 2}{V_O} = \frac{k}{2RfC_{24}} \tag{3.77}$$

3.5.3 Triple-Lift Re-Enhanced Circuit

This circuit is derived from triple-lift circuit of the main series by adding the DEC twice in each stage circuit. Its circuit and switch-on and -off equivalent circuits are shown in Figure 3.12. The voltage across capacitor C_{14} is

$$V_{C14} = \frac{4-k}{1-k} V_{in} \tag{3.78}$$

The voltage across capacitor C_{24} is

$$V_{C24} = \left(\frac{4-k}{1-k}\right)^2 V_{in} \tag{3.79}$$

By the same analysis

$$V_O = \frac{4-k}{1-k} V_{C24} = \left(\frac{4-k}{1-k}\right)^3 V_{in} \tag{3.80}$$

The voltage transfer gain is

$$G = \frac{V_O}{V_{in}} = \left(\frac{4-k}{1-k}\right)^3 \tag{3.81}$$

Analogously,

$$\Delta i_{L1} = \frac{V_{in}}{L_1}kT \qquad\qquad I_{L1} = \frac{(2-k)(3-k)}{(1-k)^3}I_O$$

$$\Delta i_{L2} = \frac{V_1}{L_2}kT \qquad\qquad I_{L2} = \frac{3-k}{(1-k)^2}I_O$$

$$\Delta i_{L3} = \frac{V_2}{L_3}kT \qquad\qquad I_{L3} = \frac{2I_O}{1-k}$$

Considering

$$\frac{V_{in}}{I_{in}} = (\frac{1-k}{2-k})^2\frac{V_O}{I_O} = (\frac{1-k}{2-k})^2 R$$

the variation ratio of current i_{L1} through inductor L_1 is

$$\begin{aligned}
\xi_1 &= \frac{\Delta i_{L1}/2}{I_{L1}} = \frac{k(1-k)^3 T V_{in}}{2(2-k)(3-k)L_1 I_O} \\
&= \frac{k(1-k)^3 T}{2(2-k)(3-k)L_1 I_O}\frac{(1-k)^3}{(2-k)^2(3-k)}V_O = \frac{k(1-k)^6}{2(2-k)^3(3-k)^2}\frac{R}{fL_1}
\end{aligned} \tag{3.82}$$

The variation ratio of current i_{L2} through inductor L_2 is

$$\begin{aligned}
\xi_2 &= \frac{\Delta i_{L2}/2}{I_{L2}} = \frac{k(1-k)^2 T V_1}{2(3-k)L_2 I_O} \\
&= \frac{k(1-k)^2 T}{2(3-k)L_2 I_O}\frac{(1-k)^2}{(2-k)(3-k)}V_O = \frac{k(1-k)^4}{2(2-k)(3-k)^2}\frac{R}{fL_2}
\end{aligned} \tag{3.83}$$

The variation ratio of current i_{L3} through inductor L_3 is

$$\xi_3 = \frac{\Delta i_{L3}/2}{I_{L3}} = \frac{k(1-k)T V_2}{4L_3 I_O} = \frac{k(1-k)T}{4L_3 I_O}\frac{1-k}{3-k}V_O = \frac{k(1-k)^2}{4(3-k)}\frac{R}{fL_3} \tag{3.84}$$

The ripple voltage of output voltage v_O is

$$\Delta v_O = \frac{\Delta Q}{C_{34}} = \frac{I_O kT}{C_{34}} = \frac{k}{fC_{34}}\frac{V_O}{R}$$

Therefore, the variation ratio of output voltage v_O is

$$\varepsilon = \frac{\Delta v_O / 2}{V_O} = \frac{k}{2RfC_{34}} \tag{3.85}$$

3.5.4 Higher Order Lift Re-Enhanced Circuit

Higher order lift additional circuits are derived from the corresponding circuit of the main series by adding DEC twice in each stage circuit. For the nth order lift additional circuit, the final output voltage is

$$V_O = (\frac{4-k}{1-k})^n V_{in}$$

The voltage transfer gain is

$$G = \frac{V_O}{V_{in}} = (\frac{4-k}{1-k})^n \tag{3.86}$$

Analogously, the variation ratio of current i_{Li} through inductor L_i ($i = 1, 2, 3, \ldots n$) is

$$\xi_i = \frac{\Delta i_{Li} / 2}{I_{Li}} = \frac{k(1-k)^{2(n-i+1)}}{2[2(2-k)]^{h(n-i)}(2-k)^{2(n-i)+1}(3-k)^{2u(n-i-1)}} \frac{R}{fL_i} \tag{3.87}$$

where

$$h(x) = \begin{cases} 0 & x > 0 \\ 1 & x \le 0 \end{cases} \quad \text{is the \textbf{Hong function}}$$

and

$$u(x) = \begin{cases} 1 & x \ge 0 \\ 0 & x < 0 \end{cases} \quad \text{is the \textbf{unit-step function}}$$

and the variation ratio of output voltage v_O is

$$\varepsilon = \frac{\Delta v_O / 2}{V_O} = \frac{k}{2RfC_{n4}} \tag{3.88}$$

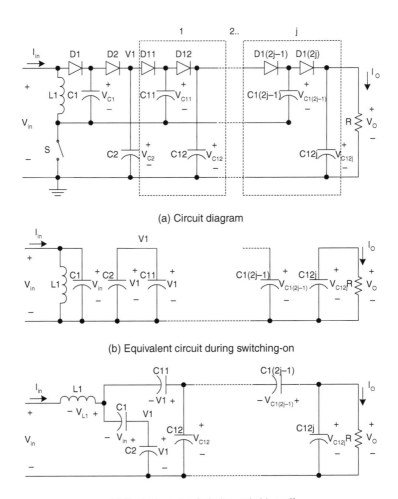

(a) Circuit diagram

(b) Equivalent circuit during switching-on

(c) Equivalent circuit during switching-off

FIGURE 3.13
Elementary multiple-enhanced circuit.

3.6   Multiple-Enhanced Series

All circuits of positive output super-lift Luo-converters — multiple-enhanced series — are derived from the corresponding circuits of the main series by adding the DEC multiple (j) times in each stage circuit. The first three stages of this series are shown in Figure 3.13 through Figure 3.15. For convenience they are called elementary multiple-enhanced circuits, re-lift multiple-enhanced circuits, and triple-lift multiple-enhanced circuits respectively, and numbered as $n = 1, 2,$ and 3.

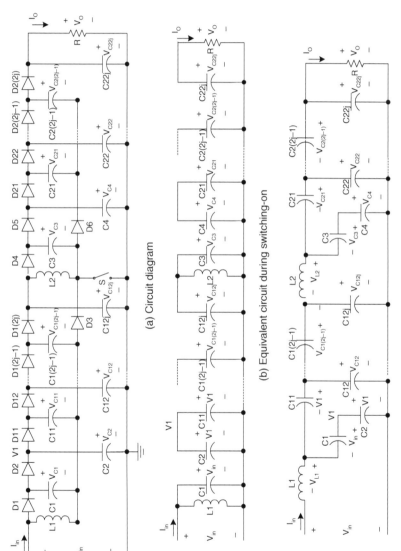

(a) Circuit diagram

(b) Equivalent circuit during switching-on

(c) Equivalent circuit during switching-off

FIGURE 3.14
Re-lift multiple-enhanced circuit.

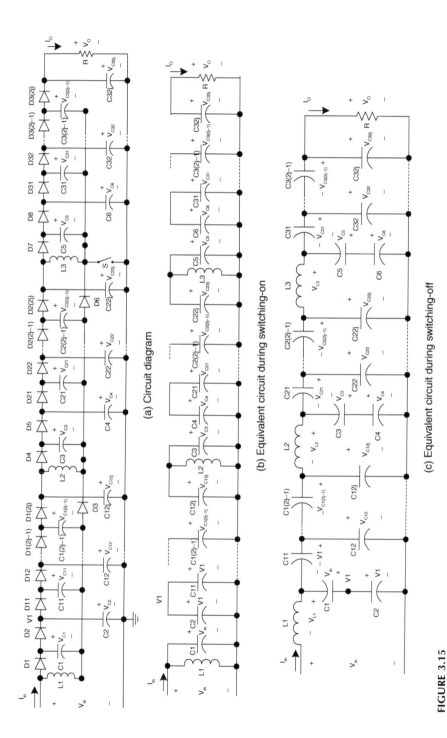

FIGURE 3.15

Triple-lift multiple-enhanced circuit.

3.6.1 Elementary Multiple-Enhanced Circuit

This circuit is derived from the elementary circuit of the main series by adding the DEC multiple (*j*) times. Its circuit and switch-on and -off equivalent circuits are shown in Figure 3.13. The output voltage is

$$V_O = \frac{j+2-k}{1-k}V_{in} \tag{3.89}$$

The voltage transfer gain is

$$G = \frac{V_O}{V_{in}} = \frac{j+2-k}{1-k} \tag{3.90}$$

Following relations are obtained:

$$i_{in-off} = I_{L1} = i_{C11-off} + i_{C1-off} = \frac{2I_O}{1-k} \qquad i_{in-on} = i_{L1-on} + i_{C1-on} = I_{L1} + \frac{I_O}{k}$$

$$i_{C1-on} = \frac{1-k}{k}i_{C1-off} = \frac{I_O}{k} \qquad i_{C1-off} = i_{C2-off} = \frac{I_O}{1-k}$$

$$i_{C2-off} = \frac{k}{1-k}i_{C2-on} = \frac{k}{1-k}i_{C11-on} = \frac{I_O}{1-k} \qquad i_{C11-on} = \frac{1-k}{k}i_{C11-off} = \frac{I_O}{k}$$

$$i_{C11-off} = I_O + i_{C12-off} = I_O + \frac{k}{1-k}i_{C12-on} = \frac{I_O}{1-k} \qquad i_{C12-off} = \frac{k}{1-k}i_{C12-on} = \frac{kI_O}{1-k}$$

If inductance L_1 is large enough, i_{L1} is nearly equal to its average current I_{L1}. Therefore,

$$i_{in-off} = I_{L1} = \frac{2I_O}{1-k} \qquad i_{in-on} = I_{L1} + \frac{I_O}{k} = (\frac{2}{1-k} + \frac{1}{k})I_O = \frac{1+k}{k(1-k)}I_O$$

Verification:

$$I_{in} = ki_{in-on} + (1-k)i_{in-off} = (\frac{1+k}{1-k} + 2)I_O = \frac{3-k}{1-k}I_O$$

Considering

$$\frac{V_{in}}{I_{in}} = (\frac{1-k}{2-k})^2\frac{V_O}{I_O} = (\frac{1-k}{2-k})^2 R$$

the variation of current i_{L1} is

$$\Delta i_{L1} = \frac{kTV_{in}}{L_1}$$

Therefore, the variation ratio of current i_{L1} through inductor L_1 is

$$\xi_1 = \frac{\Delta i_{L1}/2}{I_{L1}} = \frac{k(1-k)TV_{in}}{4L_1 I_O} = \frac{k(1-k)^2}{4(3-k)} \frac{R}{fL_1} \tag{3.91}$$

The ripple voltage of output voltage v_O is

$$\Delta v_O = \frac{\Delta Q}{C_{12j}} = \frac{I_O kT}{C_{12j}} = \frac{k}{fC_{12j}} \frac{V_O}{R}$$

Therefore, the variation ratio of output voltage v_O is

$$\varepsilon = \frac{\Delta v_O/2}{V_O} = \frac{k}{2RfC_{12j}} \tag{3.92}$$

3.6.2 Re-Lift Multiple-Enhanced Circuit

This circuit is derived from the re-lift circuit of the main series by adding the DEC multiple (j) times in each stage circuit. Its circuit diagram and switch-on and switch-off equivalent circuits are shown in Figure 3.14. The voltage across capacitor C_{12j} is

$$V_{C12j} = \frac{j+2-k}{1-k} V_{in} \tag{3.93}$$

The output voltage across capacitor C_{22j} is

$$V_O = V_{C22j} = (\frac{j+2-k}{1-k})^2 V_{in} \tag{3.94}$$

The voltage transfer gain is

$$G = \frac{V_O}{V_{in}} = (\frac{j+2-k}{1-k})^2 \tag{3.95}$$

Analogously,

$$\Delta i_{L1} = \frac{V_{in}}{L_1} kT \qquad\qquad I_{L1} = \frac{3-k}{(1-k)^2} I_O$$

$$\Delta i_{L2} = \frac{V_1}{L_2} kT \qquad\qquad I_{L2} = \frac{2I_O}{1-k}$$

Therefore, the variation ratio of current i_{L1} through inductor L_1 is

$$\xi_1 = \frac{\Delta i_{L1}/2}{I_{L1}} = \frac{k(1-k)^2 TV_{in}}{2(3-k)L_1 I_O} = \frac{k(1-k)^4}{2(2-k)(3-k)^2} \frac{R}{fL_1} \qquad (3.96)$$

and the variation ratio of current i_{L2} through inductor L_2 is

$$\xi_2 = \frac{\Delta i_{L2}/2}{I_{L2}} = \frac{k(1-k)TV_1}{4L_2 I_O} = \frac{k(1-k)^2}{4(3-k)} \frac{R}{fL_2} \qquad (3.97)$$

The ripple voltage of output voltage v_O is

$$\Delta v_O = \frac{\Delta Q}{C_{22j}} = \frac{I_O kT}{C_{22j}} = \frac{k}{fC_{22j}} \frac{V_O}{R}$$

Therefore, the variation ratio of output voltage v_O is

$$\varepsilon = \frac{\Delta v_O/2}{V_O} = \frac{k}{2RfC_{22j}} \qquad (3.98)$$

3.6.3 Triple-Lift Multiple-Enhanced Circuit

This circuit is derived from the triple-lift circuit of the main series by adding the DEC multiple (j) times in each stage circuit. Its circuit and switch-on and -off equivalent circuits are shown in Figure 3.15. The voltage across capacitor C_{12j} is

$$V_{C12j} = \frac{j+2-k}{1-k} V_{in} \qquad (3.99)$$

The voltage across capacitor C_{22j} is

$$V_{C22j} = (\frac{j+2-k}{1-k})^2 V_{in} \qquad (3.100)$$

Same analysis,

$$V_O = \frac{j+2-k}{1-k} V_{C22j} = (\frac{j+2-k}{1-k})^3 V_{in} \qquad (3.101)$$

The voltage transfer gain is

$$G = \frac{V_O}{V_{in}} = (\frac{j+2-k}{1-k})^3 \qquad (3.102)$$

Analogously,

$$\Delta i_{L1} = \frac{V_{in}}{L_1} kT \qquad\qquad I_{L1} = \frac{(2-k)(3-k)}{(1-k)^3} I_O$$

$$\Delta i_{L2} = \frac{V_1}{L_2} kT \qquad\qquad I_{L2} = \frac{3-k}{(1-k)^2} I_O$$

$$\Delta i_{L3} = \frac{V_2}{L_3} kT \qquad\qquad I_{L3} = \frac{2I_O}{1-k}$$

Considering

$$\frac{V_{in}}{I_{in}} = (\frac{1-k}{2-k})^2 \frac{V_O}{I_O} = (\frac{1-k}{2-k})^2 R$$

the variation ratio of current i_{L1} through inductor L_1 is

$$\xi_1 = \frac{\Delta i_{L1}/2}{I_{L1}} = \frac{k(1-k)^3 T V_{in}}{2(2-k)(3-k)L_1 I_O}$$

$$= \frac{k(1-k)^3 T}{2(2-k)(3-k)L_1 I_O} \frac{(1-k)^3}{(2-k)^2(3-k)} V_O = \frac{k(1-k)^6}{2(2-k)^3(3-k)^2} \frac{R}{fL_1} \qquad (3.103)$$

The variation ratio of current i_{L2} through inductor L_2 is

$$\xi_2 = \frac{\Delta i_{L2}/2}{I_{L2}} = \frac{k(1-k)^2 T V_1}{2(3-k)L_2 I_O}$$

$$= \frac{k(1-k)^2 T}{2(3-k)L_2 I_O} \frac{(1-k)^2}{(2-k)(3-k)} V_O = \frac{k(1-k)^4}{2(2-k)(3-k)^2} \frac{R}{fL_2} \qquad (3.104)$$

The variation ratio of current i_{L3} through inductor L_3 is

$$\xi_3 = \frac{\Delta i_{L3}/2}{I_{L3}} = \frac{k(1-k)TV_2}{4L_3 I_O} = \frac{k(1-k)T}{4L_3 I_O}\frac{1-k}{3-k}V_O = \frac{k(1-k)^2}{4(3-k)}\frac{R}{fL_3} \quad (3.105)$$

The ripple voltage of output voltage v_O is

$$\Delta v_O = \frac{\Delta Q}{C_{32j}} = \frac{I_O kT}{C_{32j}} = \frac{k}{fC_{32j}}\frac{V_O}{R}$$

Therefore, the variation ratio of output voltage v_O is

$$\varepsilon = \frac{\Delta v_O/2}{V_O} = \frac{k}{2RfC_{32j}} \quad (3.106)$$

3.6.4 Higher Order Lift Multiple-Enhanced Circuit

Higher order lift multiple-enhanced circuits can be derived from the corresponding circuit of the main series converters by adding the DEC multiple (j) times in each stage circuit. For the nth order lift additional circuit, the final output voltage is

$$V_O = \left(\frac{j+2-k}{1-k}\right)^n V_{in}$$

The voltage transfer gain is

$$G = \frac{V_O}{V_{in}} = \left(\frac{j+2-k}{1-k}\right)^n \quad (3.107)$$

Analogously, the variation ratio of current i_{Li} through inductor L_i ($i = 1, 2, 3, \dots n$) is

$$\xi_i = \frac{\Delta i_{Li}/2}{I_{Li}} = \frac{k(1-k)^{2(n-i+1)}}{2[2(2-k)]^{h(n-i)}(2-k)^{2(n-i)+1}(3-k)^{2u(n-i-1)}}\frac{R}{fL_i} \quad (3.108)$$

where

$$h(x) = \begin{cases} 0 & x > 0 \\ 1 & x \le 0 \end{cases} \quad \text{is the } \textbf{Hong function}$$

and

$$u(x) = \begin{cases} 1 & x \geq 0 \\ 0 & x < 0 \end{cases} \quad \text{is the \textbf{unit-step function}}$$

The variation ratio of output voltage v_O is

$$\varepsilon = \frac{\Delta v_O / 2}{V_O} = \frac{k}{2RfC_{n2j}} \qquad (3.109)$$

3.7 Summary of Positive Output Super-Lift Luo-Converters

All circuits of positive output super-lift Luo-converters can be shown in Figure 3.16 as the family tree. From the analysis in previous sections, the common formula to calculate the output voltage is presented:

$$V_O = \begin{cases} (\dfrac{2-k}{1-k})^n V_{in} & main\_series \\[2mm] (\dfrac{2-k}{1-k})^{n-1}(\dfrac{3-k}{1-k})V_{in} & additional\_series \\[2mm] (\dfrac{3-k}{1-k})^n V_{in} & enhanced\_series \\[2mm] (\dfrac{4-k}{1-k})^n V_{in} & re\text{-}enhanced\_series \\[2mm] (\dfrac{j+2-k}{1-k})^n V_{in} & multiple\text{-}enhanced\_series \end{cases} \qquad (3.110)$$

The voltage transfer gain is

$$G = \frac{V_O}{V_{in}} = \begin{cases} (\dfrac{2-k}{1-k})^n & main\_series \\[2mm] (\dfrac{2-k}{1-k})^{n-1}(\dfrac{3-k}{1-k}) & additional\_series \\[2mm] (\dfrac{3-k}{1-k})^n & enhanced\_series \\[2mm] (\dfrac{4-k}{1-k})^n & re\text{-}enhanced\_series \\[2mm] (\dfrac{j+2-k}{1-k})^n & multiple\text{-}enhanced\_series \end{cases} \qquad (3.111)$$

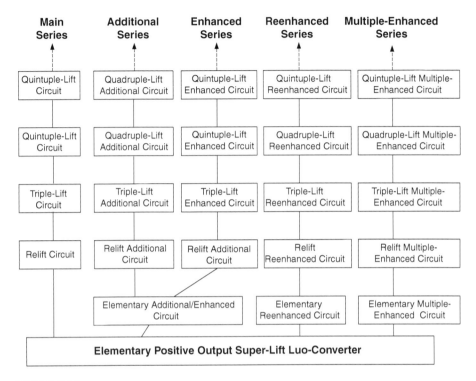

FIGURE 3.16
The family of positive output super-lift Luo-converters.

In order to show the advantages of super-lift Luo-converters, their voltage transfer gains can be compared to that of a buck converter,

$$G = \frac{V_O}{V_{in}} = k$$

forward converter,

$$G = \frac{V_O}{V_{in}} = kN \quad N \text{ is the transformer turns ratio}$$

Cúk-converter,

$$G = \frac{V_O}{V_{in}} = \frac{k}{1-k}$$

fly-back converter,

TABLE 3.1

Voltage Transfer Gains of Converters in the Condition $k = 0.2$

Stage No. (n)	1	2	3	4	5	n
Buck converter			0.2			
Forward converter			0.2 N (N is the transformer turns ratio)			
Cúk-converter			0.25			
Fly-back converter			0.25 N (N is the transformer turns ratio)			
Boost converter			1.25			
Positive output Luo-converters	1.25	2.5	3.75	5	6.25	$1.25n$
Positive output super-lift Luo-converters — main series	2.25	5.06	11.39	25.63	57.67	2.25^n
Positive output super-lift Luo-converters — additional series	3.5	7.88	17.72	39.87	89.7	$3.5*2.25^{(n-1)}$
Positive output super-lift Luo-converters — enhanced series	3.5	12.25	42.88	150	525	3.5^n
Positive output super-lift Luo-converters — re-enhanced series	4.75	22.56	107.2	509	2418	4.75^n
Positive output super-lift Luo-converters — multiple ($j = 4$)-enhanced series	7.25	52.56	381	2762	20,030	7.25^n

$$G = \frac{V_O}{V_{in}} = \frac{k}{1-k} N \qquad N \text{ is the transformer turn ratio}$$

boost converter,

$$G = \frac{V_O}{V_{in}} = \frac{1}{1-k}$$

and positive output Luo-converters.

$$G = \frac{V_O}{V_{in}} = \frac{n}{1-k} \qquad\qquad (3.112)$$

If we assume that the conduction duty k is 0.2, the output voltage transfer gains are listed in Table 3.1.

If the conduction duty k is 0.5, the output voltage transfer gains are listed in Table 3.2.

If the conduction duty k is 0.8, the output voltage transfer gains are listed in Table 3.3.

TABLE 3.2

Voltage Transfer Gains of Converters in the Condition $k = 0.5$

Stage No. (n)	1	2	3	4	5	n
Buck converter				0.5		
Forward converter			0.5 N (N is the transformer turns ratio)			
Cúk-converter				1		
Fly-back converter			N (N is the transformer turns ratio)			
Boost converter				2		
Positive output Luo-converters	2	4	6	8	10	$2n$
Positive output super-lift Luo-converters — main series	3	9	27	81	243	3^n
Positive output super-lift Luo-converters — additional series	5	15	45	135	405	$5*3^{(n-1)}$
Positive output super-lift Luo-converters — enhanced series	5	25	125	625	3125	5^n
Positive output super-lift Luo-converters — re-enhanced series	7	49	343	2401	16,807	7^n
Positive output super-lift Luo-converters — multiple ($j = 4$)-enhanced series	11	121	1331	14,641	$16*10^4$	11^n

TABLE 3.3

Voltage Transfer Gains of Converters in the Condition $k = 0.8$

Stage No. (n)	1	2	3	4	5	n
Buck converter				0.8		
Forward converter			0.8 N (N is the transformer turns ratio)			
Cúk-converter				4		
Fly-back converter			4 N (N is the transformer turns ratio)			
Boost converter				5		
Positive output Luo-converters	5	10	15	20	25	$5n$
Positive output super-lift Luo-converters — main series	6	36	216	1296	7776	6^n
Positive output super-lift Luo-converters — additional series	11	66	396	2376	14,256	$11*6^{(n-1)}$
Positive output super-lift Luo-converters — enhanced series	11	121	1331	14,641	$16*10^4$	11^n
Positive output super-lift Luo-converters — re-enhanced series	16	256	4096	65,536	$104*10^4$	16^n
Positive output super-lift Luo-converters — multiple ($j = 4$)-enhanced series	26	676	17,576	$46*10^4$	$12*10^6$	26^n

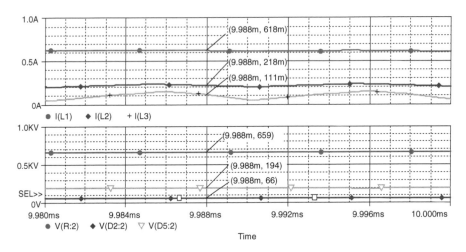

FIGURE 3.17
The simulation results of triple-lift circuit at condition $k = 0.5$ and $f = 100$ kHz.

3.8 Simulation Results

To verify the design and calculation results, the PSpice simulation package was applied to these converters. Choosing $V_{in} = 20$ V, $L_1 = L_2 = L_3 = 10$ mH, all C_1 to $C_8 = 2$ μF, and $R = 30$ kΩ, and using $k = 0.5$ and $f = 100$ kHz.

3.8.1 Simulation Results of a Triple-Lift Circuit

The voltage values V_1, V_2 and V_O of a triple-lift circuit are 66 V, 194 V, and 659 V respectively and inductor current waveforms are i_{L1} (its average value $I_{L1} = 618$ mA), i_{L2}, and i_{L3}. The simulation results are shown in Figure 3.17. The voltage values are matched to the calculated results.

3.8.2 Simulation Results of a Triple-Lift Additional Circuit

The voltage values V_1, V_2, V_3, and V_O of the triple-lift additional circuit are 57 V, 165 V, 538 V, and 910 V respectively and current waveforms are i_{L1} (its average value $I_{L1} = 1.8$ A), i_{L2}, and i_{L3}. The simulation results are shown in Figure 3.18. The voltage values are matched to the calculated results.

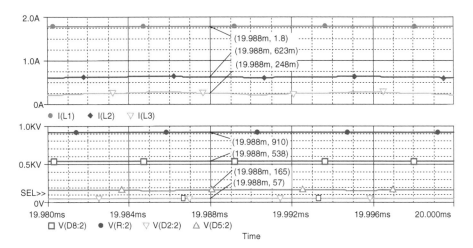

FIGURE 3.18
Simulation results of triple-lift additional circuit at condition $k = 0.5$ and $f = 100$ kHz.

3.9 Experimental Results

A test rig was constructed to verify the design and calculation results, and compare with PSpice simulation results. The testing conditions were the same: $V_{in} = 20$ V, $L_1 = L_2 = L_3 = 10$ mH, all C_1 to $C_8 = 2$ μF and $R = 30$ kΩ, and using $k = 0.5$ and $f = 100$ kHz. The component of the switch is a MOSFET device IRF950 with the rates 950 V/5 A/2 MHz. The values of the output voltage and first inductor current are measured in the following converters.

3.9.1 Experimental Results of a Triple-Lift Circuit

After careful measurement, the current value of $I_{L1} = 0.62$ A (shown in channel 1 with 1 A/Div) and voltage value of $V_O = 660$ V (shown in channel 2 with 200 V/Div). The experimental results (current and voltage values) are shown in Figure 3.19, that are identically matched to the calculated and simulation results, which are $I_{L1} = 0.618$ A and $V_O = 659$ V shown in Figure 3.17.

3.9.2 Experimental Results of a Triple-Lift Additional Circuit

The experimental results of the current value of $I_{L1} = 1.8$ A (shown in channel 1 with 1 A/Div) and voltage value of $V_O = 910$ V (shown in channel 2 with

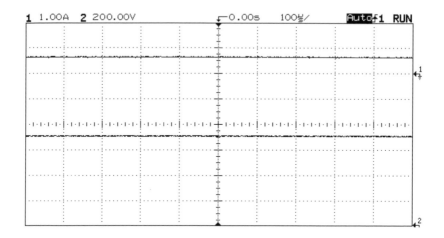

FIGURE 3.19
The experimental results of triple-lift circuit at condition $k = 0.5$ and $f = 100$ kHz.

TABLE 3.4

Comparison of Simulation and Experimental Results of a Triple-Lift Circuit

Stage No. (*n*)	I_{L1} (A)	I_{in} (A)	V_{in} (V)	P_{in} (W)	V_O (V)	P_O (W)	η (%)
Simulation results	0.618	0.927	20	18.54	659	14.47	78
Experimental results	0.62	0.93	20	18.6	660	14.52	78

TABLE 3.5

Comparison of Simulation and Experimental Results
of a Triple-Lift Additional Circuit

Stage No. (*n*)	I_{L1} (A)	I_{in} (A)	V_{in} (V)	P_{in} (W)	V_O (V)	P_O (W)	η (%)
Simulation results	1.8	2.7	20	54	910	27.6	51
Experimental results	1.8	2.7	20	54	910	27.6	51

200 V/Div) are shown in Figure 3.20 that are identically matched to the calculated and simulation results, which are $I_{L1} = 1.8$ A and $V_O = 910$ V shown in Figure 3.18.

3.9.3 Efficiency Comparison of Simulation and Experimental Results

These circuits enhanced the voltage transfer gain successfully, but efficiency, particularly, the efficiencies of the tested circuits is 51 to 78%, which is good for high voltage output equipment. Comparison of the simulation and experimental results, which are listed in the Tables 3.4 and 3.5, demonstrates that all results are well identified each other.

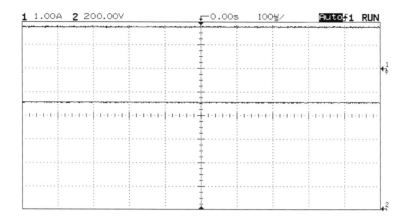

FIGURE 3.20
Experimental results of triple-lift additional circuit at condition $k = 0.5$ and $f = 100$ kHz.

Usually, there is high inrush current during the initial power-on. Therefore, the voltage across capacitors is quickly changed to certain values. The transient process is very quick in only few milliseconds.

Bibliography

Kassakian, J.G., Wolf, H-C., Miller, J.M., and Hurton, C.J., Automotive electrical systems, circa 2005, *IEEE Spectrum*, 8, 22, 1996.

Lander, C.W., *Power Electronics*, McGraw-Hill, London, 1993.

Luo, F.L., Luo-converters, a new series of positive-to-positive DC-DC step-up converters, *Power Supply Technologies and Applications*, Xi'an, China, 1, 30, 1998.

Luo, F.L., Positive output Luo-converters, the advanced voltage lift technique, *Electrical Drives*, Tianjin, China, 2, 47, 1999.

Luo, F.L. and Ye, H., Positive output super-lift Luo-converters, in *Proceedings of IEEE-PESC'2002*, Cairns, Australia, 2002, p. 425.

Luo, F.L. and Ye, H., Positive output super-lift converters, *IEEE Transactions on Power Electronics*, 18, 105, 2003.

Pelley, B.R., *Tryristor Phase Controlled Converters and Cycloconverters*, John Wiley, New York, 1971.

Pressman, A.I., *Switching Power Supply Design*, 2nd ed., McGraw-Hill, New York, 1998.

Rashid, M.H., *Power Electronics*, 2nd ed., Prentice Hall of India Pvt. Ltd., New Delhi, 1995.

Trzynadlowski, A.M., *Introduction to Modern Power Electronics*, Wiley Interscience, New York, 1998.

Ye, H., Luo, F.L., and Ye, Z.Z., Widely adjustable high-efficiency high voltage regulated power supply, *Power Supply Technologies and Applications*, Xi'an, China, 1, 18, 1998.

Ye, H., Luo, F.L., and Ye, Z.Z., DC motor Luo-converter-driver, *Power Supply World*, Guangzhou, China, 2000, p. 15.

4

Negative Output Super-Lift Luo-Converters

Along with the positive output super-lift Luo-converters, negative output (N/O) super-lift Luo-converters have also been developed. They perform super-lift technique as well.

4.1 Introduction

Negative output super-lift Luo-converters are sorted into several sub-series:

- Main series — Each circuit of the main series has one switch S, n inductors, $2n$ capacitors and $(3n - 1)$ diodes.
- Additional series — Each circuit of the additional series has one switch S, n inductors, $2(n + 1)$ capacitors and $(3n + 1)$ diodes.
- Enhanced series — Each circuit of the enhanced series has one switch S, n inductors, $4n$ capacitors and $(5n + 1)$ diodes.
- Re-enhanced series — Each circuit of the re-enhanced series has one switch S, n inductors, $6n$ capacitors and $(7n + 1)$ diodes.
- Multiple-enhanced series — Each circuit of the multiple-enhanced series has one switch S, n inductors, $2(n + j + 1)$ capacitors and $(3n + 2j + 1)$ diodes.

All analyses in this section are based on the condition of steady state operation with continuous conduction mode (CCM).

The conduction duty ratio is k, switch period $T = 1/f$ (f is the switch frequency), the load is resistive load R. The input voltage and current are V_{in} and I_{in}, output voltage and current are V_O and I_O. Assume no power losses during the conversion process, $V_{in} \times I_{in} = V_O \times I_O$. The voltage transfer gain is G:

$$G = \frac{V_O}{V_{in}}$$

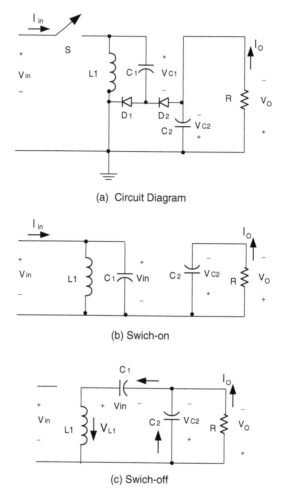

(a) Circuit Diagram

(b) Swich-on

(c) Swich-off

FIGURE 4.1
N/O elementary circuit.

4.2 Main Series

The first three stages of negative output super-lift Luo-converters — main series — are shown in Figure 4.1 to Figure 4.3. For convenience they are called elementary circuits, re-lift circuits, and triple-lift circuits respectively, and numbered as $n = 1$, 2 and 3.

4.2.1 Elementary Circuit

N/O elementary circuit and its equivalent circuits during switch-on and switch-off are shown in Figure 4.1. The voltage across capacitor C_1 is charged

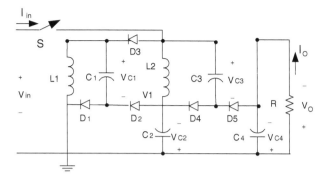

(a) Circuit Diagram

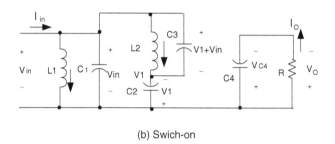

(b) Swich-on

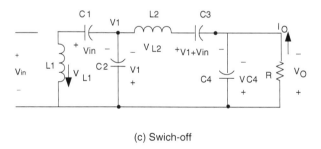

(c) Swich-off

FIGURE 4.2
N/O re-lift circuit.

to V_{in}. The current flowing through inductor L_1 increases with slop V_{in}/L_1 during switch-on period kT and decreases with slop $-(V_O - V_{in})/L_1$ during switch-off $(1 - k)T$. Therefore, the variation of current i_{L1} is

$$\Delta i_{L1} = \frac{V_{in}}{L_1}kT = \frac{V_O - V_{in}}{L_1}(1-k)T \tag{4.1}$$

$$V_O = \frac{1}{1-k}V_{in} = (\frac{2-k}{1-k} - 1)V_{in} \tag{4.2}$$

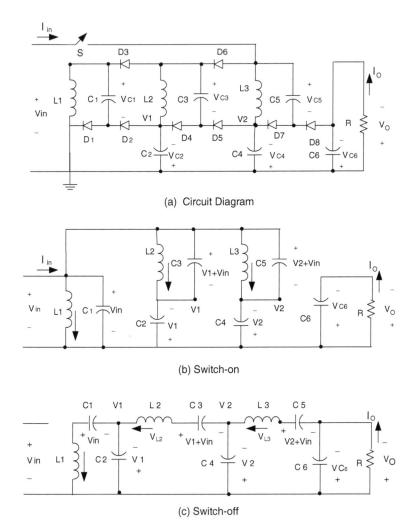

(a) Circuit Diagram

(b) Switch-on

(c) Switch-off

FIGURE 4.3
N/O triple-lift circuit.

The voltage transfer gain is

$$G_1 = \frac{V_O}{V_{in}} = \frac{2-k}{1-k} - 1 \tag{4.3}$$

In steady-state, the average charge across capacitor C_1 in a period should be zero. The relations are available:

$$kTi_{C1-on} = (1-k)Ti_{C1-off} \quad \text{and} \quad i_{C1-on} = \frac{1-k}{k}i_{C1-off}$$

This relation is available for all capacitor's current in switch-on and switch-off periods. The input current i_{in} is equal to $(i_{L1} + iC_1)$ during switch-on, and zero during switch-off. Capacitor current i_{C1} is equal to i_{L1} during switch-off.

$$i_{in-on} = i_{L1-on} + i_{C1-on} \qquad i_{L1-off} = i_{C1-off} = I_{L1}$$

If inductance L_1 is large enough, i_{L1} is nearly equal to its average current I_{L1}. Therefore,

$$i_{in-on} = i_{L1-on} + i_{C1-on} = i_{L1-on} + \frac{1-k}{k} i_{C1-off} = (1 + \frac{1-k}{k})I_{L1} = \frac{1}{k} I_{L1}$$

and

$$I_{in} = ki_{in-on} = I_{L1} \tag{4.4}$$

Further

$$i_{C2-on} = I_O \qquad\qquad i_{C2-off} = \frac{k}{1-k} I_O$$

$$I_{L1} = i_{C2-off} + I_O = \frac{k}{1-k} i_{C2-on} + I_O = \frac{1}{1-k} I_O$$

Variation ratio of inductor current i_{L1} is

$$\xi_1 = \frac{\Delta i_{L1}/2}{I_{L1}} = \frac{k(1-k)TV_{in}}{2L_1 I_O} = \frac{k(1-k)}{G_1} \frac{R}{2fL_1} \tag{4.5}$$

Usually ξ_1 is small (much lower than unity), it means this converter works in the continuous conduction mode (CCM). The ripple voltage of output voltage v_O is

$$\Delta v_O = \frac{\Delta Q}{C_2} = \frac{I_O kT}{C_2} = \frac{k}{fC_2} \frac{V_O}{R}$$

Therefore, the variation ratio of output voltage v_O is

$$\varepsilon = \frac{\Delta v_O/2}{V_O} = \frac{k}{2RfC_2} \tag{4.6}$$

Usually R is in kΩ, f in 10 kHz, and C_2 in μF, this ripple is very small.

4.2.2 N/O Re-Lift Circuit

N/O re-lift circuit is derived from N/O elementary circuit by adding the parts $(L_2$-D_3-D_4-D_5-C_3-$C_4)$. Its circuit diagram and equivalent circuits during switch-on and -off are shown in Figure 4.2. The voltage across capacitor C_1 is charged to V_{in}. As described in previous section the voltage V_1 across capacitor C_2 is

$$V_1 = \frac{1}{1-k} V_{in}$$

The voltage across capacitor C_3 is charged to $(V_1 + V_{in})$. The current flowing through inductor L_2 increases with slop $(V_1 + V_{in})/L_2$ during switch-on period kT and decreases with slop $-(V_O - 2V_1 - V_{in})/L_2$ during switch-off $(1 - k)T$. Therefore, the variation of current i_{L2} is

$$\Delta i_{L2} = \frac{V_1 + V_{in}}{L_2} kT = \frac{V_O - 2V_1 - V_{in}}{L_2}(1-k)T \qquad (4.7)$$

$$V_O = \frac{(2-k)V_1 + V_{in}}{1-k} = [(\frac{2-k}{1-k})^2 - 1]V_{in} \qquad (4.8)$$

The voltage transfer gain is

$$G_2 = \frac{V_O}{V_{in}} = (\frac{2-k}{1-k})^2 - 1 \qquad (4.9)$$

The input current i_{in} is equal to $(i_{L1} + i_{C1} + i_{L2} + i_{C3})$ during switch-on, and zero during switch-off. In steady-state, the following relations are available:

$$i_{in-on} = i_{L1-on} + i_{C1-on} + i_{L2-on} + i_{C3-on}$$

$$i_{C4-on} = I_O \qquad\qquad i_{C4-off} = \frac{k}{1-k} I_O$$

$$i_{C3-off} = I_{L2} = I_O + i_{C4-off} = \frac{I_O}{1-k} \qquad\qquad i_{C3-on} = \frac{I_O}{k}$$

$$i_{C2-on} = I_{L2} + i_{C3-on} = \frac{I_O}{1-k} + \frac{I_O}{k} = \frac{I_O}{k(1-k)} \qquad\qquad i_{C2-off} = \frac{I_O}{(1-k)^2}$$

$$i_{C1-off} = I_{L1} = I_{L2} + i_{C2-off} = \frac{I_O}{1-k} + \frac{I_O}{(1-k)^2} = \frac{2-k}{(1-k)^2} I_O \qquad i_{C1-on} = \frac{2-k}{k(1-k)} I_O$$

Thus

$$i_{in-on} = i_{L1-on} + i_{C1-on} + i_{L2-on} + i_{C3-on} = \frac{1}{k}(I_{L1} + I_{L2}) = \frac{3-2k}{k(1-k)^2} I_O$$

Therefore

$$I_{in} = ki_{in-on} = \frac{3-2k}{(1-k)^2} I_O$$

Since

$$\Delta i_{L1} = \frac{V_{in}}{L_1} kT \qquad\qquad I_{L1} = \frac{2-k}{(1-k)^2} I_O$$

$$\Delta i_{L2} = \frac{V_1 + V_{in}}{L_2} kT = \frac{2-k}{1-k} \frac{kT}{L_2} V_{in} \qquad I_{L2} = \frac{1}{1-k} I_O$$

Therefore, the variation ratio of current i_{L1} through inductor L_1 is

$$\xi_1 = \frac{\Delta i_{L1}/2}{I_{L1}} = \frac{kTV_{in}}{\frac{2-k}{(1-k)^2} 2L_1 I_O} = \frac{k(1-k)^2}{(2-k)G_2} \frac{R}{2fL_1} \qquad (4.10)$$

The variation ratio of current i_{L2} through inductor L_2 is

$$\xi_2 = \frac{\Delta i_{L2}/2}{I_{L2}} = \frac{k(2-k)TV_{in}}{2L_2 I_O} = \frac{k(2-k)}{G_2} \frac{R}{2fL_2} \qquad (4.11)$$

The ripple voltage of output voltage v_O is

$$\Delta v_O = \frac{\Delta Q}{C_4} = \frac{I_O kT}{C_4} = \frac{k}{fC_4} \frac{V_O}{R}$$

Therefore, the variation ratio of output voltage v_O is

$$\varepsilon = \frac{\Delta v_O/2}{V_O} = \frac{k}{2RfC_4} \qquad (4.12)$$

4.2.3 N/O Triple-Lift Circuit

N/O triple-lift circuit is derived from N/O re-lift circuit by double adding the parts (L_2-D_3-D_4-D_5-C_3-C_4). Its circuit diagram and equivalent circuits during switch-on and -off are shown in Figure 4.3. The voltage across capacitor C_1 is charged to V_{in}. As described in previous section the, voltage V_1 across capacitor C_2 is $V_1 = ((2-k)/(1-k)-1)V_{in} = (1/1-k)V_{in}$, and voltage V_2 across capacitor C_4 is $V_2 = [(2-k/1-k)^2 - 1]V_{in} = (3-2k/(1-k)^2)V_{in}$.

The voltage across capacitor C_5 is charged to $(V_2 + V_{in})$. The current flowing through inductor L_3 increases with slop $(V_2 + V_{in})/L_3$ during switch-on period kT and decreases with slop $-(V_O - 2V_2 - V_{in})/L_3$ during switch-off $(1-k)T$. Therefore, the variation of current i_{L3} is

$$\Delta i_{L3} = \frac{V_2 + V_{in}}{L_3} kT = \frac{V_O - 2V_2 - V_{in}}{L_3}(1-k)T \tag{4.13}$$

$$V_O = \frac{(2-k)V_2 + V_{in}}{1-k} = [(\frac{2-k}{1-k})^3 - 1]V_{in} \tag{4.14}$$

The voltage transfer gain is

$$G_3 = \frac{V_O}{V_{in}} = (\frac{2-k}{1-k})^3 - 1 \tag{4.15}$$

The input current i_{in} is equal to $(i_{L1} + i_{C1} + i_{L2} + i_{C3} + i_{L3} + i_{C5})$ during switch-on, and zero during switch-off. In steady state, the following relations are available:

$$i_{in-on} = i_{L1-on} + i_{C1-on} + i_{L2-on} + i_{C3-on} + i_{L3-on} + i_{C5-on}$$

$$i_{C6-on} = I_O \qquad\qquad\qquad\qquad i_{C6-off} = \frac{k}{1-k}I_O$$

$$i_{C5-off} = I_{L3} = I_O + i_{C6-off} = \frac{I_O}{1-k} \qquad\qquad i_{C5-on} = \frac{I_O}{k}$$

$$i_{C4-on} = I_{L3} + i_{C5-on} = \frac{I_O}{1-k} + \frac{I_O}{k} = \frac{I_O}{k(1-k)} \qquad\qquad i_{C4-off} = \frac{I_O}{(1-k)^2}$$

$$i_{C3-off} = I_{L2} = I_{L3} + i_{C4-off} = \frac{2-k}{(1-k)^2}I_O \qquad\qquad i_{C3-on} = \frac{2-k}{k(1-k)}I_O$$

$$i_{C2-on} = I_{L2} + i_{C3-on} = \frac{2-k}{k(1-k)^2} I_O \qquad\qquad i_{C2-off} = \frac{2-k}{(1-k)^3} I_O$$

$$i_{C1-off} = I_{L1} = I_{L2} + i_{C2-off} = \frac{(2-k)^2}{(1-k)^3} I_O \qquad\qquad i_{C1-on} = \frac{(2-k)^2}{k(1-k)^2} I_O$$

Thus

$$i_{in-on} = i_{L1-on} + i_{C1-on} + i_{L2-on} + i_{C3-on} + i_{L3-on} + i_{C5-on}$$

$$= \frac{1}{k}(I_{L1} + I_{L2} + I_{L3}) = \frac{7-9k+3k^2}{k(1-k)^3} I_O$$

Therefore

$$I_{in} = k i_{in-on} = \frac{7-9k+3k^2}{(1-k)^3} I_O = [(\frac{2-k}{1-k})^3 - 1]I_O$$

Analogously,

$$\Delta i_{L1} = \frac{V_{in}}{L_1} kT \qquad\qquad I_{L1} = \frac{(2-k)^2}{(1-k)^3} I_O$$

$$\Delta i_{L2} = \frac{V_1 + V_{in}}{L_2} kT = \frac{2-k}{(1-k)L_2} kTV_{in} \qquad\qquad I_{L2} = \frac{2-k}{(1-k)^2} I_O$$

$$\Delta i_{L3} = \frac{V_2 + V_{in}}{L_3} kT = (\frac{2-k}{1-k})^2 \frac{kT}{L_3} V_{in} \qquad\qquad I_{L3} = \frac{I_O}{1-k}$$

Therefore, the variation ratio of current i_{L1} through inductor L_1 is

$$\xi_1 = \frac{\Delta i_{L1}/2}{I_{L1}} = \frac{k(1-k)^3 TV_{in}}{2(2-k)^2 L_1 I_O} = \frac{k(1-k)^3}{(2-k)^2 G_3} \frac{R}{2fL_1} \qquad (4.16)$$

The variation ratio of current i_{L2} through inductor L_2 is

$$\xi_2 = \frac{\Delta i_{L2}/2}{I_{L2}} = \frac{k(1-k)TV_{in}}{2L_2 I_O} = \frac{k(1-k)}{G_3} \frac{R}{2fL_2} \qquad (4.17)$$

The variation ratio of current i_{L3} through inductor L_3 is

$$\xi_3 = \frac{\Delta i_{L3}/2}{I_{L3}} = \frac{k(2-k)^2 T V_{in}}{2(1-k)L_3 I_O} = \frac{k(2-k)^2}{(1-k)G_3}\frac{R}{2fL_3} \tag{4.18}$$

The ripple voltage of output voltage v_O is

$$\Delta v_O = \frac{\Delta Q}{C_6} = \frac{I_O kT}{C_6} = \frac{k}{fC_6}\frac{V_O}{R}$$

Therefore, the variation ratio of output voltage v_O is

$$\varepsilon = \frac{\Delta v_O/2}{V_O} = \frac{k}{2RfC_6} \tag{4.19}$$

4.2.4 N/O Higher Order Lift Circuit

N/O higher order lift circuits can be designed by repeating the parts (L_2-D_3-D_4-D_5-C_3-C_4) multiple times. For nth order lift circuit, the final output voltage across capacitor C_{2n} is

$$V_O = [(\frac{2-k}{1-k})^n - 1]V_{in} \tag{4.20}$$

The voltage transfer gain is

$$G_n = \frac{V_O}{V_{in}} = (\frac{2-k}{1-k})^n - 1 \tag{4.21}$$

The variation ratio of current i_{Li} through inductor L_i ($i = 1, 2, 3, \ldots n$) is

$$\xi_1 = \frac{\Delta i_{L1}/2}{I_{L1}} = \frac{k(1-k)^n}{(2-k)^{(n-1)}G_n}\frac{R}{2fL_i} \tag{4.22}$$

$$\xi_2 = \frac{\Delta i_{L2}/2}{I_{L2}} = \frac{k(2-k)^{(3-n)}}{(1-k)^{(n-3)}G_n}\frac{R}{2fL_i} \tag{4.23}$$

$$\xi_3 = \frac{\Delta i_{L3}/2}{I_{L3}} = \frac{k(2-k)^{(n-i+2)}}{(1-k)^{(n-i+1)}G_n}\frac{R}{2fL_3} \tag{4.24}$$

The variation ratio of output voltage v_O is

$$\varepsilon = \frac{\Delta v_O / 2}{V_O} = \frac{k}{2RfC_{2n}} \tag{4.25}$$

4.3 Additional Series

All circuits of negative output super-lift Luo-converters — additional series — are derived from the corresponding circuits of the main series by adding a double/enhanced circuit (DEC). The first three stages of this series are shown in Figure 4.4 to Figure 4.6. For convenience they are called elementary additional circuits, re-lift additional circuit, and triple-lift additional circuit respectively, and numbered as $n = 1, 2$ and 3.

4.3.1 N/O Elementary Additional Circuit

This circuit is derived from the N/O elementary circuit by adding a DEC. Its circuit and switch-on and switch-off equivalent circuits are shown in Figure 4.4. The voltage across capacitor C_1 is charged to V_{in}. The voltage across capacitor C_2 is charged to V_1 and C_{11} is charged to $(V_1 + V_{in})$. The current i_{L1} flowing through inductor L_1 increases with slope V_{in}/L_1 during switch-on period kT and decreases with slope $-(V_1 - V_{in})/L_1$ during switch-off $(1 - k)T$.
Therefore,

$$\Delta i_{L1} = \frac{V_{in}}{L_1} kT = \frac{V_1 - V_{in}}{L_1} (1 - k)T \tag{4.26}$$

$$V_1 = \frac{1}{1-k} V_{in} = (\frac{2-k}{1-k} - 1)V_{in}$$

$$V_{L1\text{-}off} = \frac{k}{1-k} V_{in}$$

The output voltage is

$$V_O = V_{in} + V_{L1} + V_1 = \frac{2}{1-k} V_{in} = [\frac{3-k}{1-k} - 1]V_{in} \tag{4.27}$$

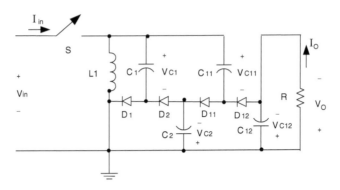

(a) Circuit Diagram

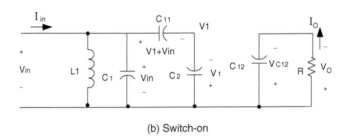

(b) Switch-on

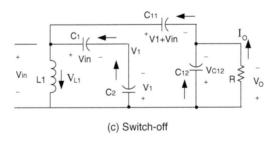

(c) Switch-off

FIGURE 4.4
N/O elementary additional (enhanced) circuit.

The voltage transfer gain is

$$G_1 = \frac{V_O}{V_{in}} = \frac{3-k}{1-k} - 1 \qquad (4.28)$$

Following relations are obtained:

$$i_{C12-on} = I_O \qquad\qquad i_{C12-off} = \frac{kI_O}{1-k}$$

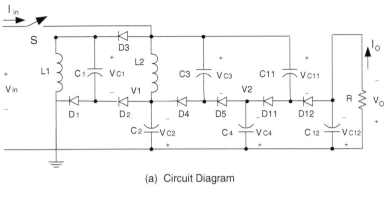

(a) Circuit Diagram

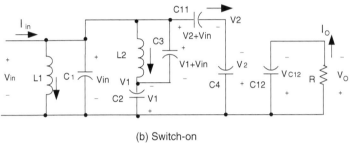

(b) Switch-on

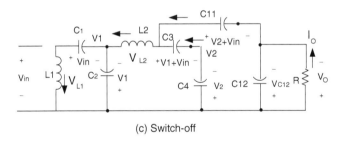

(c) Switch-off

FIGURE 4.5
N/O re-lift additional circuit.

$$i_{C11-off} = I_O + i_{C12-off} = \frac{I_O}{1-k} \qquad i_{C11-on} = i_{C2-on} = \frac{I_O}{k}$$

$$i_{C2-off} = i_{C1-off} = \frac{I_O}{1-k} \qquad i_{C1-on} = \frac{I_O}{k}$$

$$I_{L1} = i_{C1-off} + i_{C11-on} = \frac{2I_O}{1-k}$$

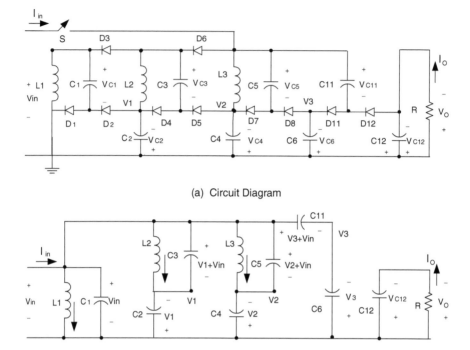

(a) Circuit Diagram

(b) Switch-on

(c) Switch-off

FIGURE 4.6
N/O triple-lift additional circuit.

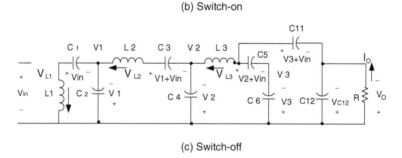

$$i_{in} = I_{L1} + i_{C1-on} + i_{C11-on} = (\frac{2}{1-k} + \frac{1}{k} + \frac{1}{k})I_O = \frac{2}{k(1-k)}I_O$$

Therefore,

$$I_{in} = ki_{in} = \frac{2}{1-k}I_O = [\frac{3-k}{1-k} - 1]I_O$$

The variation ratio of current i_{L1} through inductor L_1 is

$$\xi_1 = \frac{\Delta i_{L1}/2}{I_{L1}} = \frac{k(1-k)TV_{in}}{4L_1 I_O} = \frac{k(1-k)}{2G_1}\frac{R}{2fL_1} \tag{4.29}$$

The ripple voltage of output voltage v_O is

$$\Delta v_O = \frac{\Delta Q}{C_{12}} = \frac{I_O kT}{C_{12}} = \frac{k}{fC_{12}}\frac{V_O}{R}$$

Therefore, the variation ratio of output voltage v_O is

$$\varepsilon = \frac{\Delta v_O/2}{V_O} = \frac{1-k}{2RfC_{12}} \tag{4.30}$$

4.3.2 N/O Re-Lift Additional Circuit

The N/O re-lift additional circuit is derived from the N/O re-lift circuit by adding a DEC. Its circuit diagram and switch-on and switch-off equivalent circuits are shown in Figure 4.5. The voltage across capacitor C_1 is charged to v_{in}. As described in a previous section the voltage across C_2 is

$$V_1 = \frac{1}{1-k}V_{in}$$

The voltage across capacitor C_3 is charged to $(V_1 + V_{in})$, voltage across capacitor C_4 is charged to V_2 and voltage across capacitor C_{11} is charged to $(V_2 + V_{in})$. The current flowing through inductor L_2 increases with voltage $(V_1 + V_{in})$ during switch-on kT and decreases with voltage $-(V_2 - 2V_1 - V_{in})$ during switch-off $(1 - k)T$. Therefore,

$$\Delta i_{L2} = \frac{V_1 + V_{in}}{L_2}kT = \frac{V_2 - 2V_1 - V_{in}}{L_2}(1-k)T \tag{4.31}$$

$$V_2 = \frac{(2-k)V_1 + V_{in}}{1-k} = \frac{3-2k}{(1-k)^2} = [(\frac{2-k}{1-k})^2 - 1]V_{in}$$

and

$$V_{L2-off} = V_2 - 2V_1 - V_{in} = \frac{k(2-k)}{(1-k)^2}V_{in} \tag{4.32}$$

The output voltage is

$$V_O = V_2 + V_{in} + V_{L2} + V_1 = \frac{5-3k}{(1-k)^2}V_{in} = [\frac{3-k}{1-k}\frac{2-k}{1-k} - 1]V_{in} \qquad (4.33)$$

The voltage transfer gain is

$$G_2 = \frac{V_O}{V_{in}} = \frac{2-k}{1-k}\frac{3-k}{1-k} - 1 \qquad (4.34)$$

Following relations are obtained:

$$i_{C12-on} = I_O \qquad\qquad\qquad\qquad i_{C12-off} = \frac{kI_O}{1-k}$$

$$i_{C11-off} = I_O + i_{C12-off} = \frac{I_O}{1-k} \qquad\qquad i_{C11-on} = i_{C4-on} = \frac{I_O}{k}$$

$$i_{C4-off} = i_{C3-off} = \frac{I_O}{1-k} \qquad\qquad\qquad i_{C3-on} = \frac{I_O}{k}$$

$$I_{L2} = i_{C11-off} + i_{C3-off} = \frac{2I_O}{1-k}$$

$$i_{C2-on} = I_{L2} + i_{C3-on} = \frac{1+k}{k(1-k)}I_O \qquad\qquad i_{C2-off} = \frac{1+k}{(1-k)^2}I_O$$

$$I_{L1} = i_{C1-off} = I_{L2} + i_{C2-off} = \frac{3-k}{(1-k)^2}I_O \qquad i_{C1-on} = \frac{3-k}{k(1-k)}I_O$$

$$i_{in} = I_{L1} + i_{C1-on} + i_{C2-on} + i_{C4-on} = [\frac{3-k}{(1-k)^2} + \frac{3-k}{k(1-k)} + \frac{1+k}{k(1-k)} + \frac{1}{k}]I_O = \frac{5-3k}{k(1-k)^2}I_O$$

Therefore,

$$I_{in} = ki_{in} = \frac{5-3k}{(1-k)^2}I_O = [\frac{3-k}{1-k}\frac{2-k}{1-k} - 1]I_O$$

Analogously,

$$\Delta i_{L1} = \frac{V_{in}}{L_2}kT \qquad\qquad\qquad I_{L1} = \frac{3-k}{(1-k)^2}I_O$$

$$\Delta i_{L2} = \frac{V_1 + V_{in}}{L_2} kT = \frac{2-k}{(1-k)L_2} kTV_{in} \qquad I_{L2} = \frac{2I_O}{1-k}$$

Therefore, the variation ratio of current i_{L1} through inductor L_1 is

$$\xi_1 = \frac{\Delta i_{L1}/2}{I_{L1}} = \frac{k(1-k)^2 TV_{in}}{2(3-k)L_1 I_O} = \frac{k(1-k)^2}{(3-k)G_2} \frac{R}{2fL_1} \tag{4.35}$$

The variation ratio of current i_{L2} through inductor L_2 is

$$\xi_2 = \frac{\Delta i_{L2}/2}{I_{L2}} = \frac{k(2-k)TV_{in}}{4L_2 I_O} = \frac{k(2-k)}{2G_2} \frac{R}{2fL_2} \tag{4.36}$$

The ripple voltage of output voltage v_O is

$$\Delta v_O = \frac{\Delta Q}{C_{12}} = \frac{I_O kT}{C_{12}} = \frac{k}{fC_{12}} \frac{V_O}{R}$$

Therefore, the variation ratio of output voltage v_O is

$$\varepsilon = \frac{\Delta v_O/2}{V_O} = \frac{k}{2RfC_{12}} \tag{4.37}$$

4.3.3 N/O Triple-Lift Additional Circuit

This circuit is derived from the N/O triple-lift circuit by adding a DEC. Its circuit diagram and equivalent circuits during switch-on and switch-off are shown in Figure 4.6. The voltage across capacitor C_1 is charged to V_{in}. As described in a previous section the voltage across C_2 is

$$V_1 = \frac{1}{1-k} V_{in}$$

and voltage across C_4 is

$$V_2 = \frac{3-2k}{1-k} V_1 = \frac{3-2k}{(1-k)^2} V_{in}$$

The voltage across capacitor C_5 is charged to $(V_2 + V_{in})$, voltage across capacitor C_6 is charged to V_3 and voltage across capacitor C_{11} is charged to $(V_3 + V_{in})$. The current flowing through inductor L_3 increases with voltage

$(V_2 + V_{in})$ during switch-on period kT and decreases with voltage $-(V_3 - 2V_2 - V_{in})$ during switch-off $(1 - k)T$. Therefore,

$$\Delta i_{L3} = \frac{V_2 + V_{in}}{L_3} kT = \frac{V_3 - 2V_2 - V_{in}}{L_3}(1-k)T \tag{4.38}$$

$$V_3 = \frac{(2-k)V_2 + V_{in}}{1-k} = \frac{7-9k+3k^2}{(1-k)^3}V_{in} = [(\frac{2-k}{1-k})^3 - 1]V_{in}$$

and

$$V_{L3-off} = V_3 - 2V_2 - V_{in} = \frac{k(2-k)^2}{(1-k)^3}V_{in} \tag{4.39}$$

The output voltage is

$$V_O = V_3 + V_{in} + V_{L3} + V_2 = \frac{11-13k+4k^2}{(1-k)^3}V_{in} = [\frac{3-k}{1-k}(\frac{2-k}{1-k})^2 - 1]V_{in} \tag{4.40}$$

The voltage transfer gain is

$$G_3 = \frac{V_O}{V_{in}} = (\frac{2-k}{1-k})^2 \frac{3-k}{1-k} - 1 \tag{4.41}$$

Following relations are available:

$$i_{C12-on} = I_O \qquad\qquad i_{C12-off} = \frac{kI_O}{1-k}$$

$$i_{C11-off} = I_O + i_{C12-off} = \frac{I_O}{1-k} \qquad\qquad i_{C11-on} = i_{C6-on} = \frac{I_O}{k}$$

$$i_{C6-off} = i_{C5-off} = \frac{I_O}{1-k} \qquad\qquad i_{C5-on} = \frac{I_O}{k}$$

$$I_{L3} = i_{C11-off} + i_{C5-off} = \frac{2I_O}{1-k}$$

$$i_{C4-on} = I_{L3} + i_{C5-on} = \frac{1+k}{k(1-k)}I_O \qquad\qquad i_{C4-off} = \frac{1+k}{(1-k)^2}I_O$$

$$I_{L2} = i_{C3-off} = I_{L3} + i_{C4-off} = \frac{3-k}{(1-k)^2} I_O \qquad i_{C3-on} = \frac{3-k}{k(1-k)} I_O$$

$$i_{C2-on} = I_{L2} + i_{C3-on} = \frac{3-k}{k(1-k)^2} I_O \qquad i_{C2-off} = \frac{3-k}{(1-k)^3} I_O$$

$$I_{L1} = i_{C1-off} = I_{L2} + i_{C2-off} = \frac{(3-k)(2-k)}{(1-k)^3} I_O \qquad i_{C1-on} = \frac{(3-k)(2-k)}{k(1-k)^2} I_O$$

$$i_{in} = I_{L1} + i_{C1-on} + i_{C2-on} + i_{C4-on} + i_{C6-on}$$

$$= [\frac{(3-k)(2-k)}{(1-k)^3} + \frac{(3-k)(2-k)}{k(1-k)^2} + \frac{3-k}{k(1-k)^2} + \frac{1+k}{k(1-k)} + \frac{1}{k}]I_O$$

$$= \frac{11-13k+4k^2}{k(1-k)^3} I_O$$

Therefore,

$$I_{in} = ki_{in} = \frac{11-13k+4k^2}{(1-k)^3} I_O = [\frac{3-k}{1-k}(\frac{2-k}{1-k})^2 - 1]I_O$$

Analogously,

$$\Delta i_{L1} = \frac{V_{in}}{L_2} kT \qquad I_{L1} = \frac{(2-k)(3-k)}{(1-k)^3} I_O$$

$$\Delta i_{L2} = \frac{V_1 + V_{in}}{L_2} kT = \frac{2-k}{(1-k)L_2} kTV_{in} \qquad I_{L2} = \frac{3-k}{(1-k)^2} I_O$$

$$\Delta i_{L3} = \frac{V_2 + V_{in}}{L_3} kT = \frac{(2-k)^2}{(1-k)^2 L_3} kTV_{in} \qquad I_{L3} = \frac{2I_O}{1-k}$$

Therefore, the variation ratio of current i_{L1} through inductor L_1 is

$$\xi_1 = \frac{\Delta i_{L1}/2}{I_{L1}} = \frac{k(1-k)^3 TV_{in}}{2(2-k)(3-k)L_1 I_O} = \frac{k(1-k)^3}{(2-k)(3-k)G_3} \frac{R}{2fL_1} \qquad (4.42)$$

and the variation ratio of current i_{L2} through inductor L_2 is

$$\xi_2 = \frac{\Delta i_{L2}/2}{I_{L2}} = \frac{k(1-k)(2-k)TV_1}{2(3-k)L_2 I_O} = \frac{k(1-k)(2-k)}{(3-k)G_3} \frac{R}{2fL_2} \qquad (4.43)$$

and the variation ratio of current i_{L3} through inductor L_3 is

$$\xi_3 = \frac{\Delta i_{L3}/2}{I_{L3}} = \frac{k(2-k)^2 TV_{in}}{4(1-k)L_3 I_O} = \frac{k(2-k)^2}{2(1-k)G_3} \frac{R}{2fL_3} \tag{4.44}$$

The ripple voltage of output voltage v_O is

$$\Delta v_O = \frac{\Delta Q}{C_{12}} = \frac{I_O kT}{C_{12}} = \frac{k}{fC_{12}} \frac{V_O}{R}$$

Therefore, the variation ratio of output voltage v_O is

$$\varepsilon = \frac{\Delta v_O/2}{V_O} = \frac{k}{2RfC_{12}} \tag{4.45}$$

4.3.4 N/O Higher Order Lift Additional Circuit

Higher order N/O lift additional circuits can be derived from the corresponding circuits of the main series by adding a DEC. Each stage voltage V_i ($i = 1, 2, \ldots n$) is

$$V_i = [(\frac{2-k}{1-k})^i - 1]V_{in} \tag{4.46}$$

This means V_1 is the voltage across capacitor C_2, V_2 is the voltage across capacitor C_4 and so on. For n^{th} order lift additional circuit, the final output voltage is

$$V_O = [\frac{3-k}{1-k}(\frac{2-k}{1-k})^{n-1} - 1]V_{in} \tag{4.47}$$

The voltage transfer gain is

$$G_n = \frac{V_O}{V_{in}} = \frac{3-k}{1-k}(\frac{2-k}{1-k})^{n-1} - 1 \tag{4.48}$$

Analogously, the variation ratio of current i_{Li} through inductor L_i ($i = 1, 2, 3, \ldots n$) is

$$\xi_1 = \frac{\Delta i_{L1}/2}{I_{L1}} = \frac{k(1-k)^n}{2^{h(1-n)}[(2-k)^{(n-2)}(3-k)]^{u(n-2)}G_n} \frac{R}{fL_1} \tag{4.49}$$

$$\xi_2 = \frac{\Delta i_{L2}/2}{I_{L2}} = \frac{k(1-k)^{(n-2)}(2-k)}{2^{h(n-2)}(3-k)^{(n-2)}G_n} \frac{R}{2fL_2} \qquad (4.50)$$

$$\xi_3 = \frac{\Delta i_{L3}/2}{I_{L3}} = \frac{k(2-k)^{(n-1)}}{2^{h(n-3)}(1-k)^{(n-2)}G_n} \frac{R}{2fL_3} \qquad (4.51)$$

where

$$h(x) = \begin{cases} 0 & x > 0 \\ 1 & x \leq 0 \end{cases} \quad \text{is the } \textbf{Hong function}$$

and

$$u(x) = \begin{cases} 1 & x \geq 0 \\ 0 & x < 0 \end{cases} \quad \text{is the } \textbf{unit-step function}$$

and the variation ratio of output voltage v_O is

$$\varepsilon = \frac{\Delta v_O/2}{V_O} = \frac{k}{2RfC_{12}} \qquad (4.52)$$

4.4 Enhanced Series

All circuits of the negative output super-lift Luo-converters — enhanced series — are derived from the corresponding circuits of the main series by adding the DEC into each stage circuit of all series converters.

The first three stages of this series are shown in Figures 4.4, 4.7, and 4.8. For convenience they are called elementary enhanced circuits, re-lift enhanced circuits, and triple-lift enhanced circuits respectively, and numbered as $n = 1, 2$ and 3.

4.4.1 N/O Elementary Enhanced Circuit

This circuit is derived from N/O elementary circuit with adding a DEC. Its circuit and switch-on and switch-off equivalent circuits are shown in Figure 4.4.

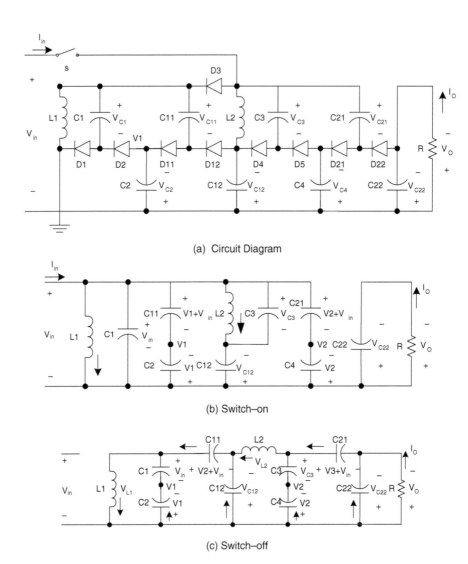

(a) Circuit Diagram

(b) Switch–on

(c) Switch–off

FIGURE 4.7
N/O re-lift enhanced circuit.

The output voltage is

$$V_O = V_{in} + V_{L1} + V_1 = \frac{2}{1-k}V_{in} = [\frac{3-k}{1-k} - 1]V_{in} \qquad (4.27)$$

The voltage transfer gain is

$$G_1 = \frac{V_O}{V_{in}} = \frac{3-k}{1-k} - 1 \qquad (4.28)$$

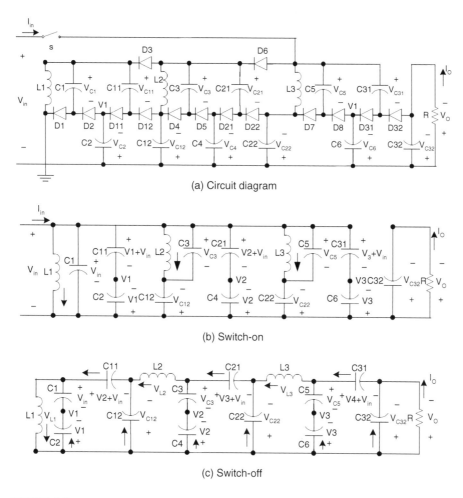

(a) Circuit diagram

(b) Switch-on

(c) Switch-off

FIGURE 4.8
N/O triple-lift enhanced circuit.

4.4.2 N/O Re-Lift Enhanced Circuit

The N/O re-lift enhanced circuit is derived from N/O re-lift circuit of the main series by adding the DEC into each stage. Its circuit diagram and switch-on and switch-off equivalent circuits are shown in Figure 4.7. The voltage across capacitor C_{12} is charged to

$$V_{C12} = \frac{3}{1-k} V_{in} \qquad (4.53)$$

The voltage across capacitor C_3 is charged to V_{C12}, and the voltage across capacitor C_4 and C_{12} is charged to V_{C4}

$$V_{C4} = \frac{2-k}{1-k} V_{C12} = \frac{2-k}{1-k} \frac{3-k}{1-k} V_{in} \qquad (4.54)$$

The current flowing through inductor L_2 increases with voltage V_{C12} during switch-on kT and decreases with voltage $-(V_{C21} - V_{C4} - V_{C12})$ during switch-off $(1-k)T$.

Therefore,

$$\Delta i_{L2} = \frac{kT}{L_2}(V_{C12} - V_{in}) = \frac{V_{C21} - V_{C4} - V_{C12}}{L_2}(1-k)T \qquad (4.55)$$

$$V_{C21} = (\frac{3-k}{1-k})^2 V_{in}$$

The output voltage is

$$V_O = V_{C21} - V_{in} = [(\frac{3-k}{1-k})^2 - 1]V_{in} \qquad (4.56)$$

The voltage transfer gain is

$$G_2 = \frac{V_O}{V_{in}} = (\frac{3-k}{1-k})^2 - 1 \qquad (4.57)$$

Following relations are obtained:

$$i_{C22-on} = I_O \qquad\qquad\qquad i_{C22-off} = \frac{kI_O}{1-k}$$

$$i_{C21-off} = I_O + i_{C22-off} = \frac{I_O}{1-k} \qquad\qquad i_{C21-on} = i_{C4-on} = \frac{I_O}{k}$$

$$i_{C4-off} = i_{C3-off} = \frac{I_O}{1-k} \qquad\qquad i_{C3-on} = \frac{I_O}{k}$$

$$I_{L2} = i_{C21-off} + i_{C3-off} = \frac{2I_O}{1-k}$$

$$i_{C12-on} = I_{L2} + i_{C3-on} = \frac{1+k}{k(1-k)} I_O \qquad\qquad i_{C12-off} = \frac{1+k}{(1-k)^2} I_O$$

$$i_{C11-off} = I_{L2} + i_{C12-off} = \frac{3-k}{(1-k)^2} I_O \qquad\qquad i_{C2-off} = \frac{3-k}{(1-k)^2} I_O$$

$$i_{C11-on} = i_{C2-on} = \frac{3-k}{k(1-k)} I_O$$

$$I_{L1} = i_{C11-off} + i_{C2-off} = 2\frac{3-k}{(1-k)^2} I_O \qquad\qquad i_{C1-on} = \frac{3-k}{k(1-k)} I_O$$

$$i_{in} = I_{L1} + i_{C1-on} + i_{C11-on} + i_{C12-on} + i_{C21-on} = \frac{4(2-k)}{k(1-k)^2} I_O$$

Therefore,

$$I_{in} = k i_{in} = \frac{4(2-k)}{(1-k)^2} I_O = [\frac{(3-k)^2}{(1-k)^2} - 1] I_O$$

Analogously,

$$\Delta i_{L1} = \frac{V_{in}}{L_2} kT \qquad\qquad I_{L1} = 2\frac{3-k}{(1-k)^2} I_O$$

$$\Delta i_{L2} = \frac{V_{C12} - V_{in}}{L_2} kT = \frac{2+k}{(1-k)L_2} kTV_{in} \qquad I_{L2} = \frac{2I_O}{1-k}$$

Therefore, the variation ratio of current i_{L1} through inductor L_1 is

$$\xi_1 = \frac{\Delta i_{L1}/2}{I_{L1}} = \frac{k(1-k)^2 TV_{in}}{4(3-k)L_1 I_O} = \frac{k(1-k)^2}{2(3-k)G_2} \frac{R}{2fL_1} \qquad (4.58)$$

The variation ratio of current i_{L2} through inductor L_2 is

$$\xi_2 = \frac{\Delta i_{L2}/2}{I_{L2}} = \frac{k(2+k)TV_{in}}{4L_2 I_O} = \frac{k(2+k)}{2G_2} \frac{R}{2fL_2} \qquad (4.59)$$

The ripple voltage of output voltage v_O is

$$\Delta v_O = \frac{\Delta Q}{C_{22}} = \frac{I_O kT}{C_{22}} = \frac{k}{fC_{22}} \frac{V_O}{R}$$

Therefore, the variation ratio of output voltage v_O is

$$\varepsilon = \frac{\Delta v_O / 2}{V_O} = \frac{k}{2RfC_{22}} \qquad (4.60)$$

4.4.3 N/O Triple-Lift Enhanced Circuit

This circuit is derived from the N/O triple-lift circuit of main series by adding the DEC into each stage. Its circuit diagram and equivalent circuits during switch-on and switch-off are shown in Figure 4.8. The voltage across capacitor C_{12} is charged to V_{C12}. As described in the previous section the voltage across C_{C12} is

$$V_{C12} = \frac{3-k}{1-k} V_{in}$$

and voltage across C_4 and C_{C22} is

$$V_{C22} = \frac{3-k}{1-k} V_{C12} = \left(\frac{3-k}{1-k}\right)^2 V_{in}$$

The voltage across capacitor C_5 is charged to V_{C22}, voltage across capacitor C_6 is charged to V_{C6}

$$V_{C6} = \frac{2-k}{1-k} V_{C22} = \frac{2-k}{1-k} \left(\frac{3-k}{1-k}\right)^2 V_{in}$$

The current flowing through inductor L_3 increases with voltage V_{C22} during switch-on period kT and decreases with voltage $-(V_{C32} - V_{C6} - V_{C22})$ during switch-off $(1-k)T$.
Therefore,

$$\Delta i_{L3} = \frac{kT}{L_3}(V_{C22} - V_{in}) = \frac{V_{C31} - V_{C6} - V_{C22}}{L_3}(1-k)T \qquad (4.61)$$

$$V_{C31} = \left(\frac{3-k}{1-k}\right)^3 V_{in}$$

and

$$V_O = V_{C31} - V_{in} = \left[\left(\frac{3-k}{1-k}\right)^3 - 1\right]V_{in} \qquad (4.62)$$

The voltage transfer gain is

$$G_3 = \frac{V_O}{V_{in}} = (\frac{3-k}{1-k})^2 - 1 \tag{4.63}$$

The following relations are obtained:

$$i_{C32-on} = I_O \qquad\qquad i_{C32-off} = \frac{kI_O}{1-k}$$

$$i_{C31-off} = I_O + i_{C32-off} = \frac{I_O}{1-k} \qquad\qquad i_{C31-on} = i_{C6-on} = \frac{I_O}{k}$$

$$i_{C6-off} = i_{C5-off} = \frac{I_O}{1-k} \qquad\qquad i_{C6-on} = i_{C5-on} = \frac{I_O}{k}$$

$$I_{L3} = i_{C31-off} + i_{C5-off} = \frac{2I_O}{1-k}$$

$$i_{C22-on} = I_{L3} + i_{C5-on} = \frac{1+k}{k(1-k)}I_O \qquad\qquad i_{C22-off} = \frac{1+k}{(1-k)^2}I_O$$

$$i_{C21-off} = i_{C4-off} = I_{L3} + i_{C22-off} = \frac{3-k}{(1-k)^2}I_O \qquad i_{C4-on} = \frac{3-k}{k(1-k)}I_O$$

$$I_{L2} = i_{C4-off} + i_{C21-off} = 2\frac{3-k}{(1-k)^2}I_O \qquad\qquad i_{C3-on} = \frac{3-k}{k(1-k)}I_O$$

$$i_{C12-on} = I_{L2} + i_{C3-on} = \frac{(3-k)(2-k)}{k(1-k)^2}I_O \qquad\qquad i_{C12-off} = \frac{(3-k)(2-k)}{(1-k)^3}I_O$$

$$i_{C11-off} = I_{L2} + i_{C12-off} = \frac{(3-k)(4-3k)}{(1-k)^3}I_O \qquad\qquad i_{C11-on} = \frac{(3-k)(4-k)}{k(1-k)^2}I_O$$

$$I_{L1} = i_{C11-off} + i_{C1-off} = 2\frac{(3-k)(4-k)}{(1-k)^3}I_O \qquad\qquad i_{C1-on} = i_{C2-on} = \frac{(3-k)(4-k)}{k(1-k)^2}I_O$$

$$i_{in} = I_{L1} + i_{C1-on} + i_{C2-on} + i_{C12-on} + i_{C4-on} + i_{C22-on} + i_{C6-on} = \frac{2(13-12k+3k^2)}{k(1-k)^3}I_O$$

Therefore,

$$I_{in} = ki_{in} = 2\frac{13 - 12k + 3k^2}{(1-k)^3}I_O = [(\frac{3-k}{1-k})^3 - 1]I_O$$

Analogously:

$$\Delta i_{L1} = \frac{V_{in}}{L_2}kT \qquad\qquad I_{L1} = \frac{2(4-k)(3-k)}{(1-k)^3}I_O$$

$$\Delta i_{L2} = \frac{V_1 + V_{in}}{L_2}kT = \frac{2-k}{(1-k)L_2}kTV_{in} \qquad I_{L2} = 2\frac{3-k}{(1-k)^2}I_O$$

$$\Delta i_{L3} = \frac{V_2 + V_{in}}{L_3}kT = \frac{(2-k)^2}{(1-k)^2 L_3}kTV_{in} \qquad I_{L3} = \frac{2I_O}{1-k}$$

Therefore, the variation ratio of current i_{L1} through inductor L_1 is

$$\xi_1 = \frac{\Delta i_{L1}/2}{I_{L1}} = \frac{k(1-k)^3 TV_{in}}{4(4-k)(3-k)L_1 I_O} = \frac{k(1-k)^3}{2(4-k)(3-k)G_3}\frac{R}{2fL_1} \tag{4.64}$$

and the variation ratio of current i_{L2} through inductor L_2 is

$$\xi_2 = \frac{\Delta i_{L2}/2}{I_{L2}} = \frac{k(1-k)(2-k)TV_1}{4(3-k)L_2 I_O} = \frac{k(1-k)(2-k)}{2(3-k)G_3}\frac{R}{2fL_2} \tag{4.65}$$

and the variation ratio of current i_{L3} through inductor L_3 is

$$\xi_3 = \frac{\Delta i_{L3}/2}{I_{L3}} = \frac{k(2-k)^2 TV_{in}}{4(1-k)L_3 I_O} = \frac{k(2-k)^2}{2(1-k)G_3}\frac{R}{2fL_3} \tag{4.66}$$

The ripple voltage of output voltage v_O is

$$\Delta v_O = \frac{\Delta Q}{C_{32}} = \frac{I_O kT}{C_{32}} = \frac{k}{fC_{32}}\frac{V_O}{R}$$

Therefore, the variation ratio of output voltage v_O is

$$\varepsilon = \frac{\Delta v_O/2}{V_O} = \frac{k}{2RfC_{32}} \tag{4.67}$$

4.4.4 N/O Higher Order Lift Enhanced Circuit

Higher order N/O lift enhanced circuit is derived from the corresponding circuit of the main series by adding the DEC in each stage. Each stage final voltage V_{Ci1} (i = 1, 2, ... n) is

$$V_{Ci1} = (\frac{3-k}{1-k})^i V_{in} \tag{4.68}$$

For nth order lift enhanced circuit, the final output voltage is

$$V_O = [(\frac{3-k}{1-k})^n - 1]V_{in} \tag{4.69}$$

The voltage transfer gain is

$$G_n = \frac{V_O}{V_{in}} = (\frac{3-k}{1-k})^n - 1 \tag{4.70}$$

The variation ratio of output voltage v_O is

$$\varepsilon = \frac{\Delta v_O / 2}{V_O} = \frac{k}{2RfC_{n2}} \tag{4.71}$$

4.5 Re-Enhanced Series

All circuits of negative output super-lift Luo-converters — re-enhanced series — are derived from the corresponding circuits of the main series by adding the DEC twice in each stage circuit.

The first three stages of this series are shown in Figure 4.9 through Figure 4.11. For convenience they are called elementary re-enhanced circuits, re-lift re-enhanced circuits, and triple-lift re-enhanced circuits respectively, and numbered as n = 1, 2, and 3.

4.5.1 N/O Elementary Re-Enhanced Circuit

This circuit is derived from the N/O elementary circuit by adding the DEC twice. Its circuit and switch-on and switch-off equivalent circuits are shown in Figure 4.9. The voltage across capacitor C_1 is charged to V_{in}. The voltage across capacitor C_{12} is charged to V_{C12}. The voltage across capacitor C_{13} is charged to V_{C13}.

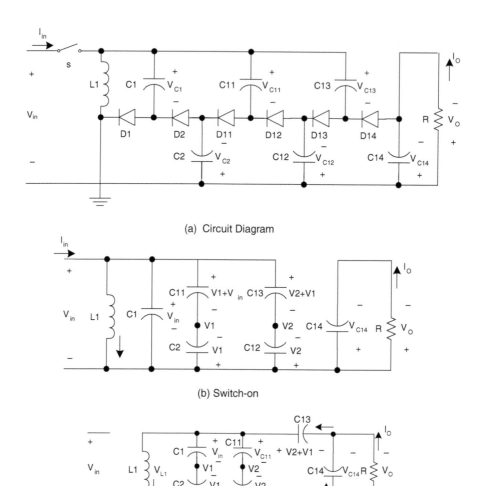

(a) Circuit Diagram

(b) Switch-on

(c) Switch-off

FIGURE 4.9
N/O elementary re-enhanced circuit.

$$V_{C13} = \frac{4-k}{1-k} V_{in} \qquad (4.72)$$

The output voltage is

$$V_O = V_{C13} - V_{in} = [\frac{4-k}{1-k} - 1]V_{in} \qquad (4.73)$$

The voltage transfer gain is

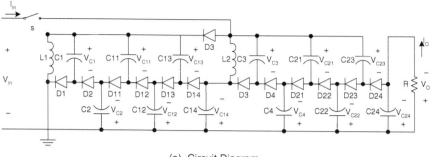

(a) Circuit Diagram

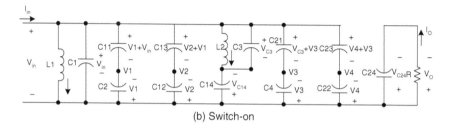

(b) Switch-on

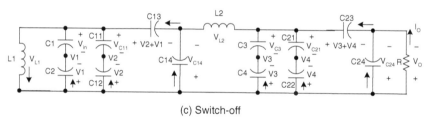

(c) Switch-off

FIGURE 4.10
N/O re-lift re-enhanced circuit.

$$G_1 = \frac{V_O}{V_{in}} = \frac{4-k}{1-k} - 1 \qquad (4.74)$$

The ripple voltage of output voltage v_O is

$$\Delta v_O = \frac{\Delta Q}{C_{14}} = \frac{I_O kT}{C_{14}} = \frac{k}{fC_{14}} \frac{V_O}{R}$$

Therefore, the variation ratio of output voltage v_O is

$$\varepsilon = \frac{\Delta v_O / 2}{V_O} = \frac{k}{2RfC_{14}} \qquad (4.75)$$

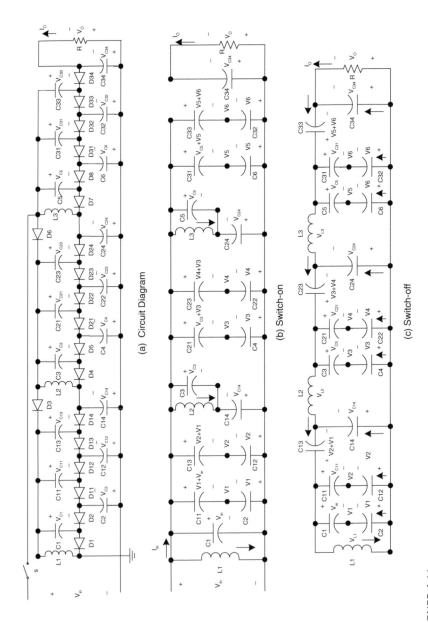

FIGURE 4.11

N/O triple-lift re-enhanced circuit.

4.5.2 N/O Re-Lift Re-Enhanced Circuit

The N/O re-lift re-enhanced circuit is derived from the N/O re-lift circuit by adding the DEC twice in each stage. Its circuit diagram and switch-on and switch-off equivalent circuits are shown in Figure 4.10. The voltage across capacitor C_{13} is charged to V_{C13}. As described in the previous section the voltage across C_{13} is

$$V_{C13} = \frac{4-k}{1-k} V_{in}$$

Analogously,

$$V_{C23} = (\frac{4-k}{1-k})^2 V_{in} \qquad (4.76)$$

The output voltage is

$$V_O = V_{C23} - V_{in} = [(\frac{4-k}{1-k})^2 - 1]V_{in} \qquad (4.77)$$

The voltage transfer gain is

$$G_2 = \frac{V_O}{V_{in}} = (\frac{4-k}{1-k})^2 - 1 \qquad (4.78)$$

The ripple voltage of output voltage v_O is

$$\Delta v_O = \frac{\Delta Q}{C_{24}} = \frac{I_O kT}{C_{24}} = \frac{k}{fC_{24}} \frac{V_O}{R}$$

Therefore, the variation ratio of output voltage v_O is

$$\varepsilon = \frac{\Delta v_O / 2}{V_O} = \frac{k}{2RfC_{24}} \qquad (4.79)$$

4.5.3 N/O Triple-Lift Re-Enhanced Circuit

This circuit is derived from N/O triple-lift circuit by adding the DEC twice in each stage circuit. Its circuit diagram and equivalent circuits during switch-on and switch-off are shown in Figure 4.11. The voltage across capacitor C_{13} is

$$V_{C13} = \frac{4-k}{1-k} V_{in}$$

The voltage across capacitor C_{23} is

$$V_{C23} = (\frac{4-k}{1-k})^2 V_{in}$$

Analogously, the voltage across capacitor C_{33} is

$$V_{C33} = (\frac{4-k}{1-k})^3 V_{in} \tag{4.80}$$

The output voltage is

$$V_O = V_{C33} - V_{in} = [(\frac{4-k}{1-k})^3 - 1]V_{in} \tag{4.81}$$

The voltage transfer gain is

$$G_3 = \frac{V_O}{V_{in}} = (\frac{4-k}{1-k})^3 - 1 \tag{4.82}$$

The ripple voltage of output voltage v_O is

$$\Delta v_O = \frac{\Delta Q}{C_{34}} = \frac{I_O kT}{C_{34}} = \frac{k}{fC_{34}} \frac{V_O}{R}$$

Therefore, the variation ratio of output voltage v_O is

$$\varepsilon = \frac{\Delta v_O / 2}{V_O} = \frac{k}{2RfC_{34}} \tag{4.83}$$

4.5.4   N/O Higher Order Lift Re-Enhanced Circuit

Higher order N/O lift re-enhanced circuits can be derived from the corresponding circuits of the main series by adding the DEC twice in each stage circuit. Each stage final voltage V_{Ci3} ($i = 1, 2, \ldots n$) is

$$V_{Ci3} = (\frac{4-k}{1-k})^i V_{in} \tag{4.84}$$

For nth order lift additional circuit, the final output voltage is

$$V_O = V_{Cn3} - V_{in} = [(\frac{4-k}{1-k})^n - 1]V_{in} \tag{4.85}$$

The voltage transfer gain is

$$G_n = \frac{V_O}{V_{in}} = (\frac{4-k}{1-k})^n - 1 \tag{4.86}$$

The variation ratio of output voltage v_O is

$$\varepsilon = \frac{\Delta v_O / 2}{V_O} = \frac{k}{2RfC_{n4}} \tag{4.87}$$

4.6 Multiple-Enhanced Series

All circuits of negative output super-lift Luo-converters — multiple-enhanced series are derived from the corresponding circuits of the main series by adding the DEC multiple (j) times in each stage circuit.

The first three stages of this series are shown in Figure 4.12 to Figure 4.14. For convenience they are called elementary multiple-enhanced circuits, re-lift multiple-enhanced circuits, and triple-lift multiple-enhanced circuits respectively, and numbered as $n = 1$, 2, and 3.

4.6.1 N/O Elementary Multiple-Enhanced Circuit

This circuit is derived from the N/O elementary circuit by adding the DEC multiple (j) times. Its circuit and switch-on and switch-off equivalent circuits are shown in Figure 4.12. The voltage across capacitor C_{12j-1} is

$$V_{C12j-1} = \frac{j+2-k}{1-k}V_{in} \tag{4.88}$$

The output voltage is

$$V_O = V_{C12j-1} - V_{in} = [\frac{j+2-k}{1-k} - 1]V_{in} \tag{4.89}$$

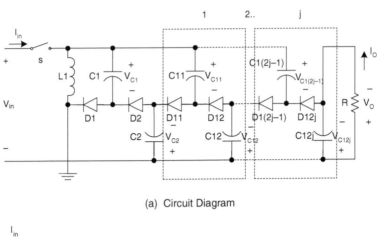

(a) Circuit Diagram

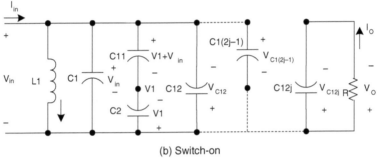

(b) Switch-on

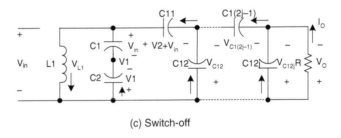

(c) Switch-off

FIGURE 4.12
N/O elementary multiple-enhanced circuit.

The voltage transfer gain is

$$G_1 = \frac{V_O}{V_{in}} = \frac{j+2-k}{1-k} - 1 \qquad (4.90)$$

The ripple voltage of output voltage v_O is

$$\Delta v_O = \frac{\Delta Q}{C_{12j}} = \frac{I_O kT}{C_{12j}} = \frac{k}{fC_{12j}} \frac{V_O}{R}$$

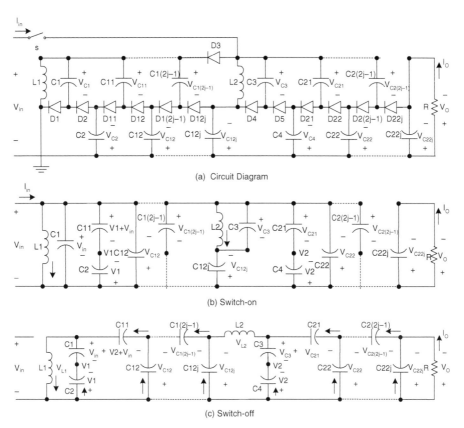

FIGURE 4.13
N/O re-lift multiple-enhanced circuit.

Therefore, the variation ratio of output voltage v_O is

$$\varepsilon = \frac{\Delta v_O / 2}{V_O} = \frac{1-k}{2RfC_{12j}} \qquad (4.91)$$

4.6.2 N/O Re-Lift Multiple-Enhanced Circuit

The N/O re-lift multiple-enhanced circuit is derived from the N/O re-lift circuit by adding the DEC multiple (j) times into each stage. Its circuit diagram and switch-on and switch-off equivalent circuits are shown in Figure 4.13. The voltage across capacitor C_{22j-1} is

$$V_{C22j-1} = (\frac{j+2-k}{1-k})^2 V_{in} \qquad (4.92)$$

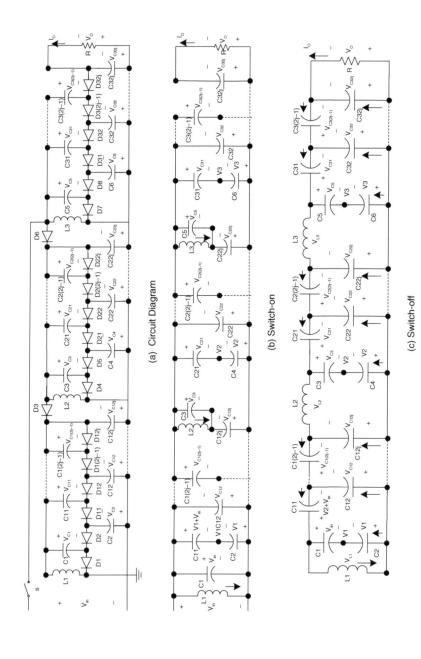

FIGURE 4.14
N/O triple-lift multiple-enhanced circuit.

The output voltage is

$$V_O = V_{C22j-1} - V_{in} = [(\frac{j+2-k}{1-k})^2 - 1]V_{in} \qquad (4.93)$$

The voltage transfer gain is

$$G_2 = \frac{V_O}{V_{in}} = (\frac{j+2-k}{1-k})^2 - 1 \qquad (4.94)$$

The ripple voltage of output voltage v_O is

$$\Delta v_O = \frac{\Delta Q}{C_{22j}} = \frac{I_O kT}{C_{22j}} = \frac{k}{fC_{22j}} \frac{V_O}{R}$$

Therefore, the variation ratio of output voltage v_O is

$$\varepsilon = \frac{\Delta v_O / 2}{V_O} = \frac{k}{2RfC_{22j}} \qquad (4.95)$$

4.6.3 N/O Triple-Lift Multiple-Enhanced Circuit

This circuit is derived from N/O triple-lift circuit by adding the DEC multiple (j) times in each stage circuit. Its circuit diagram and equivalent circuits during switch-on and switch-off are shown in Figure 4.14. The voltage across capacitor C_{32j-1} is

$$V_{C32j-1} = (\frac{j+2-k}{1-k})^3 V_{in} \qquad (4.96)$$

The output voltage is

$$V_O = V_{C32j-1} - V_{in} = [(\frac{j+2-k}{1-k})^3 - 1]V_{in} \qquad (4.97)$$

The voltage transfer gain is

$$G_3 = \frac{V_O}{V_{in}} = (\frac{j+2-k}{1-k})^3 - 1 \qquad (4.98)$$

The ripple voltage of output voltage v_O is

$$\Delta v_O = \frac{\Delta Q}{C_{32j}} = \frac{I_O kT}{C_{32j}} = \frac{k}{fC_{32j}} \frac{V_O}{R}$$

Therefore, the variation ratio of output voltage v_O is

$$\varepsilon = \frac{\Delta v_O / 2}{V_O} = \frac{k}{2RfC_{32j}} \tag{4.99}$$

4.6.4 N/O Higher Order Lift Multiple-Enhanced Circuit

The higher order N/O lift multiple-enhanced circuit is derived from the corresponding circuit of the main series by adding the DEC multiple (j) times in each stage circuit. Each stage final voltage V_{Ci2j-1} ($i = 1, 2, \dots n$) is

$$V_{Ci2j-1} = (\frac{j+2-k}{1-k})^i V_{in} \tag{4.100}$$

For nth order lift multiple-enhanced circuit, the final output voltage is

$$V_O = [(\frac{j+2-k}{1-k})^n - 1]V_{in} \tag{4.101}$$

The voltage transfer gain is

$$G_n = \frac{V_O}{V_{in}} = (\frac{j+2-k}{1-k})^n - 1 \tag{4.102}$$

The variation ratio of output voltage v_O is

$$\varepsilon = \frac{\Delta v_O / 2}{V_O} = \frac{k}{2RfC_{n2j}} \tag{4.103}$$

4.7 Summary of Negative Output Super-Lift Luo-Converters

All circuits of the negative output super-lift Luo-converters as a family can be shown in Figure 4.15. From the analysis in previous sections the common formula to calculate the output voltage can be presented:

$$V_O = \begin{cases} [(\frac{2-k}{1-k})^n - 1]V_{in} & main\_series \\[2mm] [(\frac{2-k}{1-k})^{n-1}(\frac{3-k}{1-k}) - 1]V_{in} & additional\_series \\[2mm] [(\frac{3-k}{1-k})^n - 1]V_{in} & enhanced\_series \\[2mm] [(\frac{4-k}{1-k})^n - 1]V_{in} & re\text{-}enhanced\_series \\[2mm] [(\frac{j+2-k}{1-k})^n - 1]V_{in} & multiple\text{-}enhanced\_series \end{cases} \tag{4.104}$$

The corresponding voltage transfer gain is

$$G = \frac{V_O}{V_{in}} = \begin{cases} (\frac{2-k}{1-k})^n - 1 & main\_series \\[2mm] (\frac{2-k}{1-k})^{n-1}(\frac{3-k}{1-k}) - 1 & additional\_series \\[2mm] (\frac{3-k}{1-k})^n - 1 & enhanced\_series \\[2mm] (\frac{4-k}{1-k})^n - 1 & re\text{-}enhanced\_series \\[2mm] (\frac{j+2-k}{1-k})^n - 1 & multiple\text{-}enhanced\_series \end{cases} \tag{4.105}$$

In order to show the advantages of N/O super-lift converters, their voltage transfer gains can be compared to that of buck converters,

$$G = \frac{V_O}{V_{in}} = k$$

Forward converters,

$$G = \frac{V_O}{V_{in}} = kN \quad (N \text{ is the transformer turn's ratio})$$

Cúk-converters,

$$G = \frac{V_O}{V_{in}} = \frac{k}{1-k}$$

fly-back converters,

$$G = \frac{V_O}{V_{in}} = \frac{kN}{1-k} \quad (N \text{ is the transformer turn's ratio})$$

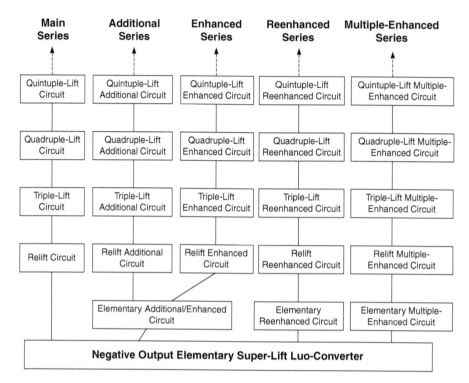

FIGURE 4.15
The family of negative output super-lift Luo-converters.

boost converters,

$$G = \frac{V_O}{V_{in}} = \frac{1}{1-k}$$

and negative output Luo-converters

$$G = \frac{V_O}{V_{in}} = \frac{n}{1-k} \qquad (4.106)$$

If we assume the conduction duty k is 0.2, the output voltage transfer gains are listed in Table 4.1, if the conduction duty k is 0.5, the output voltage transfer gains are listed in Table 4.2, and if the conduction duty k is 0.8, the output voltage transfer gains are listed in Table 4.3.

TABLE 4.1

Voltage Transfer Gains of Converters in the Condition $k = 0.2$

Stage No. (n)	1	2	3	4	5	n
Buck converter				0.2		
Forward converter			0.2N (N is the transformer turn's ratio)			
Cúk-converter				0.25		
Fly-back converter			0.25N (N is the transformer turn's ratio)			
Boost converter				1.25		
Negative output Luo-converters	1.25	2.5	3.75	5	6.25	1.25n
Negative output super-lift converters — main series	1.25	4.06	10.39	24.63	56.67	2.25^n-1
Negative output super-lift converters — additional series	2.5	6.88	16.72	38.87	88.7	$3.5*2.25^{(n-1)}$-1

TABLE 4.2

Voltage Transfer Gains of Converters in the Condition $k = 0.5$

Stage No. (n)	1	2	3	4	5	n
Buck converter				0.5		
Forward converter			0.5N (N is the transformer turn's ratio)			
Cúk-converter				1		
Fly-back converter			N (N is the transformer turn's ratio)			
Boost converter				2		
Negative output Luo-converters	2	4	6	8	10	2n
Negative output super-lift converters — main series	2	8	26	80	242	3^n-1
Negative output super-lift converters — additional series	4	14	44	134	404	$5*3^{(n-1)}$-1

TABLE 4.3

Voltage Transfer Gains of Converters in the Condition $k = 0.8$

Stage No. (n)	1	2	3	4	5	n
Buck converter				0.8		
Forward converter			0.8N (N is the transformer turn's ratio)			
Cúk-converter				4		
Fly-back converter			4N (N is the transformer turn's ratio)			
Boost converter				5		
Negative output Luo-converters	5	10	15	20	25	5n
Negative output super-lift converters — main series	5	35	215	1295	7775	6^n-1
Negative output super-lift converters — additional series	10	65	395	2375	14,255	$11*6^{(n-1)}$-1

4.8   Simulation Results

To verify the design and calculation results, PSpice simulation package was applied to these converters. Choosing V_{in} = 20 V, L_1 = L_2 = L_3 = 10 mH, all C_1 to C_8 = 2 μF, and R = 30 k, and using k = 0.5 and f = 100kHz.

4.8.1   Simulation Results of a N/O Triple-Lift Circuit

The voltage values V_1, V_2, and V_O of a N/O triple-lift circuit are –46 V, –174 V, and –639 V respectively and current waveforms i_{L1} (its average value I_{L1} = 603 mA), i_{L2}, and i_{L3}. The simulation results are shown in Figure 4.16. The voltage values are matched to the calculated results.

4.8.2   Simulation Results of a N/O Triple-Lift Additional Circuit

The voltage values V_1, V_2, V_3, and V_O of a N/O triple-lift additional circuit are –38 V, –146 V, –517 V, and –889 V respectively and current waveforms i_{L1} (its average value I_{L1} = 1.79 A), i_{L2}, and i_{L3}. The simulation results are shown in Figure 4.17. The voltage values are matched to the calculated results.

4.9   Experimental Results

A test rig was constructed to verify the design and calculation results, and compare with PSpice simulation results. The testing conditions are the same: V_{in} = 20 V, L_1 = L_2 = L_3 = 10 mH, all C_1 to C_8 = 2 μF and R = 30 k, and using k = 0.5 and f = 100kHz. The component of the switch is a MOSFET device IRF950 with the rates 950 V/5 A/2 MHz. The output voltage and the first diode current values are measured in the following converters.

4.9.1   Experimental Results of a N/O Triple-Lift Circuit

After careful measurement, the current value of I_{L1} = 0.6 A (shown in channel 1 with 1 A/Div) and voltage value of V_O = –640 V (shown in channel 2 with 200 V/Div) are obtained. The experimental results (current and voltage values) in Figure 4.18 are identically matched to the calculated and simulation results, which are I_{L1} = 0.603 A and V_O = –639 V shown in Figure 4.16.

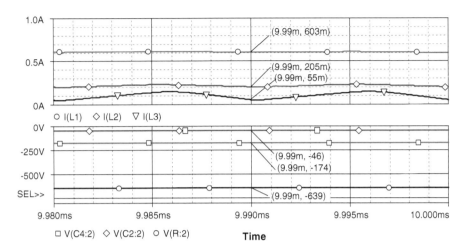

FIGURE 4.16
Simulation results of a N/O triple-lift circuit at condition $k = 0.5$ and $f = 100$ kHz.

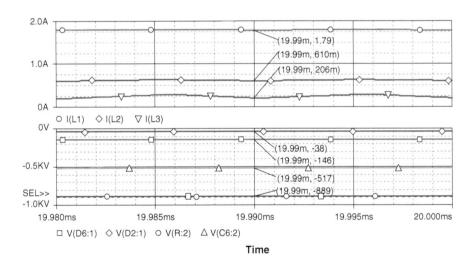

FIGURE 4.17
Simulation results of a N/O triple-lift additional circuit at condition $k = 0.5$ and $f = 100$ kHz.

4.9.2 Experimental Results of a N/O Triple-Lift Additional Circuit

The experimental results (voltage and current values) are identically matched to the calculated and simulation results as shown in Figure 4.19. The current value of $I_{L1} = 1.78$ A (shown in channel 1 with 1 A/Div) and voltage value of $V_O = -890$ V (shown in channel 2 with 200 V/Div) are obtained. The experimental results are identically matched to the calculated and simulation results, which are $I_{L1} = 1.79$ A and $V_O = -889$ V shown in Figure 4.17.

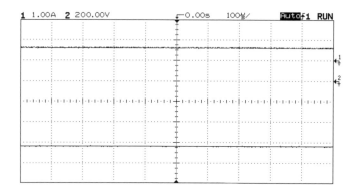

FIGURE 4.18
Experimental results of a N/O triple-lift circuit at condition $k = 0.5$ and $f = 100$ kHz.

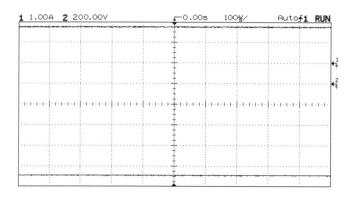

FIGURE 4.19
Experimental results of a N/O triple-lift additional circuit at $k = 0.5$ and $f = 100$ kHz.

4.9.3 Efficiency Comparison of Simulation and Experimental Results

These circuits enhanced the voltage transfer gain successfully, but efficiency, particularly the efficiencies of the tested circuits are 51 to 78%, which is good for high voltage output equipment. Comparison of the simulation and experimental results, which are listed in the Table 4.4 and Table 4.5, demonstrates that all results are well identified with each other.

4.9.4 Transient Process and Stability Analysis

Usually, there is high inrush current during the first power-on. Therefore, the voltage across capacitors is quickly changed to certain values. The transient process is very quick lasting only a few milliseconds.

TABLE 4.4

Comparison of Simulation and Experimental Results of a N/O Triple-Lift Circuit

Stage No. (*n*)	I_{L1} (A)	I_{in} (A)	V_{in} (V)	P_{in} (W)	$|V_O|$ (V)	P_O (W)	η (%)
Simulation results	0.603	0.871	20	17.42	639	13.61	78.12
Experimental results	0.6	0.867	20	17.33	640	13.65	78.75

TABLE 4.5

Comparison of Simulation and Experimental Results of a N/O Triple-Lift Additional Circuit

Stage No. (*n*)	I_{L1} (A)	I_{in} (A)	V_{in} (V)	P_{in} (W)	$|V_O|$ (V)	P_O (W)	η (%)
Simulation results	1.79	2.585	20	51.7	889	26.34	51
Experimental results	1.78	2.571	20	51.4	890	26.4	51

Bibliography

Baliga, B.J., *Modern Power Devices*, New York, John Wiley & Sons, 1987.

Luo, F.L., Advanced voltage lift technique — negative output Luo-converters, *Power Supply Technologies and Applications*, Xi'an, China, 3, 112, 1998.

Luo, F.L. and Ye, H., Negative output super-lift converters, *IEEE Transactions on Power Electronics*, 18, 268, 2003.

Luo, F.L. and Ye, H., Negative output super-lift Luo-converters, in *Proceedings of IEEE-PESC'2003*, Acapulco, Mexico, 2003, p. 1361.

Mitchell, D.M., *DC-to-DC Switching Regulator Analysis*, New York: McGraw-Hill, 1988.

Ye, H., Luo, F.L., and Ye, Z.Z., Practical circuits of Luo-converters, *Power Supply Technologies and Applications*, Xi'an, China, 2, 19, 1999.

5

Positive Output Cascade Boost Converters

Super-lift technique increases the voltage transfer gain in geometric progression. However, these circuits are a bit complex. This chapter introduces a novel approach — the positive output cascade boost converter — that implements the output voltage increasing in geometric progression, but with simpler structure. They also effectively enhance the voltage transfer gain in power-law.

5.1 Introduction

In order to sort these converters differently from existing voltage-lift (VL) and super-lift (SL) converters, these converters are entitled *positive output cascade boost converters*. There are several subseries:

- Main series — Each circuit of the main series has one switch S, n inductors, n capacitors, and $(2n - 1)$ diodes.
- Additional series — Each circuit of the additional series has one switch S, n inductors, $(n + 2)$ capacitors, and $(2n + 1)$ diodes.
- Double series — Each circuit of the double series has one switch S, n inductors, $3n$ capacitors, and $(3n - 1)$ diodes.
- Triple series — Each circuit of the triple series has one switch S, n inductors, $5n$ capacitors, and $(5n - 1)$ diodes.
- Multiple series — Each multiple series circuit has one switch S and a higher number of capacitors and diodes.

In order to concentrate the super-lift function, these converters work in the steady state with the condition of continuous conduction mode (CCM).

The conduction duty ratio is k, switching frequency is f, switching period is $T = 1/f$, the load is resistive load R. The input voltage and current are V_{in} and I_{in}, output voltage and current are V_O and I_O. Assume no power losses during the conversion process, $V_{in} \times I_{in} = V_O \times I_O$. The voltage transfer gain is G:

$$G = \frac{V_O}{V_{in}}$$

5.2   Main Series

The first three stages of positive output cascade boost converters — main series — are shown in Figure 5.1 to Figure 5.3. For convenience they are called elementary boost converter, two-stage circuit, and three-stage circuit respectively, and numbered as $n = 1$, 2, and 3.

5.2.1   Elementary Boost Circuit

The elementary boost converter is the fundamental boost converter introduced in Chapter 1 (see Figure 1.24). Its circuit diagram and its equivalent circuits during switch-on and switch-off are shown in Figure 5.1. The voltage across capacitor C_1 is charged to V_O. The current i_{L1} flowing through inductor L_1 increases with voltage V_{in} during switch-on period kT and decreases with voltage $-(V_O - V_{in})$ during switch-off period $(1 - k)T$. Therefore, the ripple of the inductor current i_{L1} is

$$\Delta i_{L1} = \frac{V_{in}}{L_1} kT = \frac{V_O - V_{in}}{L_1}(1-k)T \tag{5.1}$$

$$V_O = \frac{1}{1-k} V_{in} \tag{5.2}$$

The voltage transfer gain is

$$G = \frac{V_O}{V_{in}} = \frac{1}{1-k} \tag{5.3}$$

The inductor average current is

$$I_{L1} = (1-k)\frac{V_O}{R} \tag{5.4}$$

The variation ratio of current i_{L1} through inductor L_1 is

$$\xi_1 = \frac{\Delta i_{L1}/2}{I_{L1}} = \frac{kTV_{in}}{(1-k)2L_1V_O / R} = \frac{k}{2}\frac{R}{fL_1} \tag{5.5}$$

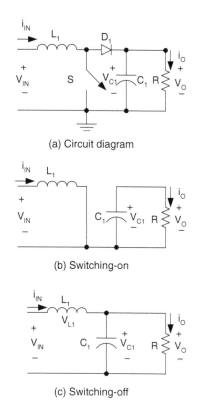

(a) Circuit diagram

(b) Switching-on

(c) Switching-off

FIGURE 5.1
Elementary boost converter.

Usually ξ_1 is small (much lower than unity), which means this converter works in the continuous mode. The ripple voltage of output voltage v_O is

$$\Delta v_O = \frac{\Delta Q}{C_1} = \frac{I_O kT}{C_1} = \frac{k}{fC_1} \frac{V_O}{R}$$

Therefore, the variation ratio of output voltage v_O is

$$\varepsilon = \frac{\Delta v_O / 2}{V_O} = \frac{k}{2RfC_1} \tag{5.6}$$

Usually R is in kΩ, f in 10 kHz, and C_1 in μF, the ripple is smaller than 1%.

5.2.2 Two-Stage Boost Circuit

The two-stage boost circuit is derived from elementary boost converter by adding the parts (L_2-D_2-D_3-C_2). Its circuit diagram and equivalent circuits

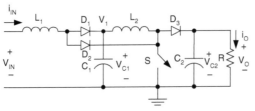

(a) Circuit diagram

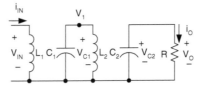

(b) Equivalent circuit during switching-on

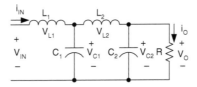

(c) Equivalent circuit during switching-off

FIGURE 5.2
Two-stage boost circuit.

during switch-on and switch-off are shown in Figure 5.2. The voltage across capacitor C_1 is charged to V_1. As described in previous section the voltage V_1 across capacitor C_1 is

$$V_1 = \frac{1}{1-k} V_{in}$$

The voltage across capacitor C_2 is charged to V_O. The current flowing through inductor L_2 increases with voltage V_1 during switching-on period kT and decreases with voltage $-(V_O - V_1)$ during switch-off period $(1-k)T$. Therefore, the ripple of the inductor current i_{L2} is

$$\Delta i_{L2} = \frac{V_1}{L_2} kT = \frac{V_O - V_1}{L_2}(1-k)T \tag{5.7}$$

$$V_O = \frac{1}{1-k} V_1 = (\frac{1}{1-k})^2 V_{in} \tag{5.8}$$

The voltage transfer gain is

$$G = \frac{V_O}{V_{in}} = (\frac{1}{1-k})^2 \tag{5.9}$$

Analogously,

$$\Delta i_{L1} = \frac{V_{in}}{L_1} kT \qquad I_{L1} = \frac{I_O}{(1-k)^2}$$

$$\Delta i_{L2} = \frac{V_1}{L_2} kT \qquad I_{L2} = \frac{I_O}{1-k}$$

Therefore, the variation ratio of current i_{L1} through inductor L_1 is

$$\xi_1 = \frac{\Delta i_{L1}/2}{I_{L1}} = \frac{k(1-k)^2 TV_{in}}{2L_1 I_O} = \frac{k(1-k)^4}{2} \frac{R}{fL_1} \tag{5.10}$$

the variation ratio of current i_{L2} through inductor L_2 is

$$\xi_2 = \frac{\Delta i_{L2}/2}{I_{L2}} = \frac{k(1-k)TV_1}{2L_2 I_O} = \frac{k(1-k)^2}{2} \frac{R}{fL_2} \tag{5.11}$$

and the variation ratio of output voltage v_O is

$$\varepsilon = \frac{\Delta v_O/2}{V_O} = \frac{k}{2RfC_2} \tag{5.12}$$

5.2.3 Three-Stage Boost Circuit

The three-stage boost circuit is derived from the two-stage boost circuit by double adding the parts (L_2-D_2-D_3-C_2). Its circuit diagram and equivalent circuits during switch-on and switch-off are shown in Figure 5.3. The voltage across capacitor C_1 is charged to V_1. As described previously, the voltage V_1 across capacitor C_1 is

$$V_1 = \frac{1}{1-k} V_{in}$$

and voltage V_2 across capacitor C_2 is

$$V_2 = (\frac{1}{1-k})^2 V_{in}$$

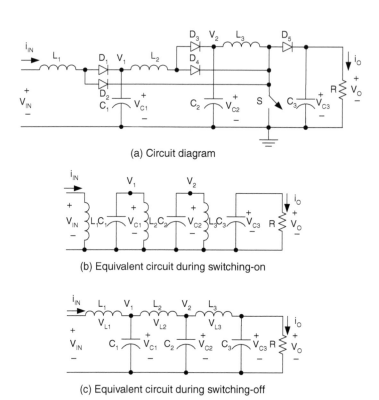

(a) Circuit diagram

(b) Equivalent circuit during switching-on

(c) Equivalent circuit during switching-off

FIGURE 5.3
Three-stage boost circuit.

The voltage across capacitor C_3 is charged to V_O. The current flowing through inductor L_3 increases with voltage V_2 during switching-on period kT and decreases with voltage $-(V_O - V_2)$ during switch-off $(1 - k)T$. Therefore, the ripple of the inductor current i_{L3} is

$$\Delta i_{L3} = \frac{V_2}{L_3} kT = \frac{V_O - V_2}{L_3} (1 - k)T \qquad (5.13)$$

$$V_O = \frac{1}{1 - k} V_2 = (\frac{1}{1 - k})^2 V_1 = (\frac{1}{1 - k})^3 V_{in} \qquad (5.14)$$

The voltage transfer gain is

$$G = \frac{V_O}{V_{in}} = (\frac{1}{1 - k})^3 \qquad (5.15)$$

Analogously,

$$\Delta i_{L1} = \frac{V_{in}}{L_1} kT \qquad I_{L1} = \frac{I_O}{(1-k)^3}$$

$$\Delta i_{L2} = \frac{V_1}{L_2} kT \qquad I_{L2} = \frac{I_O}{(1-k)^2}$$

$$\Delta i_{L3} = \frac{V_2}{L_3} kT \qquad I_{L3} = \frac{I_O}{1-k}$$

Therefore, the variation ratio of current i_{L1} through inductor L_1 is

$$\xi_1 = \frac{\Delta i_{L1}/2}{I_{L1}} = \frac{k(1-k)^3 TV_{in}}{2L_1 I_O} = \frac{k(1-k)^6}{2} \frac{R}{fL_1} \tag{5.16}$$

The variation ratio of current i_{L2} through inductor L_2 is

$$\xi_2 = \frac{\Delta i_{L2}/2}{I_{L2}} = \frac{k(1-k)^2 TV_1}{2L_2 I_O} = \frac{k(1-k)^4}{2} \frac{R}{fL_2} \tag{5.17}$$

The variation ratio of current i_{L3} through inductor L_3 is

$$\xi_3 = \frac{\Delta i_{L3}/2}{I_{L3}} = \frac{k(1-k)TV_2}{2L_3 I_O} = \frac{k(1-k)^2}{2} \frac{R}{fL_3} \tag{5.18}$$

and the variation ratio of output voltage v_O is

$$\varepsilon = \frac{\Delta v_O/2}{V_O} = \frac{k}{2RfC_3} \tag{5.19}$$

5.2.4 Higher Stage Boost Circuit

Higher stage boost circuit can be designed by just multiple repeating of the parts (L_2-D_2-D_3-C_2). For n^{th} stage boost circuit, the final output voltage across capacitor C_n is

$$V_O = (\frac{1}{1-k})^n V_{in}$$

The voltage transfer gain is

$$G = \frac{V_O}{V_{in}} = (\frac{1}{1-k})^n \tag{5.20}$$

the variation ratio of current i_{Li} through inductor L_i ($i = 1, 2, 3, ...n$) is

$$\xi_i = \frac{\Delta i_{Li}/2}{I_{Li}} = \frac{k(1-k)^{2(n-i+1)}}{2} \frac{R}{fL_i} \tag{5.21}$$

and the variation ratio of output voltage v_O is

$$\varepsilon = \frac{\Delta v_O/2}{V_O} = \frac{k}{2RfC_n} \tag{5.22}$$

5.3   Additional Series

All circuits of positive output cascade boost converters — additional series — are derived from the corresponding circuits of the main series by adding a DEC.

The first three stages of this series are shown in Figure 5.4 to Figure 5.6. For convenience they are called elementary additional circuits, two-stage additional circuits, and three-stage additional circuits respectively, and numbered as $n = 1, 2,$ and 3.

5.3.1   Elementary Boost Additional (Double) Circuit

This elementary boost additional circuit is derived from elementary boost converter by adding a DEC. Its circuit and switch-on and switch-off equivalent circuits are shown in Figure 5.4. The voltage across capacitor C_1 and C_{11} is charged to V_1 and voltage across capacitor C_{12} is charged to $V_O = 2 V_1$. The current i_{L1} flowing through inductor L_1 increases with voltage V_{in} during switching-on period kT and decreases with voltage $-(V_1 - V_{in})$ during switching-off $(1 - k)T$. Therefore,

$$\Delta i_{L1} = \frac{V_{in}}{L_1}kT = \frac{V_1 - V_{in}}{L_1}(1-k)T \tag{5.23}$$

$$V_1 = \frac{1}{1-k}V_{in}$$

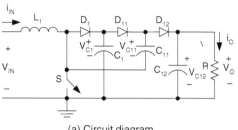

(a) Circuit diagram

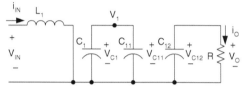

(b) Equivalent circuit during switching-on

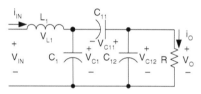

(c) Equivalent circuit during switching-off

FIGURE 5.4
Elementary boost additional (double) circuit.

The output voltage is

$$V_O = 2V_1 = \frac{2}{1-k} V_{in} \qquad (5.24)$$

The voltage transfer gain is

$$G = \frac{V_O}{V_{in}} = \frac{2}{1-k} \qquad (5.25)$$

and

$$i_{in} = I_{L1} = \frac{2}{1-k} I_O \qquad (5.26)$$

The variation ratio of current i_{L1} through inductor L_1 is

$$\xi_1 = \frac{\Delta i_{L1}/2}{I_{L1}} = \frac{k(1-k)TV_{in}}{4L_1 I_O} = \frac{k(1-k)^2}{8} \frac{R}{fL_1} \qquad (5.27)$$

The ripple voltage of output voltage v_O is

$$\Delta v_O = \frac{\Delta Q}{C_{12}} = \frac{I_O kT}{C_{12}} = \frac{k}{fC_{12}} \frac{V_O}{R}$$

Therefore, the variation ratio of output voltage v_O is

$$\varepsilon = \frac{\Delta v_O/2}{V_O} = \frac{k}{2RfC_{12}} \qquad (5.28)$$

5.3.2 Two-Stage Boost Additional Circuit

The two-stage additional boost circuit is derived from the two-stage boost circuit by adding a DEC. Its circuit diagram and switch-on and switch-off equivalent circuits are shown in Figure 5.5. The voltage across capacitor C_1 is charged to V_1. As described in the previous section the voltage V_1 across capacitor C_1 is

$$V_1 = \frac{1}{1-k} V_{in}$$

The voltage across capacitor C_2 and capacitor C_{11} is charged to V_2 and voltage across capacitor C_{12} is charged to V_O. The current flowing through inductor L_2 increases with voltage V_1 during switch-on period kT and decreases with voltage $-(V_2 - V_1)$ during switch-off period $(1 - k)T$. Therefore, the ripple of the inductor current i_{L2} is

$$\Delta i_{L2} = \frac{V_1}{L_2} kT = \frac{V_2 - V_1}{L_2}(1-k)T \qquad (5.29)$$

$$V_2 = \frac{1}{1-k} V_1 = (\frac{1}{1-k})^2 V_{in} \qquad (5.30)$$

The output voltage is

$$V_O = 2V_2 = \frac{2}{1-k} V_1 = 2(\frac{1}{1-k})^2 V_{in} \qquad (5.31)$$

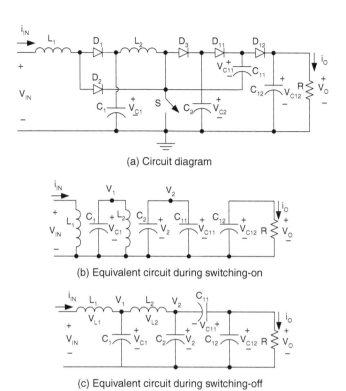

(a) Circuit diagram

(b) Equivalent circuit during switching-on

(c) Equivalent circuit during switching-off

FIGURE 5.5
Two-stage boost additional circuit.

The voltage transfer gain is

$$G = \frac{V_O}{V_{in}} = 2(\frac{1}{1-k})^2 \qquad (5.32)$$

Analogously,

$$\Delta i_{L1} = \frac{V_{in}}{L_1} kT \qquad I_{L1} = \frac{2}{(1-k)^2} I_O$$

$$\Delta i_{L2} = \frac{V_1}{L_2} kT \qquad I_{L2} = \frac{2I_O}{1-k}$$

Therefore, the variation ratio of current i_{L1} through inductor L_1 is

$$\xi_1 = \frac{\Delta i_{L1}/2}{I_{L1}} = \frac{k(1-k)^2 TV_{in}}{4L_1 I_O} = \frac{k(1-k)^4}{8}\frac{R}{fL_1} \tag{5.33}$$

and the variation ratio of current i_{L2} through inductor L_2 is

$$\xi_2 = \frac{\Delta i_{L2}/2}{I_{L2}} = \frac{k(1-k)TV_1}{4L_2 I_O} = \frac{k(1-k)^2}{8}\frac{R}{fL_2} \tag{5.34}$$

The ripple voltage of output voltage v_O is

$$\Delta v_O = \frac{\Delta Q}{C_{12}} = \frac{I_O kT}{C_{12}} = \frac{k}{fC_{12}}\frac{V_O}{R}$$

Therefore, the variation ratio of output voltage v_O is

$$\varepsilon = \frac{\Delta v_O/2}{V_O} = \frac{k}{2RfC_{12}} \tag{5.35}$$

5.3.3 Three-Stage Boost Additional Circuit

This circuit is derived from the three-stage boost circuit by adding a DEC. Its circuit diagram and equivalent circuits during switch-on and switch-off are shown in Figure 5.6. The voltage across capacitor C_1 is charged to V_1. As described previously the voltage V_1 across capacitor C_1 is $V_1 = (1/1-k)V_{in}$, and voltage V_2 across capacitor C_2 is $V_2 = (1/1-k)^2 V_{in}$.

The voltage across capacitor C_3 and capacitor C_{11} is charged to V_3. The voltage across capacitor C_{12} is charged to V_O. The current flowing through inductor L_3 increases with voltage V_2 during switch-on period kT and decreases with voltage $-(V_3 - V_2)$ during switch-off $(1-k)T$. Therefore,

$$\Delta i_{L3} = \frac{V_2}{L_3}kT = \frac{V_3 - V_2}{L_3}(1-k)T \tag{5.36}$$

and

$$V_3 = \frac{1}{1-k}V_2 = \left(\frac{1}{1-k}\right)^2 V_1 = \left(\frac{1}{1-k}\right)^3 V_{in} \tag{5.37}$$

The output voltage is

$$V_O = 2V_3 = 2\left(\frac{1}{1-k}\right)^3 V_{in} \tag{5.38}$$

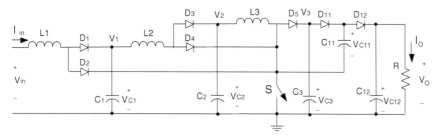

(a) Circuit diagram

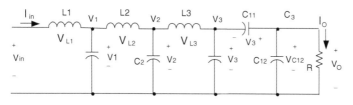

(b) Equivalent circuit during switch-on

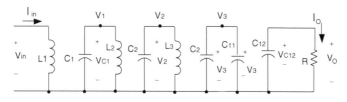

(b) Equivalent circuit during switch-off

FIGURE 5.6
Three-stage boost additional circuit.

The voltage transfer gain is

$$G = \frac{V_O}{V_{in}} = 2(\frac{1}{1-k})^3 \tag{5.39}$$

Analogously:

$$\Delta i_{L1} = \frac{V_{in}}{L_1} kT \qquad I_{L1} = \frac{2}{(1-k)^3} I_O$$

$$\Delta i_{L2} = \frac{V_1}{L_2} kT \qquad I_{L2} = \frac{2}{(1-k)^2} I_O$$

$$\Delta i_{L3} = \frac{V_2}{L_3} kT \qquad I_{L3} = \frac{2I_O}{1-k}$$

Therefore, the variation ratio of current i_{L1} through inductor L_1 is

$$\xi_1 = \frac{\Delta i_{L1}/2}{I_{L1}} = \frac{k(1-k)^3 TV_{in}}{4L_1 I_O} = \frac{k(1-k)^6}{8}\frac{R}{fL_1} \tag{5.40}$$

and the variation ratio of current i_{L2} through inductor L_2 is

$$\xi_2 = \frac{\Delta i_{L2}/2}{I_{L2}} = \frac{k(1-k)^2 TV_1}{4L_2 I_O} = \frac{k(1-k)^4}{8}\frac{R}{fL_2} \tag{5.41}$$

and the variation ratio of current i_{L3} through inductor L_3 is

$$\xi_3 = \frac{\Delta i_{L3}/2}{I_{L3}} = \frac{k(1-k)TV_2}{4L_3 I_O} = \frac{k(1-k)^2}{8}\frac{R}{fL_3} \tag{5.42}$$

The ripple voltage of output voltage v_O is

$$\Delta v_O = \frac{\Delta Q}{C_{12}} = \frac{I_O kT}{C_{12}} = \frac{k}{fC_{12}}\frac{V_O}{R}$$

Therefore, the variation ratio of output voltage v_O is

$$\varepsilon = \frac{\Delta v_O/2}{V_O} = \frac{k}{2RfC_{12}} \tag{5.43}$$

5.3.4 Higher Stage Boost Additional Circuit

Higher stage boost additional circuits can be designed by repeating the parts $(L_2\text{-}D_2\text{-}D_3\text{-}C_2)$ multiple times. For nth stage additional circuit, the final output voltage is

$$V_O = 2(\frac{1}{1-k})^n V_{in}$$

The voltage transfer gain is

$$G = \frac{V_O}{V_{in}} = 2(\frac{1}{1-k})^n \tag{5.44}$$

Analogously, the variation ratio of current i_{Li} through inductor L_i ($i = 1, 2, 3, \ldots n$) is

$$\xi_i = \frac{\Delta i_{Li} / 2}{I_{Li}} = \frac{k(1-k)^{2(n-i+1)}}{8} \frac{R}{fL_i} \qquad (5.45)$$

and the variation ratio of output voltage v_O is

$$\varepsilon = \frac{\Delta v_O / 2}{V_O} = \frac{k}{2RfC_{12}} \qquad (5.46)$$

5.4 Double Series

All circuits of the positive output cascade boost converter — double series — are derived from the corresponding circuits of the main series by adding a DEC in each stage circuit. The first three stages of this series are shown in Figures 5.4, 5.7, and 5.8. For convenience to explain, they are called elementary double circuits, two-stage double circuits, and three-stage double circuits respectively, and numbered as $n = 1, 2,$ and 3.

5.4.1 Elementary Double Boost Circuit

From the construction principle, the elementary double boost circuit is derived from the elementary boost converter by adding a DEC. Its circuit and switch-on and switch-off equivalent circuits are shown in Figure 5.4, which is the same as the elementary boost additional circuit.

5.4.2 Two-Stage Double Boost Circuit

The two-stage double boost circuit is derived from the two-stage boost circuit by adding a DEC in each stage circuit. Its circuit diagram and switch-on and switch-off equivalent circuits are shown in Figure 5.7. The voltage across capacitor C_1 and capacitor C_{11} is charged to V_1. As described in the previous section, the voltage V_1 across capacitor C_1 and capacitor C_{11} is $V_1 = (1/1-k)V_{in}$. The voltage across capacitor C_{12} is charged to $2V_1$.

The current flowing through inductor L_2 increases with voltage $2V_1$ during switch-on period kT and decreases with voltage $-(V_2 - 2V_1)$ during switch-off period $(1 - k)T$. Therefore, the ripple of the inductor current i_{L2} is

$$\Delta i_{L2} = \frac{2V_1}{L_2} kT = \frac{V_2 - 2V_1}{L_2}(1-k)T \qquad (5.47)$$

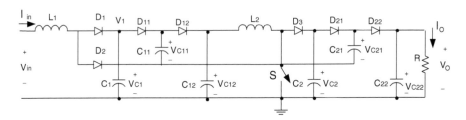

(a) Circuit diagram

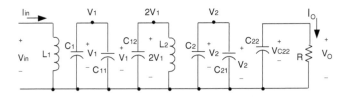

(b) Equivalent circuit during switch-on

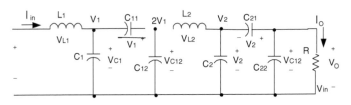

(c) Equivalent circuit during switch-off

FIGURE 5.7
Two-stage boost double circuit.

$$V_2 = \frac{2}{1-k}V_1 = 2(\frac{1}{1-k})^2 V_{in} \qquad (5.48)$$

The output voltage is

$$V_O = 2V_2 = (\frac{2}{1-k})^2 V_{in} \qquad (5.49)$$

The voltage transfer gain is

$$G = \frac{V_O}{V_{in}} = (\frac{2}{1-k})^2 \qquad (5.50)$$

Analogously,

$$\Delta i_{L1} = \frac{V_{in}}{L_1} kT \qquad\qquad I_{L1} = (\frac{2}{1-k})^2 I_O$$

$$\Delta i_{L2} = \frac{V_1}{L_2} kT \qquad\qquad I_{L2} = \frac{2I_O}{1-k}$$

Therefore, the variation ratio of current i_{L1} through inductor L_1 is

$$\xi_1 = \frac{\Delta i_{L1}/2}{I_{L1}} = \frac{k(1-k)^2 T V_{in}}{8 L_1 I_O} = \frac{k(1-k)^4}{16} \frac{R}{f L_1} \qquad (5.51)$$

and the variation ratio of current i_{L2} through inductor L_2 is

$$\xi_2 = \frac{\Delta i_{L2}/2}{I_{L2}} = \frac{k(1-k)T V_1}{4 L_2 I_O} = \frac{k(1-k)^2}{8} \frac{R}{f L_2} \qquad (5.52)$$

The ripple voltage of output voltage v_O is

$$\Delta v_O = \frac{\Delta Q}{C_{22}} = \frac{I_O kT}{C_{22}} = \frac{k}{f C_{22}} \frac{V_O}{R}$$

Therefore, the variation ratio of output voltage v_O is

$$\varepsilon = \frac{\Delta v_O/2}{V_O} = \frac{k}{2 R f C_{22}} \qquad (5.53)$$

5.4.3 Three-Stage Double Boost Circuit

This circuit is derived from the three-stage boost circuit by adding a DEC in each stage circuit. Its circuit diagram and equivalent circuits during switch-on and -off are shown in Figure 5.8. The voltage across capacitor C_1 and capacitor C_{11} is charged to V_1. As described earlier the voltage V_1 across capacitor C_1 and capacitor C_{11} is $V_1 = (1/1-k)V_{in}$, and voltage V_2 across capacitor C_2 and capacitor C_{12} is $V_2 = 2(1/1-k)^2 V_{in}$.

The voltage across capacitor C_{22} is $2V_2 = (2/1-k)^2 V_{in}$. The voltage across capacitor C_3 and capacitor C_{31} is charged to V_3. The voltage across capacitor C_{12} is charged to V_O. The current flowing through inductor L_3 increases with

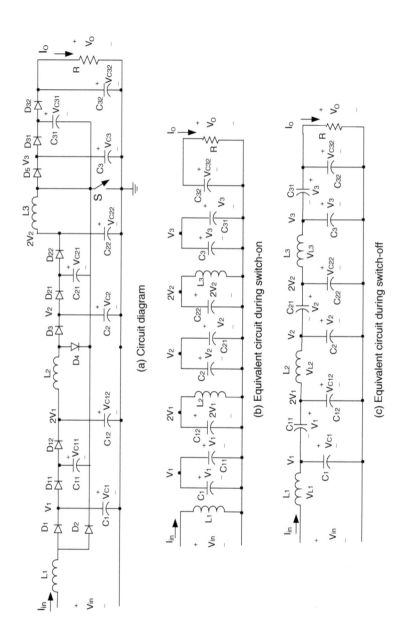

(a) Circuit diagram

(b) Equivalent circuit during switch-on

(c) Equivalent circuit during switch-off

FIGURE 5.8
Three-stage boost double circuit.

voltage V_2 during switch-on period kT and decreases with voltage $-(V_3 - 2V_2)$ during switch-off $(1 - k)T$. Therefore,

$$\Delta i_{L3} = \frac{2V_2}{L_3} kT = \frac{V_3 - 2V_2}{L_3} (1-k)T \tag{5.54}$$

and

$$V_3 = \frac{2V_2}{(1-k)} = \frac{4}{(1-k)^3} V_{in} \tag{5.55}$$

The output voltage is

$$V_O = 2V_3 = (\frac{2}{1-k})^3 V_{in} \tag{5.56}$$

The voltage transfer gain is

$$G = \frac{V_O}{V_{in}} = (\frac{2}{1-k})^3 \tag{5.57}$$

Analogously,

$$\Delta i_{L1} = \frac{V_{in}}{L_1} kT \qquad\qquad I_{L1} = \frac{8}{(1-k)^3} I_O$$

$$\Delta i_{L2} = \frac{V_1}{L_2} kT \qquad\qquad I_{L2} = \frac{4}{(1-k)^2} I_O$$

$$\Delta i_{L3} = \frac{V_2}{L_3} kT \qquad\qquad I_{L3} = \frac{2I_O}{1-k}$$

Therefore, the variation ratio of current i_{L1} through inductor L_1 is

$$\xi_1 = \frac{\Delta i_{L1}/2}{I_{L1}} = \frac{k(1-k)^3 TV_{in}}{16L_1 I_O} = \frac{k(1-k)^6}{128} \frac{R}{fL_1} \tag{5.58}$$

and the variation ratio of current i_{L2} through inductor L_2 is

$$\xi_2 = \frac{\Delta i_{L2}/2}{I_{L2}} = \frac{k(1-k)^2 TV_1}{8L_2 I_O} = \frac{k(1-k)^4}{32} \frac{R}{fL_2} \tag{5.59}$$

and the variation ratio of current i_{L3} through inductor L_3 is

$$\xi_3 = \frac{\Delta i_{L3}/2}{I_{L3}} = \frac{k(1-k)TV_2}{4L_3 I_O} = \frac{k(1-k)^2}{8}\frac{R}{fL_3} \tag{5.60}$$

The ripple voltage of output voltage v_O is

$$\Delta v_O = \frac{\Delta Q}{C_{32}} = \frac{I_O(1-k)T}{C_{32}} = \frac{1-k}{fC_{32}}\frac{V_O}{R}$$

Therefore, the variation ratio of output voltage v_O is

$$\varepsilon = \frac{\Delta v_O/2}{V_O} = \frac{k}{2RfC_{32}} \tag{5.61}$$

5.4.4 Higher Stage Double Boost Circuit

The higher stage double boost circuits can be derived from the corresponding main series circuits by adding a DEC in each stage circuit. For nth stage additional circuit, the final output voltage is

$$V_O = (\frac{2}{1-k})^n V_{in}$$

The voltage transfer gain is

$$G = \frac{V_O}{V_{in}} = (\frac{2}{1-k})^n \tag{5.62}$$

Analogously, the variation ratio of current i_{Li} through inductor L_i ($i = 1, 2, 3, \ldots n$) is

$$\xi_i = \frac{\Delta i_{Li}/2}{I_{Li}} = \frac{k(1-k)^{2(n-i+1)}}{2*2^{2n}}\frac{R}{fL_i} \tag{5.63}$$

The variation ratio of output voltage v_O is

$$\varepsilon = \frac{\Delta v_O/2}{V_O} = \frac{k}{2RfC_{n2}} \tag{5.64}$$

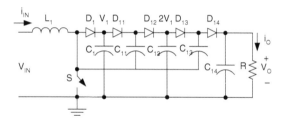

(a) Circuit Diagram

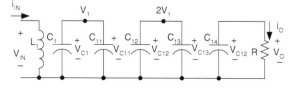

(b) Equivalent circuit during switching-on

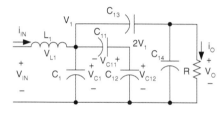

(c) Equivalent circuit during switching-off

FIGURE 5.9
Cascade boost re-double circuit.

5.5 Triple Series

All circuits of P/O cascade boost converters — triple series — are derived from the corresponding circuits of the double series by adding the DEC twice in each stage circuit. The first three stages of this series are shown in Figure 5.9 to Figure 5.11. For convenience they are called elementary triple boost circuit, two-stage triple boost circuit, and three-stage triple boost circuit respectively, and numbered as $n = 1$, 2 and 3.

5.5.1 Elementary Triple Boost Circuit

From the construction principle, the elementary triple boost circuit is derived from the elementary double boost circuit by adding another DEC. Its circuit and switch-on and -off equivalent circuits are shown in Figure 5.9. The output voltage of first stage boost circuit is V_1, $V_1 = V_{in}/(1-k)$.

The voltage across capacitors C_1 and C_{11} is charged to V_1 and voltage across capacitors C_{12} and C_{13} is charged to $V_{C13} = 2V_1$. The current i_{L1} flowing through inductor L_1 increases with voltage V_{in} during switch-on period kT and decreases with voltage $-(V_1 - V_{in})$ during switch-off $(1-k)T$. Therefore,

$$\Delta i_{L1} = \frac{V_{in}}{L_1} kT = \frac{V_1 - V_{in}}{L_1}(1-k)T \tag{5.65}$$

$$V_1 = \frac{1}{1-k} V_{in}$$

The output voltage is

$$V_O = V_{C1} + V_{C13} = 3V_1 = \frac{3}{1-k} V_{in} \tag{5.66}$$

The voltage transfer gain is

$$G = \frac{V_O}{V_{in}} = \frac{3}{1-k} \tag{5.67}$$

5.5.2 Two-Stage Triple Boost Circuit

The two-stage triple boost circuit is derived from the two-stage double boost circuit by adding another DEC in each stage circuit. Its circuit diagram and switch-on and -off equivalent circuits are shown in Figure 5.10. As described in the previous section the voltage V_1 across capacitors C_1 and C_{11} is $V_1 = (1/1-k)V_{in}$. The voltage across capacitor C_{14} is charged to $3V_1$.

The voltage across capacitors C_2 and C_{21} is charged to V_2 and voltage across capacitors C_{22} and C_{23} is charged to $V_{C23} = 2V_2$. The current flowing through inductor L_2 increases with voltage $3V_1$ during switch-on period kT, and decreases with voltage $-(V_2 - 3V_1)$ during switch-off period $(1-k)T$. Therefore, the ripple of the inductor current i_{L2} is

$$\Delta i_{L2} = \frac{3V_1}{L_2} kT = \frac{V_2 - 3V_1}{L_2}(1-k)T \tag{5.68}$$

$$V_2 = \frac{3}{1-k} V_1 = 3(\frac{1}{1-k})^2 V_{in} \tag{5.69}$$

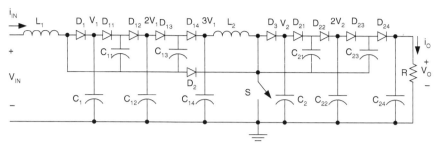

(a) Circuit diagram

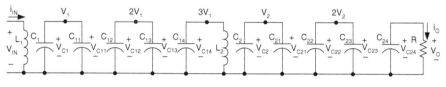

(b) Equivalent circuit during switching-on

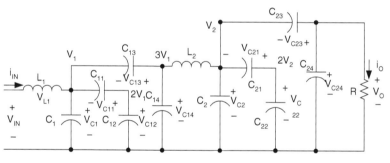

(c) Equivalent circuit during switching-off

FIGURE 5.10
Two-stage boost re-double circuit.

The output voltage is

$$V_O = V_{C2} + V_{C23} = 3V_2 = (\frac{3}{1-k})^2 V_{in} \tag{5.70}$$

The voltage transfer gain is

$$G = \frac{V_O}{V_{in}} = (\frac{3}{1-k})^2 \tag{5.71}$$

Analogously,

$$\Delta i_{L1} = \frac{V_{in}}{L_1}kT \qquad\qquad I_{L1} = (\frac{2}{1-k})^2 I_O$$

$$\Delta i_{L2} = \frac{V_1}{L_2}kT \qquad\qquad I_{L2} = \frac{2I_O}{1-k}$$

Therefore, the variation ratio of current i_{L1} through inductor L_1 is

$$\xi_1 = \frac{\Delta i_{L1}/2}{I_{L1}} = \frac{k(1-k)^2 TV_{in}}{8L_1 I_O} = \frac{k(1-k)^4}{16}\frac{R}{fL_1} \qquad (5.72)$$

and the variation ratio of current i_{L2} through inductor L_2 is

$$\xi_2 = \frac{\Delta i_{L2}/2}{I_{L2}} = \frac{k(1-k)TV_1}{4L_2 I_O} = \frac{k(1-k)^2}{8}\frac{R}{fL_2} \qquad (5.73)$$

The ripple voltage of output voltage v_O is

$$\Delta v_O = \frac{\Delta Q}{C_{24}} = \frac{I_O kT}{C_{24}} = \frac{k}{fC_{24}}\frac{V_O}{R}$$

Therefore, the variation ratio of output voltage v_O is

$$\varepsilon = \frac{\Delta v_O/2}{V_O} = \frac{k}{2RfC_{24}} \qquad (5.74)$$

5.5.3   Three-Stage Triple Boost Circuit

This circuit is derived from the three-stage double boost circuit by adding another DEC in each stage circuit. Its circuit diagram and equivalent circuits during switch-on and -off are shown in Figure 5.11. As described earlier the voltage V_2 across capacitors C_2 and C_{11} is $V_2 = 3V_1 = (3/1-k)V_{in}$, and voltage across capacitor C_{24} is charged to $3V_2$.

The voltage across capacitors C_3 and C_{31} is charged to V_3 and voltage across capacitors C_{32} and C_{33} is charged to $V_{C33} = 2V_3$. The current flowing through inductor L_3 increases with voltage $3V_2$ during switch-on period kT and decreases with voltage $-(V_3 - 3V_2)$ during switch-off $(1-k)T$. Therefore, the ripple of the inductor current i_{L3} is

$$\Delta i_{L3} = \frac{3V_2}{L_3}kT = \frac{V_3 - 3V_2}{L_3}(1-k)T \qquad (5.75)$$

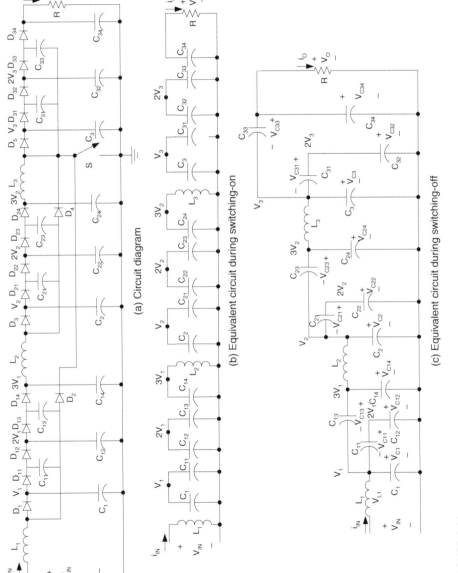

(a) Circuit diagram

(b) Equivalent circuit during switching-on

(c) Equivalent circuit during switching-off

FIGURE 5.11

Three-stage boost re-double circuit.

and

$$V_3 = \frac{3}{1-k}V_2 = 9(\frac{1}{1-k})^3 V_{in} \tag{5.76}$$

The output voltage is

$$V_O = V_{C3} + V_{C33} = 3V_3 = (\frac{3}{1-k})^3 V_{in} \tag{5.77}$$

The voltage transfer gain is

$$G = \frac{V_O}{V_{in}} = (\frac{3}{1-k})^3 \tag{5.78}$$

Analogously,

$$\Delta i_{L1} = \frac{V_{in}}{L_1}kT \qquad I_{L1} = \frac{32}{(1-k)^3}I_O$$

$$\Delta i_{L2} = \frac{V_1}{L_2}kT \qquad I_{L2} = \frac{8}{(1-k)^2}I_O$$

$$\Delta i_{L3} = \frac{V_2}{L_3}kT \qquad I_{L3} = \frac{2}{1-k}I_O$$

Therefore, the variation ratio of current i_{L1} through inductor L_1 is

$$\xi_1 = \frac{\Delta i_{L1}/2}{I_{L1}} = \frac{k(1-k)^3 TV_{in}}{64L_1 I_O} = \frac{k(1-k)^6}{12^3}\frac{R}{fL_1} \tag{5.79}$$

and the variation ratio of current i_{L2} through inductor L_2 is

$$\xi_2 = \frac{\Delta i_{L2}/2}{I_{L2}} = \frac{k(1-k)^2 TV_1}{16L_2 I_O} = \frac{k(1-k)^4}{12^2}\frac{R}{fL_2} \tag{5.80}$$

and the variation ratio of current i_{L3} through inductor L_3 is

$$\xi_3 = \frac{\Delta i_{L3}/2}{I_{L3}} = \frac{k(1-k)TV_2}{4L_3 I_O} = \frac{k(1-k)^2}{12}\frac{R}{fL_3} \tag{5.81}$$

The ripple voltage of output voltage v_O is

$$\Delta v_O = \frac{\Delta Q}{C_{34}} = \frac{I_O kT}{C_{34}} = \frac{k}{fC_{34}} \frac{V_O}{R}$$

Therefore, the variation ratio of output voltage v_O is

$$\varepsilon = \frac{\Delta v_O / 2}{V_O} = \frac{k}{2RfC_{34}} \tag{5.82}$$

5.5.4 Higher Stage Triple Boost Circuit

Higher stage triple boost circuits can be derived from the corresponding circuit of double boost series by adding another DEC in each stage circuit. For nth stage additional circuit, the final output voltage is

$$V_O = (\frac{3}{1-k})^n V_{in}$$

The voltage transfer gain is

$$G = \frac{V_O}{V_{in}} = (\frac{3}{1-k})^n \tag{5.83}$$

Analogously, the variation ratio of current i_{Li} through inductor L_i ($i = 1, 2, 3, \dots n$) is

$$\xi_i = \frac{\Delta i_{Li} / 2}{I_{Li}} = \frac{k(1-k)^{2(n-i+1)}}{12^{(n-i+1)}} \frac{R}{fL_i} \tag{5.84}$$

and the variation ratio of output voltage v_O is

$$\varepsilon = \frac{\Delta v_O / 2}{V_O} = \frac{k}{2RfC_{n2}} \tag{5.85}$$

5.6 Multiple Series

All circuits of P/O cascade boost converters — multiple series — are derived from the corresponding circuits of the main series by adding DEC multiple

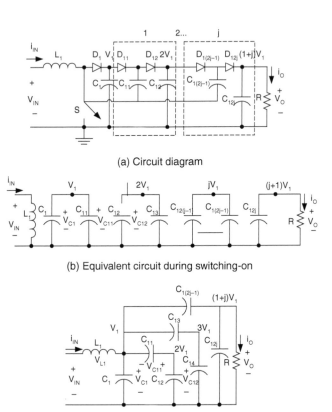

(a) Circuit diagram

(b) Equivalent circuit during switching-on

(c) Equivalent circuit during switching-off

FIGURE 5.12
Cascade boost multiple-double circuit.

(j) times in each stage circuit. The first three stages of this series are shown in Figure 5.12 to Figure 5.14. For convenience they are called elementary multiple boost circuits, two-stage multiple boost circuits, and three-stage multiple boost circuits respectively, and numbered as $n = 1, 2$, and 3.

5.6.1 Elementary Multiple Boost Circuit

From the construction principle, the elementary multiple boost circuit is derived from the elementary boost converter by adding DEC multiple (j) times in the circuit. Its circuit and switch-on and -off equivalent circuits are shown in Figure 5.12.

The voltage across capacitors C_1 and C_{11} is charged to V_1 and voltage across capacitors C_{12} and C_{13} is charged to $V_{C13} = 2V_1$. The voltage across capacitors $C_{1(2j-2)}$ and $C_{1(2j-1)}$ is charged to $V_{C1(2j-1)} = jV_1$. The current i_{L1} flowing through inductor L_1 increases with voltage V_{in} during switch-on period kT and decreases with voltage $-(V_1 - V_{in})$ during switch-off $(1 - k)T$. Therefore,

$$\Delta i_{L1} = \frac{V_{in}}{L_1} kT = \frac{V_1 - V_{in}}{L_1} (1-k)T \qquad (5.86)$$

$$V_1 = \frac{1}{1-k} V_{in} \qquad (5.87)$$

The output voltage is

$$V_O = V_{C1} + V_{C1(2j-1)} = (1+j)V_1 = \frac{1+j}{1-k} V_{in} \qquad (5.88)$$

The voltage transfer gain is

$$G = \frac{V_O}{V_{in}} = \frac{1+j}{1-k} \qquad (5.89)$$

5.6.2 Two-Stage Multiple Boost Circuit

The two-stage multiple boost circuit is derived from the two-stage boost circuit by adding multiple (j) DECs in each stage circuit. Its circuit diagram and switch-on and -off equivalent circuits are shown in Figure 5.13. The voltage across capacitor C_1 and capacitor C_{11} is charged to $V_1 = (1/1-k)V_{in}$. The voltage across capacitor $C_{1(2j)}$ is charged to $(1 + j)V_1$.

The current flowing through inductor L_2 increases with voltage $(1 + j)V_1$ during switch-on period kT and decreases with voltage $-[V_2 - (1 + j)V_1]$ during switch-off period $(1 - k)T$. Therefore, the ripple of the inductor current i_{L2} is

$$\Delta i_{L2} = \frac{1+j}{L_2} kTV_1 = \frac{V_2 - (1+j)V_1}{L_2} (1-k)T \qquad (5.90)$$

$$V_2 = \frac{1+j}{1-k} V_1 = (1+j)(\frac{1}{1-k})^2 V_{in} \qquad (5.91)$$

The output voltage is

$$V_O = V_{C1} + V_{C1(2j-1)} = (1+j)V_2 = (\frac{1+j}{1-k})^2 V_{in} \qquad (5.92)$$

The voltage transfer gain is

$$G = \frac{V_O}{V_{in}} = (\frac{1+j}{1-k})^2 \qquad (5.93)$$

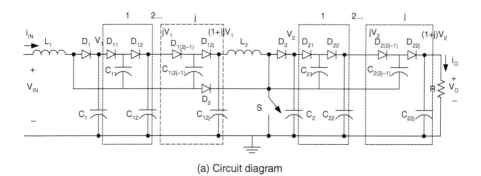

(a) Circuit diagram

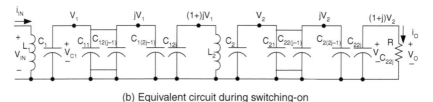

(b) Equivalent circuit during switching-on

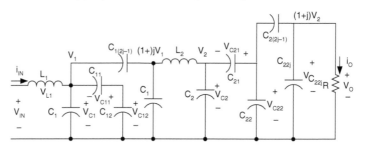

(c) Equivalent circuit during switching-off

FIGURE 5.13
Two-stage boost multiple-double circuit.

The ripple voltage of output voltage v_O is

$$\Delta v_O = \frac{\Delta Q}{C_{22j}} = \frac{I_O kT}{C_{22j}} = \frac{k}{fC_{22j}} \frac{V_O}{R}$$

Therefore, the variation ratio of output voltage v_O is

$$\varepsilon = \frac{\Delta v_O / 2}{V_O} = \frac{k}{2RfC_{22j}} \qquad (5.94)$$

5.6.3 Three-Stage Multiple Boost Circuit

This circuit is derived from the three-stage boost circuit by adding multiple (j) DECs in each stage circuit. Its circuit diagram and equivalent circuits during switch-on and -off are shown in Figure 5.14. The voltage across capacitor C_1 and capacitor C_{11} is charged to $V_1 = \left(1/1-k\right)V_{in}$. The voltage across capacitor $C_{1(2j)}$ is charged to $(1 + j)V_1$. The voltage V_2 across capacitor C_2 and capacitor $C_{2(2j)}$ is charged to $(1 + j)V_2$.

The current flowing through inductor L_3 increases with voltage $(1 + j)V_2$ during switch-on period kT and decreases with voltage $-[V_3 - (1 + j)V_2]$ during switch-off $(1 - k)T$. Therefore,

$$\Delta i_{L3} = \frac{1+j}{L_3}kTV_2 = \frac{V_3 - (1+j)V_2}{L_3}(1-k)T \tag{5.95}$$

and

$$V_3 = \frac{(1+j)V_2}{(1-k)} = \frac{(1+j)^2}{(1-k)^3}V_{in} \tag{5.96}$$

The output voltage is

$$V_O = V_{C3} + V_{C3(2j-1)} = (1+j)V_3 = (\frac{1+j}{1-k})^3 V_{in} \tag{5.97}$$

The voltage transfer gain is

$$G = \frac{V_O}{V_{in}} = (\frac{1+j}{1-k})^3 \tag{5.98}$$

The ripple voltage of output voltage v_O is

$$\Delta v_O = \frac{\Delta Q}{C_{32j}} = \frac{I_O kT}{C_{32j}} = \frac{k}{fC_{32j}}\frac{V_O}{R}$$

Therefore, the variation ratio of output voltage v_O is

$$\varepsilon = \frac{\Delta v_O / 2}{V_O} = \frac{k}{2RfC_{32j}} \tag{5.99}$$

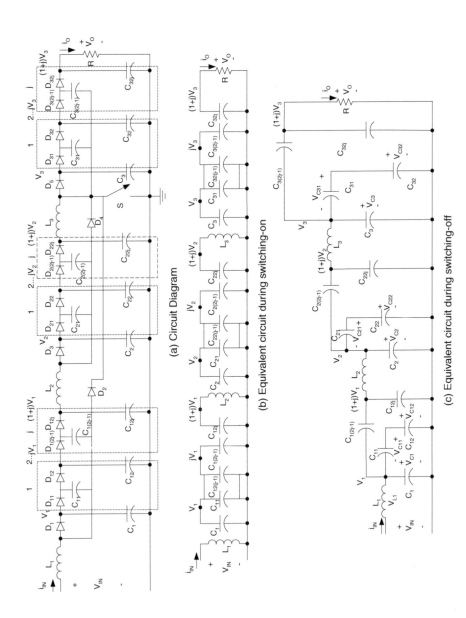

FIGURE 5.14
Three-stage boost multiple-double circuit.

5.6.4 Higher Stage Multiple Boost Circuit

Higher stage multiple boost circuit is derived from the corresponding circuit of the main series by adding multiple (*j*) DECs in each stage circuit. For n^{th} stage additional circuit, the final output voltage is

$$V_O = (\frac{1+j}{1-k})^n V_{in}$$

The voltage transfer gain is

$$G = \frac{V_O}{V_{in}} = (\frac{1+j}{1-k})^n \qquad (5.100)$$

Analogously, the variation ratio of output voltage v_O is

$$\varepsilon = \frac{\Delta v_O / 2}{V_O} = \frac{k}{2RfC_{n2j}} \qquad (5.101)$$

5.7 Summary of Positive Output Cascade Boost Converters

All circuits of the positive output cascade boost converters as a family are shown in Figure 5.15 as the family tree. From the analysis of the previous two sections we have the common formula to calculate the output voltage:

$$V_O = \begin{cases} (\dfrac{1}{1-k})^n V_{in} & main\_series \\[2mm] 2*(\dfrac{1}{1-k})^n V_{in} & additional\_series \\[2mm] (\dfrac{2}{1-k})^n V_{in} & double\_series \\[2mm] (\dfrac{3}{1-k})^n V_{in} & triple\_series \\[2mm] (\dfrac{j+1}{1-k})^n V_{in} & multiple(j)\_series \end{cases} \qquad (5.102)$$

The voltage transfer gain is

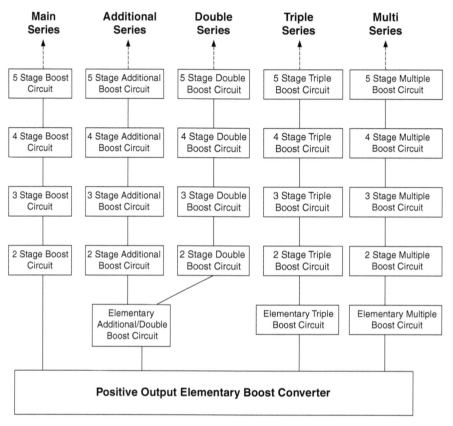

FIGURE 5.15
The family of positive output cascade boost converters

$$G = \frac{V_O}{V_{in}} = \begin{cases} (\frac{1}{1-k})^n & main\_series \\ 2*(\frac{1}{1-k})^n & additional\_series \\ (\frac{2}{1-k})^n & double\_series \\ (\frac{3}{1-k})^n & triple\_series \\ (\frac{j+1}{1-k})^n & multiple(j)\_series \end{cases} \qquad (5.103)$$

In order to show the advantages of the positive output cascade boost converters, we compare their voltage transfer gains to that of the buck converter,

$$G = \frac{V_O}{V_{in}} = k$$

forward converter,

$$G = \frac{V_O}{V_{in}} = kN \quad N \text{ is the transformer turn ratio}$$

Cúk-converter,

$$G = \frac{V_O}{V_{in}} = \frac{k}{1-k}$$

fly-back converter,

$$G = \frac{V_O}{V_{in}} = \frac{k}{1-k} N \quad N \text{ is the transformer turn ratio}$$

boost converter,

$$G = \frac{V_O}{V_{in}} = \frac{1}{1-k}$$

and positive output Luo-converters

$$G = \frac{V_O}{V_{in}} = \frac{n}{1-k} \tag{5.104}$$

If we assume that the conduction duty k is 0.2, the output voltage transfer gains are listed in Table 5.1; if the conduction duty k is 0.5, the output voltage transfer gains are listed in Table 5.2; if the conduction duty k is 0.8, the output voltage transfer gains are listed in Table 5.3.

5.8 Simulation and Experimental Results

5.8.1 Simulation Results of a Three-Stage Boost Circuit

To verify the design and calculation results, the PSpice simulation package was applied to a three-stage boost converter. Choosing $V_{in} = 20$ V, $L_1 = L_2 = L_3 = 10$ mH, all C_1 to $C_8 = 2$ µF, and $R = 30$ kΩ, $k = 0.7$ and $f = 100$ kHz. The

TABLE 5.1

Voltage Transfer Gains of Converters in the Condition $k = 0.2$

Stage No. (n)	1	2	3	4	5	n
Buck converter			0.2			
Forward converter		0.2N (N is the transformer turn ratio)				
Cúk-converter			0.25			
Fly-back converter		0.25N (N is the transformer turn ratio)				
Boost converter			1.25			
Positive output Luo-converters	1.25	2.5	3.75	5	6.25	$1.25n$
Positive output cascade boost converters — main series	1.25	1.563	1.953	2.441	3.052	1.25^n
Positive output cascade boost converters — additional series	2.5	3.125	3.906	4.882	6.104	$2*1.25^n$
Positive output cascade boost converters — double series	2.5	6.25	15.625	39.063	97.66	$(2*1.25)^n$
Positive output cascade boost converters — triple series	3.75	14.06	52.73	197.75	741.58	$(3*1.25)^n$
Positive output cascade boost ($j = 3$) converters — multiple series	5	25	125	625	3125	$(4*1.25)^n$

TABLE 5.2

Voltage Transfer Gains of Converters in the Condition $k = 0.5$

Stage No. (n)	1	2	3	4	5	n
Buck converter			0.5			
Forward converter		0.5N (N is the transformer turn ratio)				
Cúk-converter			1			
Fly-back converter		N (N is the transformer turn ratio)				
Boost converter			2			
Positive output Luo-converters	2	4	6	8	10	$2n$
Positive output cascade boost converters — main series	2	4	8	16	32	2^n
Positive output cascade boost converters — additional series	4	8	16	32	64	$2*2^n$
Positive output cascade boost converters — double series	4	16	64	256	1024	$(2*2)^n$
Positive output cascade boost converters — triple series	6	36	216	1296	7776	$(3*2)^n$
Positive output cascade boost ($j = 3$) converters — multiple series	8	64	512	4096	32,768	$(4*2)^n$

obtained voltage values V_1, V_2, and V_O of a triple-lift circuit are 66 V, 194 V, and 659 V respectively and inductor current waveforms i_{L1} (its average value $I_{L1} = 618$ mA), i_{L2}, and i_{L3}. The simulation results are shown in Figure 5.16. The voltage values match the calculated results.

TABLE 5.3

Voltage Transfer Gains of Converters in the Condition $k = 0.8$

Stage No. (n)	1	2	3	4	5	n
Buck converter			0.8			
Forward converter			0.8N (N is the transformer turn ratio)			
Cúk-converter			4			
Fly-back converter			4N (N is the transformer turn ratio)			
Boost converter			5			
Positive output Luo-converters	5	10	15	20	25	$5n$
Positive output cascade boost converters — main series	5	25	125	625	3125	5^n
Positive output cascade boost converters — additional series	10	50	250	1250	6250	$2*5^n$
Positive output cascade boost converters — double series	10	100	1000	10,000	100,000	$(2*5)^n$
Positive output cascade boost converters — triple series	15	225	3375	50,625	759,375	$(3*5)^n$
Positive output cascade boost (j=3) converters — multiple series	20	400	8000	160,000	$32*10^5$	$(4*5)^n$

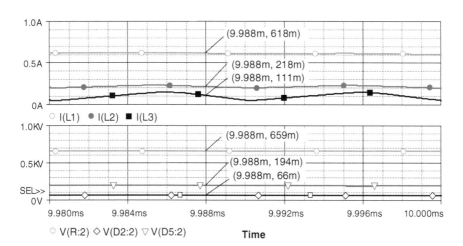

FIGURE 5.16

The simulation results of a three-stage boost circuit at condition $k = 0.7$ and $f = 100$ kHz.

5.8.2 Experimental Results of a Three-Stage Boost Circuit

A test rig was constructed to verify the design and calculation results, and compared with PSpice simulation results. The test conditions are still $V_{in} = 20$ V, $L_1 = L_2 = L_3 = 10$ mH, all C_1 to $C_8 = 2$ μF and $R = 30$ kΩ, $k = 0.7$, and $f = 100$kHz. The component of the switch is a MOSFET device IRF950 with the rate 950 V/5 A/2 MHz. The measured values of the output voltage and

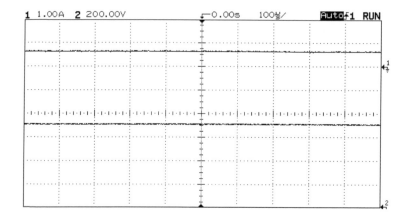

FIGURE 5.17
The experimental results of a three-stage boost circuit at condition $k = 0.7$ and $f = 100$ kHz

TABLE 5.4

Comparison of Simulation and Experimental Results of a Triple-Lift Circuit

Stage No. (n)	I_{L1} (A)	I_{in} (A)	V_{in} (V)	P_{in} (W)	V_O (V)	P_O (W)	η (%)
Simulation results	0.618	0.927	20	18.54	659	14.47	78
Experimental results	0.62	0.93	20	18.6	660	14.52	78

first inductor current in a three-stage boost converter. After careful measurement, we obtained the current value of $I_{L1} = 0.62$ A (shown in channel 1 with 1 A/Div) and voltage value of $V_O = 660$ V (shown in channel 2 with 200 V/ Div). The experimental results (current and voltage values) in Figure 5.17 match the calculated and simulation results, which are $I_{L1} = 0.618$ A and $V_O = 659$ V shown in Figure 5.16.

5.8.3 Efficiency Comparison of Simulation and Experimental Results

These circuits enhanced the voltage transfer gain successfully, and efficiently. Particularly, the efficiency of the tested circuits is 78%, which is good for high voltage output equipment. Comparison of the simulation and experimental results is shown in Table 5.4.

5.8.4 Transient Process

Usually, there is high inrush current during the first power-on. Therefore, the voltage across capacitors is quickly changed to certain values. The transient process is very quick taking only a few milliseconds.

Bibliography

Chokhawala, R.S., Catt, J., and Pelly, B.R., Gate drive considerations for IGBT modules, *IEEE Trans. on Industry Applications*, 31, 603, 1995.

Czarkowski, D. and Kazimierczuk, M.K., Phase-controlled series-parallel resonant converter, *IEEE Trans. on Power Electronics*, 8, 309, 1993.

Kassakian, J.G., Schlecht, M.F., and Verghese, G.C., *Principles of Power Electronics*, Addison-Wesley, New York, 1991.

Khan, I.A., DC-to-DC converters for electric and hybrid vehicles, in *Proceedings of Power Electronics in Transportation*, 1994, p. 113.

Luo, F.L. and Ye, H., Positive output cascade boost converters, in *Proceedings of IEEE-IPEMC'2003*, Xi'an, China, 866, 2002.

Luo, F.L. and Ye, H., Fundamentals of positive output cascade boost converters, submitted to *IEEE Transactions on Power Electronics*, 18, 1526, 2003.

Luo, F. L. and Ye, H., Development of positive output cascade boost converters, submitted to *IEEE Transactions on Power Electronics*, 18, 1536, 2003.

Poon, N.K. and Pong, M.H., Computer aided design of a crossing current resonant converter (XCRC), in *Proceedings of IECON'94*, Bologna, Italy, 1994, p. 135.

Steigerwald, R.L., High-frequency resonant transistor DC-DC converters, *IEEE Trans. on Industrial Electronics*, 31, 181, 1984.

6

Negative Output Cascade Boost Converters

6.1 Introduction

This chapter introduces negative output cascade boost converters. Just as with positive output cascade boost converters these converters use the super-lift technique. There are several sub-series:

- Main series — Each circuit of the main series has one switch S, n inductors, n capacitors, and $(2n - 1)$ diodes.
- Additional series — Each circuit of the additional series has one switch S, n inductors, $(n + 2)$ capacitors, and $(2n + 1)$ diodes.
- Double series — Each circuit of the double series has one switch S, n inductors, $3n$ capacitors, and $(3n - 1)$ diodes.
- Triple series — Each circuit of the triple series has one switch S, n inductors, $5n$ capacitors, and $(5n - 1)$ diodes.
- Multiple series — Multiple series circuits have one switch S and a higher number of capacitors and diodes.

The conduction duty ratio is k, switching frequency is f, switching period is $T = 1/f$, the load is resistive load R. The input voltage and current are V_{in} and I_{in}, output voltage and current are V_O and I_O. Assume no power losses during the conversion process, $V_{in} \times I_{in} = V_O \times I_O$. The voltage transfer gain is G:

$$G = \frac{V_O}{V_{in}}$$

6.2 Main Series

The first three stages of the negative output cascade boost converters — main series — are shown in Figure 6.1 to Figure 6.3. For convenience they are

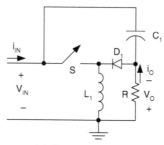

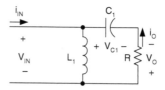

(a) Circuit diagram

(b) Equivalent circuit during switching-on

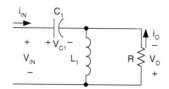

(c) Equivalent circuit during switching-off

FIGURE 6.1
Elementary boost converter.

called elementary boost converter, two-stage boost circuit and three-stage boost circuit respectively, and numbered as $n = 1, 2$ and 3.

6.2.1 N/O Elementary Boost Circuit

The N/O elementary boost converter and its equivalent circuits during switch-on and -off are shown in Figure 6.1. The voltage across capacitor C_1 is charged to V_{C1}. The current i_{L1} flowing through inductor L_1 increases with voltage V_{in} during switch-on period kT and decreases with voltage $-(V_{C1} - V_{in})$ during switch-off period $(1-k)T$. Therefore, the ripple of the inductor current i_{L1} is

$$\Delta i_{L1} = \frac{V_{in}}{L_1} kT = \frac{V_{C1} - V_{in}}{L_1} (1-k)T \tag{6.1}$$

$$V_{C1} = \frac{1}{1-k} V_{in}$$

$$V_O = V_{C1} - V_{in} = \frac{k}{1-k} V_{in} \tag{6.2}$$

The voltage transfer gain is

$$G = \frac{V_O}{V_{in}} = \frac{1}{1-k} - 1 \tag{6.3}$$

The inductor average current is

$$I_{L1} = (1-k)\frac{V_O}{R} \tag{6.4}$$

The variation ratio of current i_{L1} through inductor L_1 is

$$\xi_1 = \frac{\Delta i_{L1}/2}{I_{L1}} = \frac{kTV_{in}}{(1-k)2L_1V_O/R} = \frac{k}{2}\frac{R}{fL_1} \tag{6.5}$$

Usually ξ_1 is small (much lower than unity), it means this converter works in the continuous mode.

The ripple voltage of output voltage v_O is

$$\Delta v_O = \frac{\Delta Q}{C_1} = \frac{I_O kT}{C_1} = \frac{k}{fC_1}\frac{V_O}{R}$$

Therefore, the variation ratio of output voltage v_O is

$$\varepsilon = \frac{\Delta v_O/2}{V_O} = \frac{k}{2RfC_1} \tag{6.6}$$

Usually R is in kΩ, f in 10 kHz, and C_1 in μF, this ripple is smaller than 1%.

6.2.2 N/O Two-Stage Boost Circuit

The N/O two-stage boost circuit is derived from elementary boost converter by adding the parts (L_2-D_2-D_3-C_2). Its circuit diagram and equivalent circuits during switch-on and switch -off are shown in Figure 6.2. The voltage across capacitor C_1 is charged to V_1. As described in the previous section the voltage V_1 across capacitor C_1 is

$$V_1 = \frac{1}{1-k} V_{in}$$

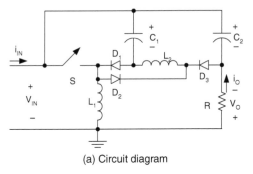

(a) Circuit diagram

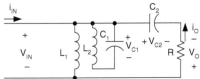

(b) Equivalent circuit during switching-on

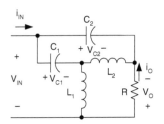

(c) Equivalent circuit during switching-off

FIGURE 6.2
Two-stage boost circuit.

The voltage across capacitor C_2 is charged to V_{C2}. The current flowing through inductor L_2 increases with voltage V_1 during switch-on period kT and decreases with voltage $-(V_{C2} - V_1)$ during switch-off period $(1 - k)T$. Therefore, the ripple of the inductor current i_{L2} is

$$\Delta i_{L2} = \frac{V_1}{L_2}kT = \frac{V_{C2} - V_1}{L_2}(1-k)T \tag{6.7}$$

$$V_{C2} = \frac{1}{1-k}V_1 = (\frac{1}{1-k})^2 V_{in}$$

$$V_O = V_{C2} - V_{in} = [(\frac{1}{1-k})^2 - 1]V_{in} \tag{6.8}$$

The voltage transfer gain is

$$G = \frac{V_O}{V_{in}} = (\frac{1}{1-k})^2 - 1 \tag{6.9}$$

Analogously,

$$\Delta i_{L1} = \frac{V_{in}}{L_1} kT \qquad I_{L1} = \frac{I_O}{(1-k)^2}$$

$$\Delta i_{L2} = \frac{V_1}{L_2} kT \qquad I_{L2} = \frac{I_O}{1-k}$$

Therefore, the variation ratio of current i_{L1} through inductor L_1 is

$$\xi_1 = \frac{\Delta i_{L1}/2}{I_{L1}} = \frac{k(1-k)^2 TV_{in}}{2L_1 I_O} = \frac{k(1-k)^4}{2} \frac{R}{fL_1} \tag{6.10}$$

the variation ratio of current i_{L2} through inductor L_2 is

$$\xi_2 = \frac{\Delta i_{L2}/2}{I_{L2}} = \frac{k(1-k)TV_1}{2L_2 I_O} = \frac{k(1-k)^2}{2} \frac{R}{fL_2} \tag{6.11}$$

and the variation ratio of output voltage v_O is

$$\varepsilon = \frac{\Delta v_O/2}{V_O} = \frac{k}{2RfC_2} \tag{6.12}$$

6.2.3 N/O Three-Stage Boost Circuit

The N/O three-stage boost circuit is derived from the two-stage boost circuit by double adding the parts (L_2-D_2-D_3-C_2). Its circuit diagram and equivalent circuits during switch-on and -off are shown in Figure 6.3. The voltage across capacitor C_1 is charged to V_1. As described previously the voltage V_{C1} across capacitor C_1 is $V_{C1} = (1/1-k)V_{in}$, and voltage V_{C2} across capacitor C_2 is $V_{C2} = (1/1-k)^2 V_{in}$.

The voltage across capacitor C_3 is charged to V_O. The current flowing through inductor L_3 increases with voltage V_{C2} during switch-on period kT and decreases with voltage $-(V_{C3} - V_{C2})$ during switch-off $(1-k)T$. Therefore, the ripple of the inductor current i_{L3} is

$$\Delta i_{L3} = \frac{V_{C2}}{L_3} kT = \frac{V_{C3} - V_{C2}}{L_3} (1-k)T \tag{6.13}$$

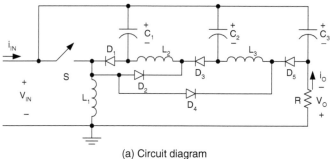

(a) Circuit diagram

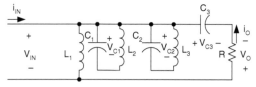

(b) Equivalent circuit during switching-on

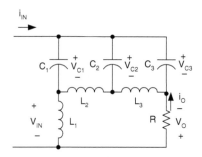

(c) Equivalent circuit during switching-off

FIGURE 6.3
Three-stage boost circuit.

$$V_{C3} = \frac{1}{1-k} V_{C2} = (\frac{1}{1-k})^2 V_{C1} = (\frac{1}{1-k})^3 V_{in}$$

$$V_O = V_{C3} - V_{in} = [(\frac{1}{1-k})^3 - 1]V_{in} \tag{6.14}$$

The voltage transfer gain is

$$G = \frac{V_O}{V_{in}} = (\frac{1}{1-k})^3 - 1 \tag{6.15}$$

Analogously,

$$\Delta i_{L1} = \frac{V_{in}}{L_1} kT \qquad\qquad I_{L1} = \frac{I_O}{(1-k)^3}$$

$$\Delta i_{L2} = \frac{V_1}{L_2} kT \qquad\qquad I_{L2} = \frac{I_O}{(1-k)^2}$$

$$\Delta i_{L3} = \frac{V_2}{L_3} kT \qquad\qquad I_{L3} = \frac{I_O}{1-k}$$

Therefore, the variation ratio of current i_{L1} through inductor L_1 is

$$\xi_1 = \frac{\Delta i_{L1}/2}{I_{L1}} = \frac{k(1-k)^3 TV_{in}}{2L_1 I_O} = \frac{k(1-k)^6}{2} \frac{R}{fL_1} \qquad (6.16)$$

the variation ratio of current i_{L2} through inductor L_2 is

$$\xi_2 = \frac{\Delta i_{L2}/2}{I_{L2}} = \frac{k(1-k)^2 TV_1}{2L_2 I_O} = \frac{k(1-k)^4}{2} \frac{R}{fL_2} \qquad (6.17)$$

the variation ratio of current i_{L3} through inductor L_3 is

$$\xi_3 = \frac{\Delta i_{L3}/2}{I_{L3}} = \frac{k(1-k)TV_2}{2L_3 I_O} = \frac{k(1-k)^2}{2} \frac{R}{fL_3} \qquad (6.18)$$

and the variation ratio of output voltage v_O is

$$\varepsilon = \frac{\Delta v_O/2}{V_O} = \frac{k}{2RfC_3} \qquad (6.19)$$

6.2.4   N/O Higher Stage Boost Circuit

N/O higher stage boost circuit can be designed by multiple repetition of the parts (L_2-D_2-D_3-C_2). For nth stage boost circuit, the final output voltage across capacitor C_n is

$$V_O = [(\frac{1}{1-k})^n - 1]V_{in}$$

The voltage transfer gain is

$$G = \frac{V_O}{V_{in}} = (\frac{1}{1-k})^n - 1 \tag{6.20}$$

the variation ratio of current i_{Li} through inductor L_i ($i = 1, 2, 3, \ldots n$) is

$$\xi_i = \frac{\Delta i_{Li}/2}{I_{Li}} = \frac{k(1-k)^{2(n-i+1)}}{2} \frac{R}{fL_i} \tag{6.21}$$

and the variation ratio of output voltage v_O is

$$\varepsilon = \frac{\Delta v_O/2}{V_O} = \frac{k}{2RfC_n} \tag{6.22}$$

6.3 Additional Series

All circuits of negative output cascade boost converters — additional series — are derived from the corresponding circuits of the main series by adding a DEC.

The first three stages of this series are shown in Figure 6.4 to Figure 6.6. For convenience they are called elementary additional boost circuit, two-stage additional boost circuit, and three-stage additional boost circuit respectively, and numbered as $n = 1, 2$ and 3.

6.3.1 N/O Elementary Additional Boost Circuit

This N/O elementary boost additional circuit is derived from N/O elementary boost converter by adding a DEC. Its circuit and switch-on and switch-off equivalent circuits are shown in Figure 6.4. The voltage across capacitor C_1 and C_{11} is charged to V_{C1} and voltage across capacitor C_{12} is charged to $V_{C12} = 2V_{C1}$. The current i_{L1} flowing through inductor L_1 increases with voltage V_{in} during switch-on period kT and decreases with voltage $-(V_{C1} - V_{in})$ during switch-off $(1 - k)T$. Therefore,

$$\Delta i_{L1} = \frac{V_{in}}{L_1} kT = \frac{V_{C1} - V_{in}}{L_1}(1-k)T \tag{6.23}$$

$$V_{C1} = \frac{1}{1-k} V_{in}$$

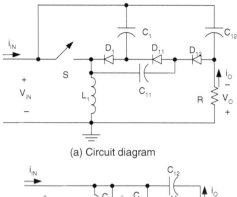

(a) Circuit diagram

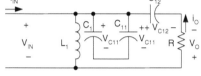

(b) Equivalent circuit during switching-on

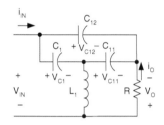

(c) Equivalent circuit during switching-off

FIGURE 6.4
Elementary boost additional (double) circuit.

The voltage V_{C12} is

$$V_{C12} = 2V_{C1} = \frac{2}{1-k}V_{in} \qquad (6.24)$$

The output voltage is

$$V_O = V_{C12} - V_{in} = [\frac{2}{1-k} - 1]V_{in} \qquad (6.25)$$

The voltage transfer gain is

$$G = \frac{V_O}{V_{in}} = \frac{2}{1-k} - 1 \qquad (6.26)$$

The variation ratio of current i_{L1} through inductor L_1 is

$$\xi_1 = \frac{\Delta i_{L1}/2}{I_{L1}} = \frac{k(1-k)TV_{in}}{4L_1 I_O} = \frac{k(1-k)^2}{8}\frac{R}{fL_1} \tag{6.27}$$

The ripple voltage of output voltage v_O is

$$\Delta v_O = \frac{\Delta Q}{C_{12}} = \frac{I_O kT}{C_{12}} = \frac{k}{fC_{12}}\frac{V_O}{R}$$

Therefore, the variation ratio of output voltage v_O is

$$\varepsilon = \frac{\Delta v_O/2}{V_O} = \frac{k}{2RfC_{12}} \tag{6.28}$$

6.3.2   N/O Two-Stage Additional Boost Circuit

The N/O two-stage additional boost circuit is derived from the N/O two-stage boost circuit by adding a DEC. Its circuit diagram and switch-on and switch-off equivalent circuits are shown in Figure 6.5. The voltage across capacitor C_1 is charged to V_{C1}. As described in the previous section the voltage V_{C1} across capacitor C_1 is

$$V_{C1} = \frac{1}{1-k}V_{in}$$

The voltage across capacitor C_2 and capacitor C_{11} is charged to V_{C2} and voltage across capacitor C_{12} is charged to V_{C12}. The current flowing through inductor L_2 increases with voltage V_{C1} during switch-on period kT and decreases with voltage $-(V_{C2} - V_{C1})$ during switch-off period $(1 - k)T$. Therefore, the ripple of the inductor current i_{L2} is

$$\Delta i_{L2} = \frac{V_{C1}}{L_2}kT = \frac{V_{C2} - V_{C1}}{L_2}(1-k)T \tag{6.29}$$

$$V_{C2} = \frac{1}{1-k}V_{C1} = (\frac{1}{1-k})^2 V_{in} \tag{6.30}$$

$$V_{C12} = 2V_{C2} = \frac{2}{1-k}V_{C1} = 2(\frac{1}{1-k})^2 V_{in}$$

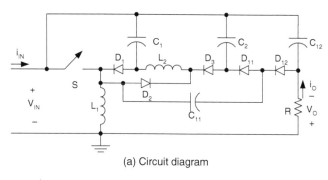

(a) Circuit diagram

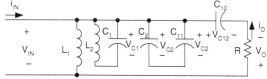

(b) Equivalent circuit during switching-on

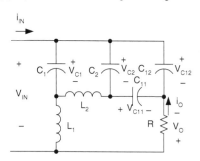

(c) Equivalent circuit during switching-off

FIGURE 6.5
Two-stage additional boost circuit.

The output voltage is

$$V_O = V_{C12} - V_{in} = [2(\frac{1}{1-k})^2 - 1]V_{in} \qquad (6.31)$$

The voltage transfer gain is

$$G = \frac{V_O}{V_{in}} = 2(\frac{1}{1-k})^2 - 1 \qquad (6.32)$$

Analogously,

$$\Delta i_{L1} = \frac{V_{in}}{L_1} kT \qquad\qquad I_{L1} = \frac{2}{(1-k)^2} I_O$$

$$\Delta i_{L2} = \frac{V_1}{L_2} kT \qquad\qquad I_{L2} = \frac{2I_O}{1-k}$$

Therefore, the variation ratio of current i_{L1} through inductor L_1 is

$$\xi_1 = \frac{\Delta i_{L1}/2}{I_{L1}} = \frac{k(1-k)^2 TV_{in}}{4L_1 I_O} = \frac{k(1-k)^4}{8} \frac{R}{fL_1} \tag{6.33}$$

and the variation ratio of current i_{L2} through inductor L_2 is

$$\xi_2 = \frac{\Delta i_{L2}/2}{I_{L2}} = \frac{k(1-k)TV_1}{4L_2 I_O} = \frac{k(1-k)^2}{8} \frac{R}{fL_2} \tag{6.34}$$

The ripple voltage of output voltage v_O is

$$\Delta v_O = \frac{\Delta Q}{C_{12}} = \frac{I_O kT}{C_{12}} = \frac{k}{fC_{12}} \frac{V_O}{R}$$

Therefore, the variation ratio of output voltage v_O is

$$\varepsilon = \frac{\Delta v_O/2}{V_O} = \frac{k}{2RfC_{12}} \tag{6.35}$$

6.3.3 N/O Three-Stage Additional Boost Circuit

This N/O circuit is derived from three-stage boost circuit by adding a DEC. Its circuit diagram and equivalent circuits during switch-on and switch-off are shown in Figure 6.6. The voltage across capacitor C_1 is charged to V_{C1}. As described previously, the voltage V_{C1} across capacitor C_1 is $V_{C1} = (1/1-k)V_{in}$, and voltage V_2 across capacitor C_2 is $V_{C2} = (1/1-k)^2 V_{in}$. The voltage across capacitor C_3 and capacitor C_{11} is charged to V_{C3}. The voltage across capacitor C_{12} is charged to V_{C12}. The current flowing through inductor L_3 increases with voltage V_{C2} during switch-on period kT and decreases with voltage $-(V_{C3} - V_{C2})$ during switch-off $(1-k)T$. Therefore,

$$\Delta i_{L3} = \frac{V_{C2}}{L_3} kT = \frac{V_{C3} - V_{C2}}{L_3} (1-k)T \tag{6.36}$$

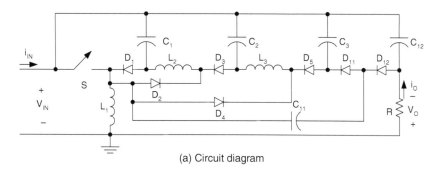

(a) Circuit diagram

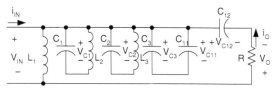

(b) Equivalent circuit during switching-on

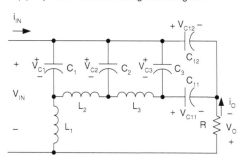

(c) Equivalent circuit during switching-off

FIGURE 6.6
Three-stage additional boost circuit.

and

$$V_{C3} = \frac{1}{1-k} V_{C2} = (\frac{1}{1-k})^2 V_{C1} = (\frac{1}{1-k})^3 V_{in} \qquad (6.37)$$

The voltage V_{C12} is

$$V_{C12} = 2V_{C3} = 2(\frac{1}{1-k})^3 V_{in}$$

The output voltage is

$$V_O = V_{C12} - V_{in} = [2(\frac{1}{1-k})^3 - 1]V_{in} \qquad (6.38)$$

The voltage transfer gain is

$$G = \frac{V_O}{V_{in}} = 2(\frac{1}{1-k})^3 - 1 \tag{6.39}$$

Analogously,

$$\Delta i_{L1} = \frac{V_{in}}{L_1} kT \qquad\qquad I_{L1} = \frac{2}{(1-k)^3} I_O$$

$$\Delta i_{L2} = \frac{V_1}{L_2} kT \qquad\qquad I_{L2} = \frac{2}{(1-k)^2} I_O$$

$$\Delta i_{L3} = \frac{V_2}{L_3} kT \qquad\qquad I_{L3} = \frac{2I_O}{1-k}$$

Therefore, the variation ratio of current i_{L1} through inductor L_1 is

$$\xi_1 = \frac{\Delta i_{L1}/2}{I_{L1}} = \frac{k(1-k)^3 TV_{in}}{4L_1 I_O} = \frac{k(1-k)^6}{8} \frac{R}{fL_1} \tag{6.40}$$

and the variation ratio of current i_{L2} through inductor L_2 is

$$\xi_2 = \frac{\Delta i_{L2}/2}{I_{L2}} = \frac{k(1-k)^2 TV_1}{4L_2 I_O} = \frac{k(1-k)^4}{8} \frac{R}{fL_2} \tag{6.41}$$

and the variation ratio of current i_{L3} through inductor L_3 is

$$\xi_3 = \frac{\Delta i_{L3}/2}{I_{L3}} = \frac{k(1-k)TV_2}{4L_3 I_O} = \frac{k(1-k)^2}{8} \frac{R}{fL_3} \tag{6.42}$$

The ripple voltage of output voltage v_O is

$$\Delta v_O = \frac{\Delta Q}{C_{12}} = \frac{I_O kT}{C_{12}} = \frac{k}{fC_{12}} \frac{V_O}{R}$$

Therefore, the variation ratio of output voltage v_O is

$$\varepsilon = \frac{\Delta v_O/2}{V_O} = \frac{k}{2RfC_{12}} \tag{6.43}$$

6.3.4 N/O Higher Stage Additional Boost Circuit

The N/O higher stage boost additional circuit is derived from the corresponding circuit of the main series by adding a DEC. For the nth stage additional circuit, the final output voltage is

$$V_O = [2(\frac{1}{1-k})^n - 1]V_{in}$$

The voltage transfer gain is

$$G = \frac{V_O}{V_{in}} = 2(\frac{1}{1-k})^n - 1 \tag{6.44}$$

Analogously, the variation ratio of current i_{Li} through inductor L_i ($i = 1, 2, 3, \dots n$) is

$$\xi_i = \frac{\Delta i_{Li}/2}{I_{Li}} = \frac{k(1-k)^{2(n-i+1)}}{8} \frac{R}{fL_i} \tag{6.45}$$

and the variation ratio of output voltage v_O is

$$\varepsilon = \frac{\Delta v_O/2}{V_O} = \frac{k}{2RfC_{12}} \tag{6.46}$$

6.4 Double Series

All circuits of N/O cascade boost converters — double series — are derived from the corresponding circuits of the main series by adding a DEC in each stage circuit. The first three stages of this series are shown in Figures 6.4, 6.7, and 6.8. For convenience they are called elementary double boost circuit, two-stage double boost circuit, and three-stage double boost circuit respectively, and numbered as $n = 1, 2$ and 3.

6.4.1 N/O Elementary Double Boost Circuit

This N/O elementary double boost circuit is derived from the elementary boost converter by adding a DEC. Its circuit and switch-on and switch-off equivalent circuits are shown in Figure 6.4, which is the same as the elementary boost additional circuit.

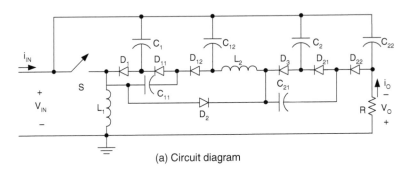

(a) Circuit diagram

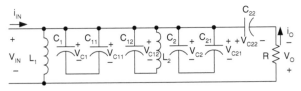

(b) Equivalent circuit during switching–on

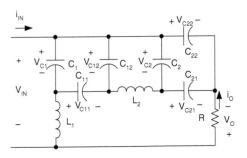

(c) Equivalent circuit during switching–off

FIGURE 6.7
Two-stage double boost circuit.

6.4.2 N/O Two-Stage Double Boost Circuit

The N/O two-stage double boost circuit is derived from two-stage boost circuit by adding a DEC in each stage circuit. Its circuit diagram and switch-on and switch-off equivalent circuits are shown in Figure 6.7. The voltage across capacitor C_1 and capacitor C_{11} is charged to V_1. As described in the previous section the voltage V_{C1} across capacitor C_1 and capacitor C_{11} is $V_{C1} = (1/1-k)V_{in}$. The voltage across capacitor C_{12} is charged to $2V_{C1}$.

The current flowing through inductor L_2 increases with voltage $2V_{C1}$ during switch-on period kT and decreases with voltage $-(V_{C2} - 2V_{C1})$ during switch-off period $(1 - k)T$. Therefore, the ripple of the inductor current i_{L2} is

$$\Delta i_{L2} = \frac{2V_{C1}}{L_2}kT = \frac{V_{C2} - 2V_{C1}}{L_2}(1-k)T \qquad (6.47)$$

$$V_{C2} = \frac{2}{1-k} V_{C1} = 2(\frac{1}{1-k})^2 V_{in} \quad (6.48)$$

The voltage V_{C22} is

$$V_{C22} = 2V_{C2} = (\frac{2}{1-k})^2 V_{in}$$

The output voltage is

$$V_O = V_{C22} - V_{in} = [(\frac{2}{1-k})^2 - 1]V_{in} \quad (6.49)$$

The voltage transfer gain is

$$G = \frac{V_O}{V_{in}} = (\frac{2}{1-k})^2 - 1 \quad (6.50)$$

Analogously,

$$\Delta i_{L1} = \frac{V_{in}}{L_1} kT \qquad I_{L1} = (\frac{2}{1-k})^2 I_O$$

$$\Delta i_{L2} = \frac{V_1}{L_2} kT \qquad I_{L2} = \frac{2I_O}{1-k}$$

Therefore, the variation ratio of current i_{L1} through inductor L_1 is

$$\xi_1 = \frac{\Delta i_{L1}/2}{I_{L1}} = \frac{k(1-k)^2 TV_{in}}{8L_1 I_O} = \frac{k(1-k)^4}{16} \frac{R}{fL_1} \quad (6.51)$$

and the variation ratio of current i_{L2} through inductor L_2 is

$$\xi_2 = \frac{\Delta i_{L2}/2}{I_{L2}} = \frac{k(1-k)TV_1}{4L_2 I_O} = \frac{k(1-k)^2}{8} \frac{R}{fL_2} \quad (6.52)$$

The ripple voltage of output voltage v_O is

$$\Delta v_O = \frac{\Delta Q}{C_{22}} = \frac{I_O kT}{C_{22}} = \frac{k}{fC_{22}} \frac{V_O}{R}$$

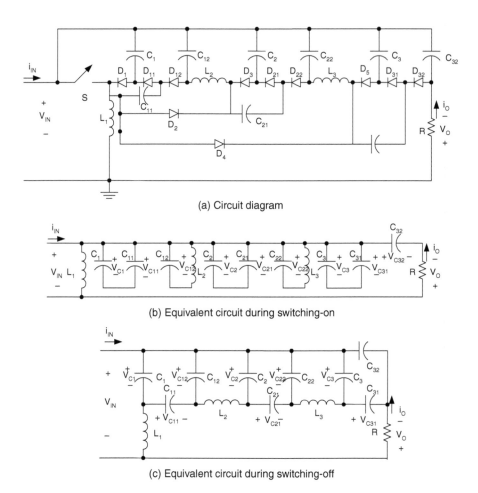

(a) Circuit diagram

(b) Equivalent circuit during switching-on

(c) Equivalent circuit during switching-off

FIGURE 6.8
Three-stage double boost circuit.

Therefore, the variation ratio of output voltage v_O is

$$\varepsilon = \frac{\Delta v_O / 2}{V_O} = \frac{k}{2RfC_{22}} \tag{6.53}$$

6.4.3 N/O Three-Stage Double Boost Circuit

This N/O circuit is derived from the three-stage boost circuit by adding DEC in each stage circuit. Its circuit diagram and equivalent circuits during switch-on and switch-off are shown in Figure 6.8. The voltage across capacitor C_1 and capacitor C_{11} is charged to V_{C1}. As described previously, the voltage V_{C1} across capacitor C_1 and capacitor C_{11} is $V_{C1} = (1/1-k)V_{in}$, and voltage V_{C2} across capacitor C_2 and capacitor C_{12} is $V_{C2} = 2(1/1-k)^2 V_{in}$.

The voltage across capacitor C_{22} is $2V_{C2} = (2/1-k)^2 V_{in}$. The voltage across capacitor C_3 and capacitor C_{31} is charged to V_3. The voltage across capacitor C_{12} is charged to V_O. The current flowing through inductor L_3 increases with voltage V_2 during switch-on period kT and decreases with voltage $-(V_{C3} - 2V_{C2})$ during switch-off $(1-k)T$. Therefore,

$$\Delta i_{L3} = \frac{2V_{C2}}{L_3} kT = \frac{V_{C3} - 2V_{C2}}{L_3} (1-k)T \qquad (6.54)$$

and

$$V_{C3} = \frac{2V_{C2}}{(1-k)} = \frac{4}{(1-k)^3} V_{in} \qquad (6.55)$$

The voltage V_{C32} is

$$V_{C32} = 2V_{C3} = (\frac{2}{1-k})^3 V_{in}$$

The output voltage is

$$V_O = V_{C32} - V_{in} = [(\frac{2}{1-k})^3 - 1]V_{in} \qquad (6.56)$$

The voltage transfer gain is

$$G = \frac{V_O}{V_{in}} = (\frac{2}{1-k})^3 - 1 \qquad (6.57)$$

Analogously,

$$\Delta i_{L1} = \frac{V_{in}}{L_1} kT \qquad I_{L1} = \frac{8}{(1-k)^3} I_O$$

$$\Delta i_{L2} = \frac{V_1}{L_2} kT \qquad I_{L2} = \frac{4}{(1-k)^2} I_O$$

$$\Delta i_{L3} = \frac{V_2}{L_3} kT \qquad I_{L3} = \frac{2I_O}{1-k}$$

Therefore, the variation ratio of current i_{L1} through inductor L_1 is

$$\xi_1 = \frac{\Delta i_{L1}/2}{I_{L1}} = \frac{k(1-k)^3 TV_{in}}{16L_1 I_O} = \frac{k(1-k)^6}{128} \frac{R}{fL_1} \tag{6.58}$$

and the variation ratio of current i_{L2} through inductor L_2 is

$$\xi_2 = \frac{\Delta i_{L2}/2}{I_{L2}} = \frac{k(1-k)^2 TV_1}{8L_2 I_O} = \frac{k(1-k)^4}{32} \frac{R}{fL_2} \tag{6.59}$$

and the variation ratio of current i_{L3} through inductor L_3 is

$$\xi_3 = \frac{\Delta i_{L3}/2}{I_{L3}} = \frac{k(1-k)TV_2}{4L_3 I_O} = \frac{k(1-k)^2}{8} \frac{R}{fL_3} \tag{6.60}$$

The ripple voltage of output voltage v_O is

$$\Delta v_O = \frac{\Delta Q}{C_{32}} = \frac{I_O kT}{C_{32}} = \frac{k}{fC_{32}} \frac{V_O}{R}$$

Therefore, the variation ratio of output voltage v_O is

$$\varepsilon = \frac{\Delta v_O/2}{V_O} = \frac{k}{2RfC_{32}} \tag{6.61}$$

6.4.4 N/O Higher Stage Double Boost Circuit

The N/O higher stage double boost circuit is derived from the corresponding circuit of the main series by adding DEC in each stage circuit. For nth stage additional circuit, the final output voltage is

$$V_O = [(\frac{2}{1-k})^n - 1]V_{in}$$

The voltage transfer gain is

$$G = \frac{V_O}{V_{in}} = (\frac{2}{1-k})^n - 1 \tag{6.62}$$

Analogously, the variation ratio of current i_{Li} through inductor L_i ($i = 1, 2, 3, \ldots n$) is

$$\xi_i = \frac{\Delta i_{Li}/2}{I_{Li}} = \frac{k(1-k)^{2(n-i+1)}}{2*2^{2n}} \frac{R}{fL_i} \tag{6.63}$$

and the variation ratio of output voltage v_O is

$$\varepsilon = \frac{\Delta v_O / 2}{V_O} = \frac{k}{2RfC_{n2}} \tag{6.64}$$

6.5 Triple Series

All circuits of N/O cascade boost converters — triple series — are derived from the corresponding circuits of the main series by adding DEC twice in each stage circuit. The first three stages of this series are shown in Figure 6.9 to Figure 6.11. For convenience they are called elementary triple boost (or additional) circuit, two-stage triple boost circuit, and three-stage triple boost circuit respectively, and numbered as $n = 1$, 2 and 3.

6.5.1 N/O Elementary Triple Boost Circuit

This N/O elementary triple boost circuit is derived from the elementary boost converter by adding DEC twice. Its circuit and switch-on and switch-off equivalent circuits are shown in Figure 6.9. The output voltage of the first stage boost circuit is V_{C1}, $V_{C1} = V_{in}/(1-k)$.

After the first DEC, the voltage (across capacitor C_{12}) increases to

$$V_{C12} = 2V_{C1} = \frac{2}{1-k} V_{in} \tag{6.65}$$

After the second DEC, the voltage (across capacitor C_{14}) increases to

$$V_{C14} = V_{C12} + V_{C1} = \frac{3}{1-k} V_{in} \tag{6.66}$$

The final output voltage V_O is equal to

$$V_O = V_{C14} - V_{in} = [\frac{3}{1-k} - 1]V_{in} \tag{6.67}$$

The voltage transfer gain is

$$G = \frac{V_O}{V_{in}} = \frac{3}{1-k} - 1 \tag{6.68}$$

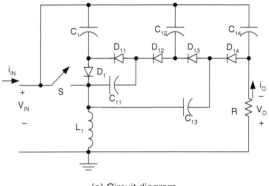

(a) Circuit diagram

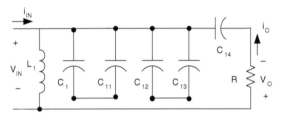

(b) Equivalent circuit during switching-on

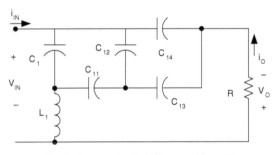

(c) Equivalent circuit during switching-off

FIGURE 6.9
Elementary triple boost circuit.

6.5.2 N/O Two-Stage Triple Boost Circuit

The N/O two-stage triple boost circuit is derived from two-stage boost circuit by adding DEC twice in each stage circuit. Its circuit diagram and switch-on and switch-off equivalent circuits are shown in Figure 6.10.

As described in the previous section the voltage across capacitor C_{14} is $V_{C14} = (3/1-k)V_{in}$. Analogously, the voltage across capacitor C_{24} is.

$$V_{C24} = (\frac{3}{1-k})^2 V_{in} \tag{6.69}$$

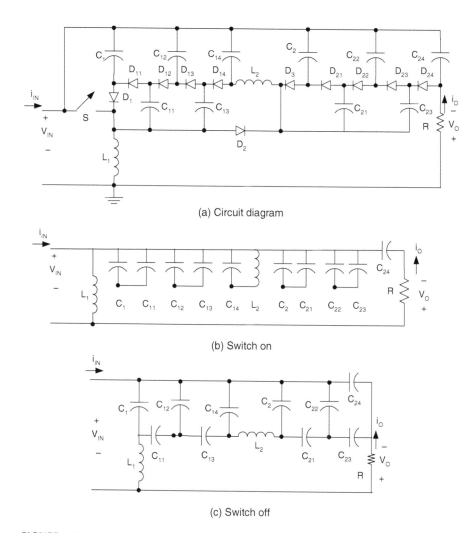

(a) Circuit diagram

(b) Switch on

(c) Switch off

FIGURE 6.10
Two-stage triple boost circuit.

The final output voltage V_O is equal to

$$V_O = V_{C24} - V_{in} = [(\frac{3}{1-k})^2 - 1]V_{in} \qquad (6.70)$$

The voltage transfer gain is

$$G = \frac{V_O}{V_{in}} = (\frac{3}{1-k})^2 - 1 \qquad (6.71)$$

Analogously,

$$\Delta i_{L1} = \frac{V_{in}}{L_1} kT \qquad\qquad I_{L1} = (\frac{2}{1-k})^2 I_O$$

$$\Delta i_{L2} = \frac{V_1}{L_2} kT \qquad\qquad I_{L2} = \frac{2I_O}{1-k}$$

Therefore, the variation ratio of current i_{L1} through inductor L_1 is

$$\xi_1 = \frac{\Delta i_{L1}/2}{I_{L1}} = \frac{k(1-k)^2 TV_{in}}{8L_1 I_O} = \frac{k(1-k)^4}{16} \frac{R}{fL_1} \qquad (6.72)$$

and the variation ratio of current i_{L2} through inductor L_2 is

$$\xi_2 = \frac{\Delta i_{L2}/2}{I_{L2}} = \frac{k(1-k)TV_1}{4L_2 I_O} = \frac{k(1-k)^2}{8} \frac{R}{fL_2} \qquad (6.73)$$

The ripple voltage of output voltage v_O is

$$\Delta v_O = \frac{\Delta Q}{C_{22}} = \frac{I_O kT}{C_{22}} = \frac{k}{fC_{22}} \frac{V_O}{R}$$

Therefore, the variation ratio of output voltage v_O is

$$\varepsilon = \frac{\Delta v_O/2}{V_O} = \frac{k}{2RfC_{22}} \qquad (6.74)$$

6.5.3   N/O Three-Stage Triple Boost Circuit

This N/O circuit is derived from the three-stage boost circuit by adding DEC twice in each stage circuit. Its circuit diagram and equivalent circuits during switch-on and switch-off are shown in Figure 6.11.

As described in the previous section the voltage across capacitor C_{14} is $V_{C14} = (3/1-k)V_{in}$, and the voltage across capacitor C_{24} is $V_{C24} = (3/1-k)^2 V_{in}$. Analogously, the voltage across capacitor C_{34} is

$$V_{C34} = (\frac{3}{1-k})^3 V_{in} \qquad (6.75)$$

The final output voltage V_O is equal to

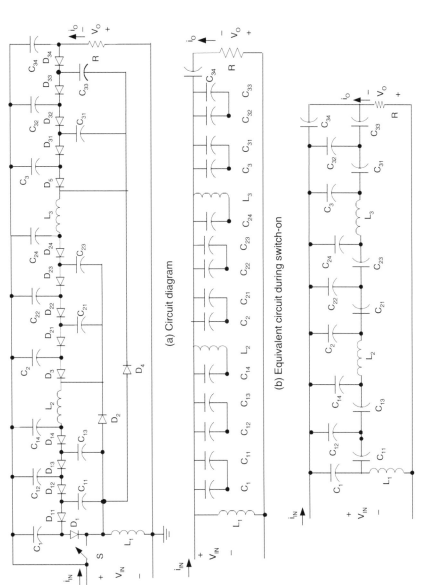

FIGURE 6.11
Three-stage triple boost circuit.

$$V_O = V_{C34} - V_{in} = [(\frac{3}{1-k})^3 - 1]V_{in} \qquad (6.76)$$

The voltage transfer gain is

$$G = \frac{V_O}{V_{in}} = (\frac{3}{1-k})^3 - 1 \qquad (6.77)$$

Analogously,

$$\Delta i_{L1} = \frac{V_{in}}{L_1} kT \qquad\qquad I_{L1} = \frac{32}{(1-k)^3} I_O$$

$$\Delta i_{L2} = \frac{V_1}{L_2} kT \qquad\qquad I_{L2} = \frac{8}{(1-k)^2} I_O$$

$$\Delta i_{L3} = \frac{V_2}{L_3} kT \qquad\qquad I_{L3} = \frac{2}{1-k} I_O$$

Therefore, the variation ratio of current i_{L1} through inductor L_1 is

$$\xi_1 = \frac{\Delta i_{L1}/2}{I_{L1}} = \frac{k(1-k)^3 TV_{in}}{64 L_1 I_O} = \frac{k(1-k)^6}{12^3} \frac{R}{f L_1} \qquad (6.78)$$

and the variation ratio of current i_{L2} through inductor L_2 is

$$\xi_2 = \frac{\Delta i_{L2}/2}{I_{L2}} = \frac{k(1-k)^2 TV_1}{16 L_2 I_O} = \frac{k(1-k)^4}{12^2} \frac{R}{f L_2} \qquad (6.79)$$

and the variation ratio of current i_{L3} through inductor L_3 is

$$\xi_3 = \frac{\Delta i_{L3}/2}{I_{L3}} = \frac{k(1-k)TV_2}{4 L_3 I_O} = \frac{k(1-k)^2}{12} \frac{R}{f L_3} \qquad (6.80)$$

Usually ξ_1, ξ_2, and ξ_3 are small, this means that this converter works in the continuous mode. The ripple voltage of output voltage v_O is

$$\Delta v_O = \frac{\Delta Q}{C_{32}} = \frac{I_O kT}{C_{32}} = \frac{k}{f C_{32}} \frac{V_O}{R}$$

Therefore, the variation ratio of output voltage v_O is

$$\varepsilon = \frac{\Delta v_O / 2}{V_O} = \frac{k}{2RfC_{32}} \tag{6.81}$$

6.5.4 N/O Higher Stage Triple Boost Circuit

The N/O higher stage triple boost circuit is derived from the corresponding circuit of the main series by adding DEC twice in each stage circuit. For nth stage additional circuit, the voltage across capacitor C_{n4} is

$$V_{Cn4} = (\frac{3}{1-k})^n V_{in}$$

The output voltage is

$$V_O = V_{Cn4} - V_{in} = [(\frac{3}{1-k})^n - 1]V_{in} \tag{6.82}$$

The voltage transfer gain is

$$G = \frac{V_O}{V_{in}} = (\frac{3}{1-k})^n - 1 \tag{6.83}$$

Analogously, the variation ratio of current i_{Li} through inductor L_i ($i = 1, 2, 3,$ ...n) is

$$\xi_i = \frac{\Delta i_{Li} / 2}{I_{Li}} = \frac{k(1-k)^{2(n-i+1)}}{12^{(n-i+1)}} \frac{R}{fL_i} \tag{6.84}$$

and the variation ratio of output voltage v_O is

$$\varepsilon = \frac{\Delta v_O / 2}{V_O} = \frac{k}{2RfC_{n2}} \tag{6.85}$$

6.6 Multiple Series

All circuits of N/O cascade boost converters — multiple series — are derived from the corresponding circuits of the main series by adding DEC multiple

(j) times in each stage circuit. The first three stages of this series are shown in Figure 6.12 to Figure 6.14. For convenience they are called elementary multiple boost circuit, two-stage multiple boost circuit, and three-stage multiple boost circuit respectively, and numbered as $n = 1, 2$ and 3.

6.6.1 N/O Elementary Multiple Boost Circuit

This N/O elementary multiple boost circuit is derived from elementary boost converter by adding a DEC multiple (j) times. Its circuit and switch-on and switch-off equivalent circuits are shown in Figure 6.12.

The output voltage of the first DEC (across capacitor C_{12j}) increases to

$$V_{C12j} = \frac{j+1}{1-k} V_{in} \qquad (6.86)$$

The final output voltage V_O is equal to

$$V_O = V_{C12j} - V_{in} = [\frac{j+1}{1-k} - 1]V_{in} \qquad (6.87)$$

The voltage transfer gain is

$$G = \frac{V_O}{V_{in}} = \frac{j+1}{1-k} - 1 \qquad (6.88)$$

6.6.2 N/O Two-Stage Multiple Boost Circuit

The N/O two-stage multiple boost circuit is derived from the two-stage boost circuit by adding DEC multiple (j) times in each stage circuit. Its circuit diagram and switch-on and switch-off equivalent circuits are shown in Figure 6.13.

As described in the previous section the voltage across capacitor C_{12j} is

$$V_{C12j} = \frac{j+1}{1-k} V_{in}$$

Analogously, the voltage across capacitor C_{22j} is.

$$V_{C22j} = (\frac{j+1}{1-k})^2 V_{in} \qquad (6.89)$$

The final output voltage V_O is equal to

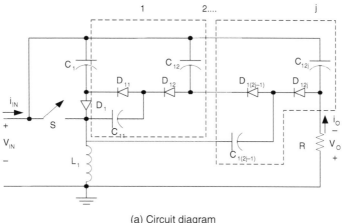

(a) Circuit diagram

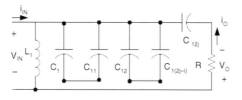

(b) Equivalent circuit during switching–on

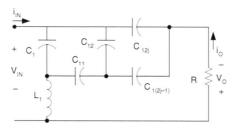

(c) Equivalent circuit during switching–off

FIGURE 6.12
Elementary multiple boost circuit.

$$V_O = V_{C22j} - V_{in} = [(\frac{j+1}{1-k})^2 - 1]V_{in} \qquad (6.90)$$

The voltage transfer gain is

$$G = \frac{V_O}{V_{in}} = (\frac{j+1}{1-k})^2 - 1 \qquad (6.91)$$

The ripple voltage of output voltage v_O is

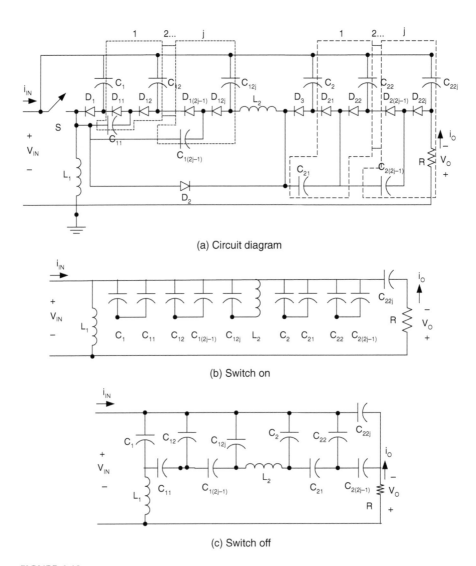

(a) Circuit diagram

(b) Switch on

(c) Switch off

FIGURE 6.13
Two-stage multiple boost circuit.

$$\Delta v_O = \frac{\Delta Q}{C_{22j}} = \frac{I_O kT}{C_{22j}} = \frac{k}{fC_{22j}} \frac{V_O}{R}$$

Therefore, the variation ratio of output voltage v_O is

$$\varepsilon = \frac{\Delta v_O / 2}{V_O} = \frac{k}{2RfC_{22j}} \qquad (6.92)$$

6.6.3 N/O Three-Stage Multiple Boost Circuit

This N/O circuit is derived from the three-stage boost circuit by adding DEC multiple (j) times in each stage circuit. Its circuit diagram and equivalent circuits during switch-on and switch-off are shown in Figure 6.14.

As described in the previous section the voltage across capacitor C_{12j} is $V_{C12j} = (j + 1/1-k)V_{in}$, and the voltage across capacitor C_{22j} is $V_{C22j} = (j + 1/1-k)^2 V_{in}$. Analogously, the voltage across capacitor C_{32j} is

$$V_{C32j} = (\frac{j+1}{1-k})^3 V_{in} \qquad (6.93)$$

The final output voltage V_O is equal to

$$V_O = V_{C32j} - V_{in} = [(\frac{j+1}{1-k})^3 - 1]V_{in} \qquad (6.94)$$

The voltage transfer gain is

$$G = \frac{V_O}{V_{in}} = (\frac{j+1}{1-k})^3 - 1 \qquad (6.95)$$

The ripple voltage of output voltage v_O is

$$\Delta v_O = \frac{\Delta Q}{C_{32j}} = \frac{I_O kT}{C_{32j}} = \frac{k}{fC_{32j}} \frac{V_O}{R}$$

Therefore, the variation ratio of output voltage v_O is

$$\varepsilon = \frac{\Delta v_O / 2}{V_O} = \frac{k}{2RfC_{32j}} \qquad (6.96)$$

6.6.4 N/O Higher Stage Multiple Boost Circuit

The N/O higher stage multiple boost circuit is derived from the corresponding circuit of the main series by adding DEC multiple (j) times in each stage circuit. For nth stage multiple boost circuit, the voltage across capacitor C_{n2j} is

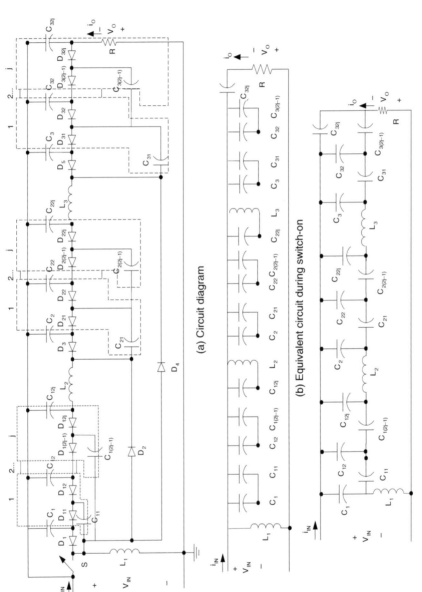

(a) Circuit diagram

(b) Equivalent circuit during switch-on

(c) Equivalent circuit during switch-off

FIGURE 6.14
Three-stage multiple boost circuit.

$$V_{Cn2j} = (\frac{j+1}{1-k})^n V_{in}$$

The output voltage is

$$V_O = V_{Cn2j} - V_{in} = [(\frac{j+1}{1-k})^n - 1]V_{in} \qquad (6.97)$$

The voltage transfer gain is

$$G = \frac{V_O}{V_{in}} = (\frac{j+1}{1-k})^n - 1 \qquad (6.98)$$

The variation ratio of output voltage v_O is

$$\varepsilon = \frac{\Delta v_O / 2}{V_O} = \frac{k}{2RfC_{n2j}} \qquad (6.99)$$

6.7 Summary of Negative Output Cascade Boost Converters

All the circuits of the N/O cascade boost converters as a family can be shown in Figure 6.15. From the analysis of the previous two sections we have the common formula to calculate the output voltage:

$$V_O = \begin{cases} [(\frac{1}{1-k})^n - 1]V_{in} & main\_series \\ [2*(\frac{1}{1-k})^n - 1]V_{in} & additional\_series \\ [(\frac{2}{1-k})^n - 1]V_{in} & double\_series \\ [(\frac{3}{1-k})^n - 1]V_{in} & triple\_series \\ [(\frac{j+1}{1-k})^n - 1]V_{in} & multiple(j)\_series \end{cases} \qquad (6.100)$$

The voltage transfer gain is

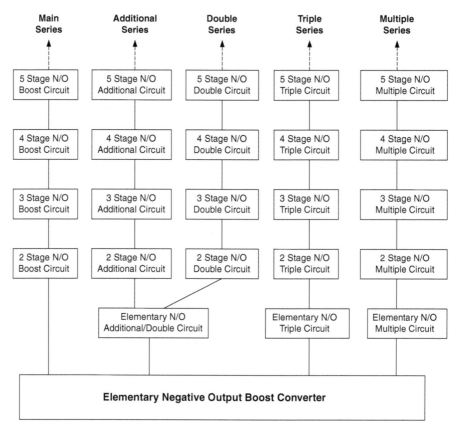

FIGURE 6.15
The family of negative output cascade boost converters.

$$G = \frac{V_O}{V_{in}} = \begin{cases} (\frac{1}{1-k})^n - 1 & main\_series \\ 2*(\frac{1}{1-k})^n - 1 & additional\_series \\ (\frac{2}{1-k})^n - 1 & double\_series \\ (\frac{3}{1-k})^n - 1 & triple\_series \\ (\frac{j+1}{1-k})^n - 1 & multiple(j)\_series \end{cases} \qquad (6.101)$$

In order to show the advantages of N/O cascade boost converters, we compare their voltage transfer gains to that of the buck converter,

$$G = \frac{V_O}{V_{in}} = k$$

forward converter,

$$G = \frac{V_O}{V_{in}} = kN \qquad N \text{ is the transformer turn ratio}$$

Cúk-converter,

$$G = \frac{V_O}{V_{in}} = \frac{k}{1-k}$$

fly-back converter,

$$G = \frac{V_O}{V_{in}} = \frac{k}{1-k} N \qquad N \text{ is the transformer turn ratio}$$

boost converter,

$$G = \frac{V_O}{V_{in}} = \frac{1}{1-k}$$

and negative output Luo-converters

$$G = \frac{V_O}{V_{in}} = \frac{n}{1-k} \tag{6.102}$$

If we assume that the conduction duty k is 0.2, the output voltage transfer gains are listed in Table 6.1, if the conduction duty k is 0.5, the output voltage transfer gains are listed in Table 6.2, if the conduction duty k is 0.8, the output voltage transfer gains are listed in Table 6.3.

6.8 Simulation and Experimental Results

6.8.1 Simulation Results of a Three-Stage Boost Circuit

To verify the design and calculation results, PSpice simulation package was applied to a three-stage boost circuit. Choosing $V_{in} = 20$ V, $L_1 = L_2 = L_3 = 10$ mH, all C_1 to $C_8 = 2$ μF and $R = 30$ kΩ, and using $k = 0.7$ and $f = 100$ kHz. The voltage values V_1, V_2, and V_O of a triple-lift circuit are 66 V, 194 V, and 659 V respectively, and inductor current waveforms i_{L1} (its average value I_{L1} = 618 mA), i_{L2}, and i_{L3}. The simulation results are shown in Figure 6.16. The voltage values are matched to the calculated results.

TABLE 6.1

Voltage Transfer Gains of Converters in the Condition $k = 0.2$

Stage No. (n)	1	2	3	4	5	n
Buck converter			0.2			
Forward converter			0.2N (N is the transformer turn ratio)			
Cúk-converter			0.25			
Fly-back converter			0.25N (N is the transformer turn ratio)			
Boost converter			1.25			
Negative output Luo-converters	1.25	2.5	3.75	5	6.25	$1.25n$
Negative output cascade boost converters — main series	0.25	0.563	0.953	1.441	2.052	$1.25^n\text{-}1$
Negative output cascade boost converters — additional series	1.5	2.125	2.906	3.882	5.104	$2*1.25^n\text{-}1$
Negative output cascade boost converters — double series	1.5	5.25	14.625	38.063	96.66	$(2*1.25)^n\text{-}1$
Negative output cascade boost converters — triple series	2.75	13.06	51.73	196.75	740.58	$(3*1.25)^n\text{-}1$
Negative output cascade boost converters — multiple series (j=3)	4	24	124	624	3124	$(4*1.25)^n\text{-}1$

TABLE 6.2

Voltage Transfer Gains of Converters in the Condition $k = 0.5$

Stage No. (n)	1	2	3	4	5	n
Buck converter			0.5			
Forward converter			0.5N (N is the transformer turn ratio)			
Cúk-converter			1			
Fly-back converter			N (N is the transformer turn ratio)			
Boost converter			2			
Negative output Luo-converters	2	4	6	8	10	$2n$
Negative output cascade boost converters — main series	1	3	7	15	31	$2^n\text{-}1$
Negative output cascade boost converters — additional series	3	7	15	31	63	$2*2^n\text{-}1$
Negative output cascade boost converters — double series	3	15	63	255	1023	$(2*2)^n\text{-}1$
Negative output cascade boost converters — triple series	5	35	215	1295	7775	$(3*2)^n\text{-}1$
Negative output cascade boost converters — multiple series (j=3)	7	63	511	4095	32767	$(4*2)^n\text{-}1$

6.8.2 Experimental Results of a Three-Stage Boost Circuit

A test rig was constructed to verify the design and calculation results, and compare with PSpice simulation results. The test conditions are still $V_{in} = 20$ V, $L_1 = L_2 = L_3 = 10$ mH, all C_1 to $C_8 = 2$ μF and $R = 30$ kΩ, and using $k = 0.7$ and $f = 100$ kHz. The component of the switch is a MOSFET device IRF950

TABLE 6.3

Voltage Transfer Gains of Converters in the Condition $k = 0.8$

Stage No. (n)	1	2	3	4	5	n
Buck converter				0.8		
Forward converter			0.8N (N is the transformer turn ratio)			
Cúk-converter				4		
Fly-back converter			4N (N is the transformer turn ratio)			
Boost converter				5		
Negative output Luo-converters	5	10	15	20	25	$5n$
Negative output cascade boost converters — main series	4	24	124	624	3124	5^n-1
Negative output cascade boost converters — additional series	9	49	249	1249	6249	$2*5^n$-1
Negative output cascade boost converters — double series	9	99	999	9999	99999	$(2*5)^n$-1
Negative output cascade boost converters — triple series	14	224	3374	50624	759374	$(3*5)^n$-1
Negative output cascade boost converters — multiple series (j=3)	19	399	7999	15999	32×10^5	$(4*5)^n$-1

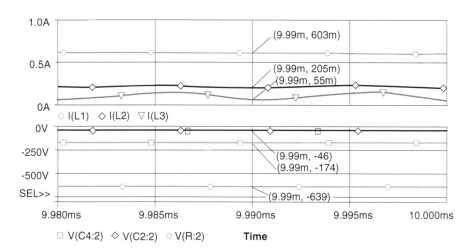

FIGURE 6.16

The simulation results of a three-stage boost circuit at condition $k = 0.7$ and $f = 100$ kHz.

with the rates 950 V/5 A/2 MHz. We measured the values of the output voltage and first inductor current in the following converters.

After careful measurement, the current value of $I_{L1} = 0.62$ A (shown in channel 1 with 1 A/Div) and voltage value of $V_O = 660$ V (shown in Channel 2 with 200 V/Div) are obtained. The experimental results (current and voltage values) in Figure 6.17 match the calculated and simulation results, which are $I_{L1} = 0.618$ A and $V_O = 659$ V shown in Figure 6.16.

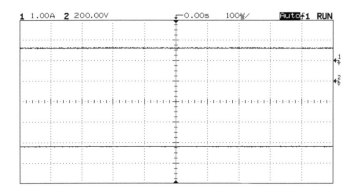

FIGURE 6.17

The experimental results of a three-stage boost circuit at condition $k = 0.7$ and $f = 100$ kHz.

TABLE 6.4

Comparison of Simulation and Experimental Results of a Triple-Lift Circuit.

Stage No. (n)	I_{L1} (A)	I_{in} (A)	V_{in} (V)	P_{in} (W)	V_O (V)	P_O (W)	η (%)
Simulation results	0.618	0.927	20	18.54	659	14.47	78
Experimental results	0.62	0.93	20	18.6	660	14.52	78

6.8.3 Efficiency Comparison of Simulation and Experimental Results

These circuits enhanced the voltage transfer gain successfully, and efficiently. Particularly, the efficiencies of the tested circuits is 78%, which is good for high output voltage equipment. To compare the simulation and experimental results, see Table 6.4. All results are well identified with each other.

6.8.4 Transient Process

Usually, there is high inrush current during the first power-on. Therefore, the voltage across capacitors is quickly changed to certain values. The transient process is very quick taking only a few milliseconds. It is difficult to demonstrate it in this section.

Bibliography

Liu, K.H. and Lee, F.C., Resonant switches — a unified approach to improved performances of switching converters, in *Proceedings of International Telecommunications Energy Conference*, New Orleans, LA, U.S., 1984, p. 344.

Luo, F.L. and Ye, H., Fundamentals of negative output cascade boost converters, in *Proceedings of IEEE-EPEMC'2003*, Xi'an, China, 2003, p. 1896.

Luo, F.L. and Ye, H., Development of negative output cascade boost converters, in *Proceedings of IEEE-EPEMC'2003*, Xi'an, China, 2003, p. 1888.

Martinez, Z.R. and Ray, B., Bidirectional DC/DC power conversion using constant-frequency multi-resonant topology, in Proceedings of Applied Power Electronics Conf. APEC'94, Orlando, FL, U.S., 1994, p. 991.

Masserant, B.J. and Stuart, T.A., A high frequency DC/DC converter for electric vehicles, in *Proceedings of Power Electronics in Transportation*, 1994, p. 123.

Pong, M.H., Ho, W.C., and Poon, N.K., Soft switching converter with power limiting feature, *IEE Proceedings on Electric Power Applications*, 146, 95, 1999.

7

Ultra-Lift Luo-Converter

Voltage-lift (VL) technique has been successfully applied to the design of power DC/DC converters. Good examples are the three-series Luo-converters. Using VL technique can produce high-voltage transfer gain. Super-lift (SL) technique has been given much attention since it yields high-voltage transfer gain. This chapter introduces the ultra-lift Luo-converter as a novel approach within the new ultra-lift (UL) technique, which produces even higher voltage transfer gain. Our analysis and calculations illustrate the advanced characteristics of this converter.

7.1 Introduction

Voltage-lift (VL) technique has been widely applied in electronic circuit design. Since the last century it has been successfully applied to the design of power DC/DC converters. Good examples are the three-series Luo-converters. Using VL technique, one can obtain the converter's voltage transfer gain stage-by-stage in arithmetical series; this gain is higher than that of other classical converters, such as the Buck converter, the Boost converter, and the Buck-Boost converter. Assume the input voltage and current of a DC/DC converter are V_1 and I_1, the output voltage and current are V_2 and I_2, and the conduction duty cycle is k. In order to compare these converters' transfer gains, we use the formulae below:

$$\text{Buck converter} \qquad G = \frac{V_2}{V_1} = k \qquad\qquad (7.1)$$

$$\text{Boost converter} \qquad G = \frac{V_2}{V_1} = \frac{1}{1-k} \qquad\qquad (7.2)$$

$$\text{Buck-Boost converter} \qquad G = \frac{V_2}{V_1} = \frac{k}{1-k} \qquad\qquad (7.3)$$

Luo-Converters $\qquad G = \dfrac{V_2}{V_1} = \dfrac{k^{h(n)}[n+h(n)]}{1-k}$ $\qquad$ (7.4)

Where n is the stage number and h(n) is the Hong Function.

$$h(n) = \begin{cases} 1 & n = 0 \\ 0 & n > 0 \end{cases} \qquad\qquad (7.5)$$

n = 0 for the elementary circuit with the voltage transfer gain.

$$G = \frac{V_2}{V_1} = \frac{k}{1-k} \qquad\qquad (7.6)$$

Super-lift (SL) technique has been paid much more attention because it yields higher voltage transfer gain. Good examples are the super-lift Luo-converters. Using this technique, one can obtain the converter's voltage transfer gain stage-by-stage in geometrical series. The gain calculation formulae are:

$$G = \frac{V_2}{V_1} = \left(\frac{j+2-k}{1-k}\right)^n \qquad\qquad (7.7)$$

where n is the stage number and j is the multiple-enhanced number. n = 1 and j = 0 for the elementary circuit, yielding

$$G = \frac{V_2}{V_1} = \frac{2-k}{1-k} \qquad\qquad (7.8)$$

This chapter introduces the ultra-lift Luo-converter as a novel approach of new technology, ultra-lift (UL) technique, which produces even higher voltage transfer gain. Simulated results verify our analysis and calculations, and illustrate the advanced characteristics of this converter.

7.2   Operation of Ultra-lift Luo-Converter

The circuit diagram is shown in Figure 7.1 (a), which consists of one switch S, two inductors L_1 and L_2, two capacitors C_1 and C_2, three diodes, and the

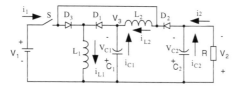

(a) Circuit diagram

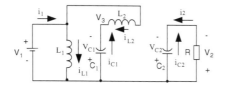

(b) Equivalent circuit during switch-on

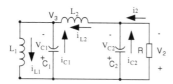

(c) Equivalent circuit during switch-off (CCM)

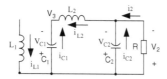

(d) Equivalent circuit during switch-off (DCM)

FIGURE 7.1
Ultra-Lift Luo-Converter.

load R. Its switch-on equivalent circuit is shown in Figure 7.1 (b). Its switch-off equivalent circuit for the continuous conduction mode (CCM) is shown in Figure 7.1 (c), and its switch-off equivalent circuit for the discontinuous conduction mode (DCM) is shown in Figure 7.1 (d).

It is a very simply structured converter in comparison to other converters. As usual, the input voltage and current of the ultra-lift Luo-converter are V_1 and I_1, the output voltage and current are V_2 and I_2, the conduction duty cycle is k and the switching frequency is f. Consequently, the repeating period $T = 1/f$, the switch-on period is kT, and the switch-off period is $(1 - k)T$. To concentrate the operation process, we assume that all components except load R are ideal ones. Therefore, no power losses are considered during power transformation, i.e., $P_{in} = P_O$ and $V_1 \times I_1 = V_2 \times I_2$.

7.2.1   Continuous Conduction Mode (CCM)

Referring to Figures 7.1 (b) and (c), we see that the current i_{L1} increases with the slope $+ V_1/L_1$ during switch-on, and decreases with the slope $- V_3/L_1$ during switch-off. In the steady state, the current increment is equal to the decrement in a whole period T. This gives relation below:

$$kT\frac{V_1}{L_1} = (1-k)T\frac{V_3}{L_1}$$

Thus,

$$V_{C1} = V_3 = \frac{k}{1-k}V_1 \tag{7.9}$$

The current i_{L2} increases with the slope $+ (V_1 - V_3)/L_2$ during switch-on, and decreases with the slope $- (V_3 - V_2)/L_2$ during switch-off. In the steady state, the current increment is equal to the decrement in a whole period T. We obtain the relationships below:

$$kT\frac{V_1+V_3}{L_2} = (1-k)T\frac{V_2-V_3}{L_2}$$

$$V_2 = V_{C2} = \frac{2-k}{1-k}V_3 = \frac{k}{1-k}\frac{2-k}{1-k}V_1 = \frac{k(2-k)}{(1-k)^2}V_1 \tag{7.10}$$

The voltage transfer gain is

$$G = \frac{V_2}{V_1} = \frac{k}{1-k}\frac{2-k}{1-k} = \frac{k(2-k)}{(1-k)^2} \tag{7.11}$$

This is much higher than the voltage transfer gains of the VL Luo-converter and SL Luo-converter in equations (7.6) and (7.8). In fact, the gain in (7.11) is the product of those in (7.6) and (7.8). Another advantage is the starting output voltage of zero. The curve of the voltage transfer gain M versus the conduction duty cycle k is shown in Figure 7.2.

The relation between input and output average currents is

$$I_2 = \frac{(1-k)^2}{k(2-k)}I_1 \tag{7.12}$$

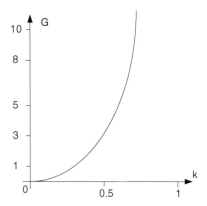

FIGURE 7.2
The voltage transfer gain M versus conduction duty cycle k.

The relation between average currents I_{L2} and I_{L1} is

$$I_{L2} = (1-k)I_{L1} \tag{7.13}$$

Other relations:

$$I_{L2} = (1 + \frac{k}{1-k})I_2 = \frac{1}{1-k}I_2 \tag{7.14}$$

$$I_{L1} = \frac{1}{1-k}I_{L2} = (\frac{1}{1-k})^2 I_2 \tag{7.15}$$

The variation of inductor current i_{L1} is

$$\Delta i_{L1} = kT\frac{V_1}{L_1} \tag{7.16}$$

and its variation ratio is

$$\xi_1 = \frac{\Delta i_{L1}/2}{I_{L1}} = \frac{k(1-k)^2 TV_1}{2L_1 I_2} = \frac{k(1-k)^2 TR}{2L_1 M} = \frac{(1-k)^4 TR}{2(2-k)fL_1} \tag{7.17}$$

The diode current i_{D1} is the same as the inductor current i_{L1} during the switching-off period. For the CCM operation, both currents do not descend to zero, i.e.,

$$\xi_1 \leq 1$$

The variation of inductor current i_{L2} is

$$\Delta i_{L2} = \frac{kTV_1}{(1-k)L_2} \tag{7.18}$$

and its variation ratio is

$$\xi_2 = \frac{\Delta i_{L2}/2}{I_{L2}} = \frac{kTV_1}{2L_2I_2} = \frac{kTR}{2L_2M} = \frac{(1-k)^2 TR}{2(2-k)fL_2} \tag{7.19}$$

The variation of capacitor voltage v_{C1} is

$$\Delta v_{C1} = \frac{\Delta Q_{C1}}{C_1} = \frac{kTI_{L2}}{C_1} = \frac{kTI_2}{(1-k)C_1} \tag{7.20}$$

and its variation ratio is

$$\sigma_1 = \frac{\Delta v_{C1}/2}{V_{C1}} = \frac{kTI_2}{2(1-k)V_3C_1} = \frac{k(2-k)}{2(1-k)^2 fC_1R} \tag{7.21}$$

The variation of capacitor voltage v_{C2} is

$$\Delta v_{C2} = \frac{\Delta Q_{C2}}{C_2} = \frac{kTI_2}{C_2} \tag{7.22}$$

and its variation ratio is

$$\varepsilon = \sigma_2 = \frac{\Delta v_{C2}/2}{V_{C2}} = \frac{kTI_2}{2V_2C_2} = \frac{k}{2fC_2R} \tag{7.23}$$

From the analysis and calculations, we can see that all variations are very small. For example, if $V_1 = 10$ V, $L_1 = L_2 = 1$ mH, $C_1 = C_2 = 1$ μF, $R = 3000$ Ω, $f = 50$ kHz, and the conduction duty cycle k varies from 0.1 to 0.9, the output voltage variation ratio ε is less than 0.003. The output voltage is very smooth DC voltage with nearly no ripple.

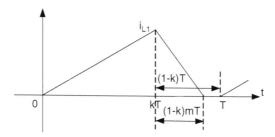

FIGURE 7.3
Discontinuous inductor current i_{L1}.

7.2.2 Discontinuous Conduction Mode (DCM)

Referring to Figures 1 (b), (c) and (d), we see that the current i_{L1} increases with the slope $+ V_1/L_1$ during switch-on, and decreases with the slope $-V_3/L_1$ during switch-off. The inductor current i_{L1} decreases to zero before $t = T$, i.e. the current becomes zero before next time the switch is turned on.

The current waveform is shown in Figure 7.3. The DCM operation condition is defined as

$$\xi_1 \geq 1$$

or

$$\xi_1 = \frac{k(1-k)^2 TR}{2L_1 M} = \frac{(1-k)^4 TR}{2(2-k)fL_1} \geq 1 \tag{7.24}$$

To obtain the boundary between the CCM and DCM operation conditions, we define the normalized impedance Z_n:

$$Z_n = \frac{R}{fL_1} \tag{7.25}$$

The boundary equation is

$$G = \frac{k(1-k)^2}{2} Z_N \tag{7.26}$$

or

$$\frac{G}{Z_N} = \frac{k(1-k)^2}{2}$$

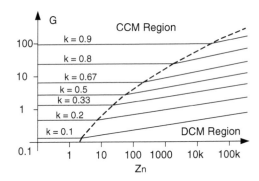

FIGURE 7.4
Boundary between CCM and DCM.

TABLE 7.1

Boundary between CCM and DCM

k	0.2	0.33	0.5	0.67	0.8	0.9
G	0.5625	1.25	3	8	24	99
G/Z_n	0.064	2/27	1/16	1/27	0.016	0.0045
Z_n	8.8	16.9	48	216	1500	22000

The corresponding Z_N is

$$Z_N = \frac{k(2-k)}{(1-k)^2} \bigg/ \frac{k(1-k)^2}{2} = \frac{2(2-k)}{(1-k)^4} \qquad (7.27)$$

The curve is shown in Figure 7.4 and Table 7.1.

We define the filling factor m to describe the current. For DCM operation,

$$0 < m \le 1$$

In the steady state, the current increment is equal to the decrement in a whole period T. This gives the relation below:

$$kT\frac{V_1}{L_1} = (1-k)mT\frac{V_3}{L_1}$$

Thus,

$$V_{C1} = V_3 = \frac{k}{(1-k)m}V_1 \qquad (7.28)$$

Comparing equations (7.9) and (7.28), we find that the voltage V_3 is higher during DCM operation since m < 1. Its expression is

$$m = \frac{1}{\xi_1} = \frac{2L_1 G}{k(1-k)^2 TR} = \frac{2(2-k)}{(1-k)^4 Z_N} \tag{7.29}$$

The current i_{L2} increases with the slope $+ (V_1-V_3)/L_2$ during switch-on, and decreases with the slope $-(V_3 - V_2)/L_2$ during switch-off. In the steady state the current increment is equal to the decrement in a whole period T. We obtain the relationship below:

$$kT\frac{V_1 + V_3}{L_2} = (1-k)T\frac{V_2 - V_3}{L_2}$$

$$V_2 = V_{C2} = \frac{2-k}{1-k}V_3 = \frac{k(2-k)}{m(1-k)^2}V_1 \tag{7.30}$$

The voltage transfer gain in DCM is

$$G_{DCM} = \frac{V_2}{V_1} = \frac{k(2-k)}{m(1-k)^2} = \frac{k(1-k)^2}{2}Z_N \tag{7.31}$$

This is higher that the voltage transfer gain during CCM operation since m < 1. We can see that the voltage transfer gain G_{DCM} increases linearly with, and is proportional to, the normalized impedance Z_N, and is shown in Figure 7.4.

7.3   Instantaneous Values

Instantaneous values of the voltage and current of each component is very important in describing the converter operation. By referring to Figure 7.1, we can obtain these values in CCM and DCM operations.

7.3.1   Continuous Conduction Mode (CCM)

Referring to Figures 7.1 (b) and (c), we obtain the instantaneous values of the voltage and current of each component in CCM operation below:

$$i_{L1}(t) = \begin{cases} I_{L1-min} + \dfrac{V_1}{L_1}t & 0 \le t \le kT \\[3ex] I_{L1-max} - \dfrac{V_3}{L_1}t & kT \le t \le T \end{cases} \tag{7.32}$$

$$i_{L2}(t) = \begin{cases} I_{L2-min} + \dfrac{V_1 - V_3}{L_2}t & 0 \le t \le kT \\[3ex] I_{L2-max} - \dfrac{V_2 - V_1}{L_2}t & kT \le t \le T \end{cases} \tag{7.33}$$

$$i_1(t) = i_s = \begin{cases} I_{1-min} + \left(\dfrac{V_1}{L_1} + \dfrac{V_1 - V_3}{L_2} \right)t & 0 \le t \le kT \\[3ex] 0 & kT \le t \le T \end{cases} \tag{7.34}$$

$$i_{D1}(t) = \begin{cases} 0 & 0 \le t \le kT \\[3ex] I_{L1-max} - \dfrac{V_3}{L_1}t & kT \le t \le T \end{cases} \tag{7.35}$$

$$i_{C1}(t) = \begin{cases} -\left(I_{L2-min} + \dfrac{V_1 - V_3}{L_2}t \right) & 0 \le t \le kT \\[3ex] I_{C1} & kT \le t \le T \end{cases} \tag{7.36}$$

$$i_{C2}(t) = \begin{cases} -I_2 & 0 \le t \le kT \\ I_{C2} & kT \le t \le T \end{cases} \tag{7.37}$$

$$v_{L1}(t) = \begin{cases} V_1 & 0 \le t \le kT \\ V_3 & kT \le t \le T \end{cases} \tag{7.38}$$

$$v_{L2}(t) = \begin{cases} V_1 - V_3 & 0 \le t \le kT \\ V_2 - V_3 & kT \le t \le T \end{cases} \tag{7.39}$$

$$v_s = \begin{cases} 0 & 0 \le t \le kT \\ V_1 - V_3 & kT \le t \le T \end{cases} \tag{7.40}$$

$$v_{D1}(t) = \begin{cases} V_1 - V_3 & 0 \le t \le kT \\ 0 & kT \le t \le T \end{cases} \tag{7.41}$$

$$v_{C1}(t) = \begin{cases} V_3 - \dfrac{I_{L2}}{C_1} t & 0 \le t \le kT \\ V_3 + \dfrac{I_{C1}}{C_1} t & kT \le t \le T \end{cases} \tag{7.42}$$

$$v_{C2}(t) = \begin{cases} V_2 - \dfrac{I_2}{C_2} t & 0 \le t \le kT \\ V_2 + \dfrac{I_{C2}}{C_2} t & kT \le t \le T \end{cases} \tag{7.43}$$

7.3.2 Discontinuous Conduction Mode (DCM)

Referring to Figures 7.1 (b), (c) and (d), we obtain the instantaneous values of the voltage and current of each component in DCM operation. Since inductor current i_{L1} is discontinuous, some parameters have three states with T′ = kT + (1 − k)mT < T.

$$i_{L1}(t) = \begin{cases} \dfrac{V_1}{L_1} t & 0 \le t \le kT \\ I_{L1-max} - \dfrac{V_3}{L_1} t & kT \le t \le T' \\ 0 & T' \le t \le kT \end{cases} \tag{7.44}$$

$$i_{L2}(t) = \begin{cases} I_{L2-min} + \dfrac{V_1 - V_3}{L_2} t & 0 \le t \le kT \\ I_{L2-max} - \dfrac{V_2 - V_1}{L_2} t & kT \le t \le T \end{cases} \tag{7.45}$$

$$i_1(t) = i_s = \begin{cases} I_{1-min} + \left(\dfrac{V_1}{L_1} + \dfrac{V_1 - V_3}{L_2} \right) t & 0 \le t \le kT \\ 0 & kT \le t \le T \end{cases} \tag{7.46}$$

$$i_{D1}(t) = \begin{cases} 0 & 0 \le t \le kT \\ I_{L1-max} - \dfrac{V_3}{L_1}t & kT \le t \le T' \\ 0 & T' \le t \le kT \end{cases} \tag{7.47}$$

$$i_{C1}(t) = \begin{cases} -\left(I_{L2-min} + \dfrac{V_1 - V_3}{L_2}t\right) & 0 \le t \le kT \\ I_{C1} & kT \le t \le T \end{cases} \tag{7.48}$$

$$i_{C2}(t) = \begin{cases} -I_2 & 0 \le t \le kT \\ I_{C2} & kT \le t \le T \end{cases} \tag{7.49}$$

$$v_{L1}(t) = \begin{cases} V_1 & 0 \le t \le kT \\ V_3 & kT \le t \le T' \\ 0 & T' \le t \le kT \end{cases} \tag{7.50}$$

$$v_{L2}(t) = \begin{cases} V_1 - V_3 & 0 \le t \le kT \\ V_2 - V_3 & kT \le t \le T \end{cases} \tag{7.51}$$

$$v_s(t) = \begin{cases} 0 & 0 \le t \le kT \\ V_1 - V_3 & kT \le t \le T' \\ V_1 & T' \le t \le kT \end{cases} \tag{7.52}$$

$$v_{D1}(t) = \begin{cases} V_1 - V_3 & 0 \le t \le kT \\ 0 & kT \le t \le T' \\ -V_3 & T' \le t \le kT \end{cases} \tag{7.53}$$

$$v_{C1}(t) = \begin{cases} V_3 - \dfrac{I_{L2}}{C_1}t & 0 \le t \le kT \\ V_3 + \dfrac{I_{C1}}{C_1}t & kT \le t \le T \end{cases} \tag{7.54}$$

$$v_{C2}(t) = \begin{cases} V_2 - \dfrac{I_2}{C_2}t & 0 \le t \le kT \\ V_2 + \dfrac{I_{C2}}{C_2}t & kT \le t \le T \end{cases} \tag{7.55}$$

TABLE 7.2

Comparison of Various Converters Gains

k	0.2	0.33	0.5	0.67	0.8	0.9
Buck	0.2	0.33	0.5	0.67	0.8	0.9
Boost	1.25	1.5	2	3	5	10
Buck-Boost	0.25	0.5	1	2	4	9
Luo-Converter	0.25	0.5	1	2	4	9
Super-Lift Luo-Converter	2.25	2.5	3	4	6	11
Ultra-Lift Luo-Converter	0.56	1.25	3	8	24	99

7.4 Comparison of the Gain to Other Converters' Gains

The ultra-lift Luo-converter has been successfully developed using the novel approach of the new technology, ultra-lift (UL). Table 7.2 lists the voltage transfer gains of various converters at k = 0.2, 0.33, 0.5, 0.67, 0.8 and 0.9. The outstanding characteristics of the ultra-lift Luo-converter are very well presented. From the comparison we can obviously see that the ultra-lift Luo-converter has very high voltage transfer gain: $G(k)|_{k=0.5} = 3$, $G(k)|_{k=0.667} = 8$, $G(k)|_{k=0.8} = 24$, $G(k)|_{k=0.9} = 99$.

7.5 Simulation Results

To verify the advantages of the ultra-lift Luo-converter, we apply the PSpice simulation method. We choose the parameter's values: $V_1 = 10$ V, $L_1 = L_2 = 1$ mH, $C_1 = C_2 = 1$ μF, R = 3 kΩ, f = 50 kHz and conduction duty cycle k = 0.6 and 0.66. The corresponding output voltage $V_2 = 52.5$ V and 78 V. The first waveform is the inductor's current i_{L1}, which flows through the inductor L_1. The second and third waveforms are the voltage V_3 and output voltage V_2. These simulation results are identical to the calculation results. These results are shown in Figures 7.5 and 7.6 respectively.

7.6 Experimental Results

To verify the advantages and design of the ultra-lift Luo-converter and compare them with the simulation results, we construct a test rig with these components: $V_1 = 10$ V, $L_1 = L_2 = 1$ mH, $C_1 = C_2 = 1$ μF, R = 3 kΩ, f = 50

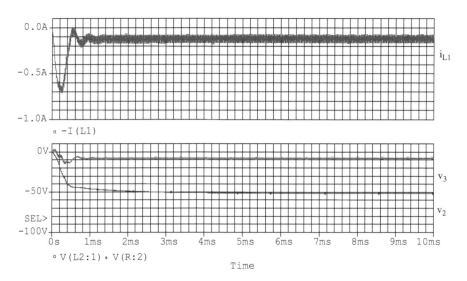

FIGURE 7.5
Simulation results for k = 0.6.

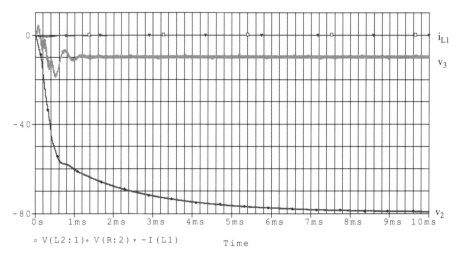

FIGURE 7.6
Simulation results for k = 0.66.

kHz and conduction duty cycle k = 0.6 and 0.66. The corresponding output voltage V_2 = 52 V and 78 V. The first waveform is the inductor's current i_{L1}, which flows through the inductor L_1. The second waveform is the output voltage V_2. The experimental results are shown in Figures 7.7 and 7.8 respectively. The test results are identical to those of the simulation results shown in Figures 7.5 and 7.6, and verify both the calculation results and our design.

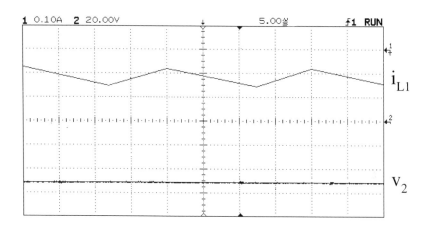

FIGURE 7.7
Experimental results for k = 0.6.

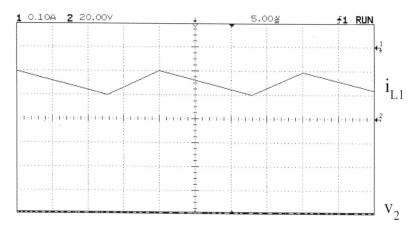

FIGURE 7.8
Experimental results for k = 0.66.

7.7 Summary

The Ultra-lift Luo-converter has been successfully developed using the novel approach of a new technology, the Ultra-lift (UL) technique, which produces especially high voltage transfer gain. It is much higher than that of Voltage-lift Luo-converters and Super-lift Luo-converters. This chapter introduces the operation and characteristics of this converter in detail. This converter will be applied in industrial applications entailing high output voltages.

Bibliography

Jozwik J. J., and Kazimerczuk M. K., Dual Sepic PWM Switching-Mode DC/DC Power Converter, *IEEE Transactions on Industrial Electronics*, 36, 1, 1989, pp. 64-70.

Luo F. L., Negative Output Luo-Converters: Voltage Lift Technique, in *IEE-EPA Proceedings*, 146, 2, 1999, pp. 208-224.

Luo F. L., Positive Output Luo-Converters: Voltage Lift Technique, in *IEE-EPA Proceedings*, 146, 4, 1999, pp. 415-432.

Luo F. L., Double Output Luo-Converters: Advanced Voltage Lift Technique, in *IEE-EPA Proceedings*, 147, 6, 2000, pp. 469-485.

Luo F. L. and Ye H., Chapter 17, "DC/DC Conversion Techniques and Nine Series Luo-Converters," in *Power Electronics Handbook*, Academic Press, San Diego, 2001, pp. 335-406.

Luo F. L. and Ye H., Positive Output Super-Lift Luo-Converters, in *Proceedings of IEEE International Conference PESC'2002*, Cairns, Australia, 2002, pp. 425-430.

Luo F. L. and Ye H., *Advanced DC/DC Converters*, CRC Press, LLC, Boca Raton, 2003.

Luo F. L. and Ye H., Negative Output Super-Lift Luo-Converters, in *Proceedings of IEEE International Conference PESC'2003*, Acapulco, Mexico, 2003, pp. 1361-1366.

Luo F. L. and Ye H., Positive Output Super-Lift Converters, in *IEEE-Transactions on PEL*, 18, 1, 2003, pp. 105-113.

Luo F. L. and Ye H., Negative Output Super-Lift Converters, in *IEEE-Transactions on PEL*, 18, 5, 2003, pp. 1113-1121.

Luo F. L. and Ye H., Ultra-Lift Luo-Converter, in *IEE-EPA Proceedings*, 152, 1, 2005, pp. 27-32.

Massey R. P. and Snyder E. C., High Voltage Single-Ended DC-DC Converter, in *IEEE PESC, 1977 Record*, pp. 156-159.

Mohan N., Undeland T. M., and Robbins W. P., *Power Electronics: Converters, Applications and Design*, 3rd Ed., John Wiley & Sons, New York, 2003.

Index

407

negative output Luo-converters; *see*
        Negative output Luo-converters
overview, 32, 41–42
positive output Luo-converters; *see*
        Positive output Luo-converters
self-lift converters; *see* Self-lift converters

Z

Zero-current-switching quasi-resonant
        converters, 35, 36
Zero-transition (ZT) converters, 35, 36
Zero-voltage-switching quasi-resonant
        converters, 35–36
ZETA pump, 15, 17

Y

Ye, Hong, 32

Printed and bound by CPI Group (UK) Ltd, Croydon, CR0 4YY

17/10/2024

01775690-0016